C0-ASN-292

REVIEWS OF INFRARED AND MILLIMETER WAVES

VOLUME 1

A Continuation Order Plan is available for this series. A continuation order will bring delivery of each new volume immediately upon publication. Volumes are billed only upon actual shipment. For further information please contact the publisher.

REVIEWS OF INFRARED AND MILLIMETER WAVES

VOLUME 1

EDITED BY KENNETH J. BUTTON

Massachusetts Institute of Technology
Cambridge, Massachusetts

PLENUM PRESS • NEW YORK AND LONDON

Library of Congress Cataloging in Publication Data

Main entry under title:

Reviews of infrared and millimeter waves.

"Papers from the XXth General Assembly of the Union-Radio-Scientifique Internationale, Washington, August, 1981"—Pref.
Includes bibliographical references and index.
1. Millimeter waves—Congresses. 2. Infra-red rays—Congresses. I. Button, Kenneth J. II. International Union of Radio Science. General Assembly (20th: 1981 Washington, D.C.)
TK7876.R48 1983 621.3813 83-2295
ISBN 0-306-41260-8

A Division of Plenum Publishing Corporation
233 Spring Street, New York, N.Y. 10013

Printed in the United States of America

PREFACE

This is the first book in the series that is being called The Reviews of Infrared and Millimeter Waves. The series will contain the manuscripts of invited papers from conferences on this subject. This first book contains some of the invited papers from the XXth General Assembly of the Union-Radio-Scientifique Internationale, Washington, August, 1981. We were asked by the URSI Committee to organize a two day symposium on millimeter and submillimeter waves This required the difficult choice of five topics which turned out to be (1) Ultra-low Noise Millimeter Wave Receivers (Detectors and Mixers), (2) Free Electron Maser and Gyrotron, (3) Measurements of Power and Noise Power, (4) Complex Dielectric Properties of Solids and Liquids, and (5) Radioastronomy. We have not yet collected all the manuscripts and perhaps we never shall because the time-consuming effort required to prepare a comprehensive review manuscript works a hardship on research scientists who are already overburdened. We are particularly grateful, therefore, to the authors who have worked so hard to contribute the chapters to this book. The first four chapters contibute to the timely topic of detectors, mixers and receivers. These authors: Tucker, Feldman, Rudner, Okamura, Hogg and their well-known colleagues have been among the leaders in this exciting emerging field for the past few years. The fifth chapter by Sakai and Genzel is the most comprehensive treatment of the metal mesh filter science that can be found in one place. Likewise, Geick and Strobel have prepared the first of Geick's comprehensive treatments on magnetic resonance phenomena. We expect another chapter shortly to expand on this theme. Finally, one of the materials measurement techniques has been treated by Gardiol and his group.

The forthcoming Volume 2 will contain additional URSI papers but will also be supplemented by manuscripts of the invited speakers at the annual Conference on Infrared and Millimeter Waves which is sponsored by the International Union of Radio Science and the IEEE Microwave Theory and Techniques Society. The current conference is being held in Marseille, February 14, 1983. The following conference will be held in Miami Beach, December 12, 1983 and the Ninth Conference will be held in Takarazuka City, Japan, October 24, 1984 under the direction of Professor Hiroshi Yoshinaga. The Proceedings of these conferences appear in the International Journal of Infrared and Millimeter Waves.

Kenneth J. Button

CONTENTS

THE QUANTUM RESPONSE OF NONLINEAR TUNNEL JUNCTIONS AS DETECTORS AND MIXERS*

John R. Tucker

Department of Electrical Engineering
and Coordinated Science Laboratory
University of Illinois
Urbana, Illinois 61801

1. INTRODUCTION

Detectors and heterodyne receivers capable of approaching quantum-limited sensitivity in the millimeter wave region are currently being developed. This new class of quasiparticle mixer utilizes the extraordinary nonlinearities that occur for single-electron tunneling into a superconductor forming one or both sides of a junction. These nonlinear tunneling devices perform the same detector functions as standard Schottky barrier diodes and other resistive mixers, but with an important physical difference. At frequencies where the photon energy exceeds the voltage scale of the dc I-V nonlinearity, a single-particle tunnel junction ceases to respond classically and becomes capable of detecting individual quanta. The results of this transformation are very dramatic, and appear certain to have a major impact on low-noise reciever technology.

Quantum effects in the tunneling of single-electron quasiparticles were first observed nearly 20 years ago by Dayem and Martin (1962) using superconductor-insulating oxide-superconductor (SIS) junctions. The well-known step structure induced on the dc I-V characteristic by applied microwave radiation is illustrated in Fig. 1a. The physical origin of these steps, displaced from the sharp current onset at the gap voltage by integral multiples of $\hbar\omega/e$, was explained shortly thereafter by Tien and Gordon (1963) in terms of photon-assisted tunneling. More than a decade elapsed,

*Supported by the Joint Services Electronics Program (U.S. Army, U.S. Navy, U.S. Air Force) under contract number N00014-C-79-0424

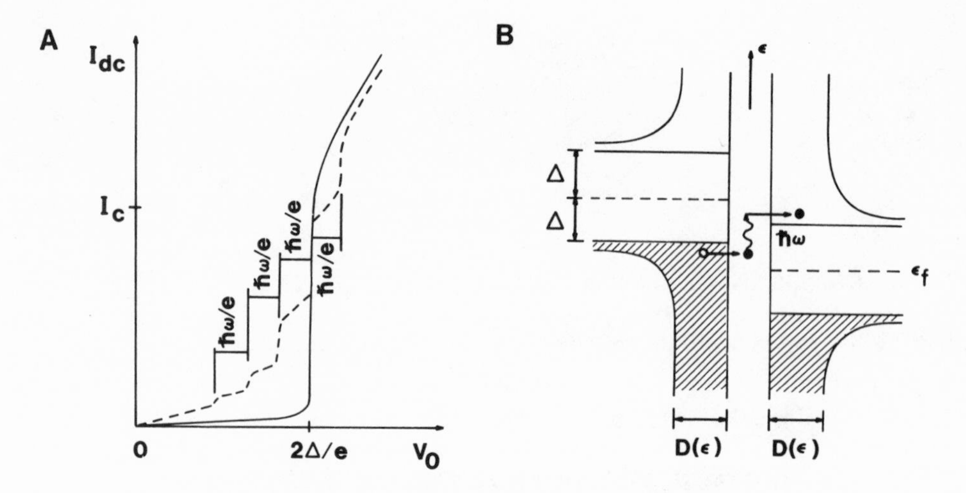

Fig. 1. (a) Photon-assisted tunneling steps (dashed curve) induced by an applied microwave field on the dc I-V characteristic (solid curve) of an SIS tunnel junction. (b) Densities of states vs. energy of single-particle excitations for identical superconductors on opposite sides of a tunnel barrier.

however, before the invention of the superconductor-semiconductor or super-Schottky diode by McColl, Millea, and Silver (1973) stimulated an inquiry (Tucker, 1975; Tucker and Millea, 1978; Tucker and Millea, 1979) into the relationship between photon-assisted tunneling and detector properties.

A full quantum generalization of classical microwave mixer theory was developed (Tucker, 1979) in order to predict the high-frequency behavior of nonlinear single-particle tunnel junctions as direct detectors and heterodyne receivers. It was found that such devices would make a smooth transition from classical rectification and mixing at low frequencies to quantum response at high frequencies. In the quantum regime, the current responsivity for direct detection represents one additional electron tunneling across the junction for each signal photon absorbed, as schematically illustrated in Fig. 1b. The mixer noise temperature for a heterodyne receiver was predicted to approach quantum limited sensitivity, $T_M \sim \hbar\omega/k$, under appropriate conditions. The frequency threshold of this transition to quantum behavior is determined by the nonlinearity of the dc I-V characteristic. When the photon energy $\hbar\omega/e$ is small on the scale of the nonlinearity, the response is classical. For frequencies $\hbar\omega/e$ larger than the voltage scale of the dc I-V nonlinearity, however, the junction responds to individual quanta and must be described in terms of photon-assisted tunneling.

The super-Schottky tunnel junction is extremely nonlinear, and it demonstrated spectacular improvements in both detector sensitivity and mixer noise temperature compared with all previous microwave devices (Vernon et al., 1977; McColl et al., 1977). It has not proved feasible, however, to extend the practical operating frequency of the super-Schottky into the region where significant quantum effects are expected (Silver et al., 1981). The quantum mixer theory nevertheless predicted a number of new and potentially significant effects, and all of these have recently been observed using small area SIS junctions. The single-electron or quasiparticle portion of the dc I-V curve for an SIS diode is distinctly more nonlinear than a super-Schottky, and so photon-assisted tunneling appears at lower frequencies. There is also no series-spreading resistance, as in a semiconductor contact, to limit the operating frequency by introducing parasitic losses which increase as the diode dimensions are reduced.

In the spring of 1979, three groups reported initial mixer experiments using SIS tunnel junctions (Richards et al., 1979; Dolan et al., 1979; Rudner and Claeson, 1979). It was immediately apparent that quantum effects were being observed at the higher frequencies, along with very low mixer noise temperatures. The response of these devices to individual photons in direct detection at 36 GHz (Richards et al., 1980) was also quickly confirmed.

A computer simulation, based on the photon-assisted tunneling theory, was developed (Tucker, 1980) in order to characterize the heterodyne mixing properties of an ideal SIS junction operating on the single-particle branch of its dc I-V curve. This simulation produced some extraordinary predictions. It showed that these devices should not only be capable of quantum-limited mixer-noise temperatures, but that conversion gain (simultaneous amplification in the process of frequency down-conversion) would also be possible. In addition, the low frequency output impedance (the dynamic resistance on the steps in Fig. 1a) was predicted to become infinite and then negative under certain conditions. The available conversion gain in this case was shown to approach infinitely large values as the output impedance increased, and to remain infinite in the negative-resistance region.

The achievement of conversion gain in these mixers is of crucial technical importance, since the photon energies are of order $\hbar\omega/k \sim 1K$ in the micro- and millimeter wave regions. Practical receivers designed to approach quantum noise-limited sensitivity at these frequencies will depend upon utilizing this internal gain mechanism to prevent amplifier noise from dominating the overall system noise temperature. This conversion gain is a new quantum phenomenon that cannot be realized in classical resistive mixers, such as Schottky barrier diodes, which are theoretically limited to a minimum conversion loss $L_c \geq 3$ db under ordinary conditions.

The first observation of conversion gain in SIS quasiparticle mixers was made only a few weeks after its prediction in experiments at 36 GHz by Shen, Richards, Harris and Lloyd (1980). Their estimate of mixer noise temperature also equalled the quantum limited value at this frequency within experimental error. Similar results for conversion efficiency was soon achieved at 73 GHz by Rudner, Feldman, Kollberg, and Claeson (1981a, 1981b) using a unique six-element series array configuration. Meanwhile, high current density single-junction SIS receivers were being developed and tested for astronomical observations at 115 GHz by Phillips, Woody, Dolan, Miller, and Linke (1981) and in collaboration with T.C.L.G. Sollner (Dolan et al., 1981). More recently, negative resistance and infinite available conversion gain have been observed at 115 GHz by Kerr, Pan, Feldman, and Davidson (1981). These effects were also seen by Smith, McGrath, Richards, van Kempen, Prober, and Santhanam (1981) at 36 GHz. In collaboration with R. A. Batchelor (McGrath et al., 1981), they report raw single-sideband conversion gains in excess of L_C^{-1} = +4db and mixer noise temperatures $T_M \simeq 9k < 4\hbar\omega/k$ within a factor of four of the quantum limit.

This field is developing so rapidly that a complete review would be inappropriate at the present time. The results achieved thus far by the Berkeley group have already been extensively summarized (Richards and Shen, 1980; Sen, 1981). This introduction has attempted only to highlight the early work on quasiparticle mixers, with particular emphasis placed on the device physics. At this point it is clear that these mixers, and SIS devices in particular, will have a major impact on the science and technology of ultra-low noise detection in the micro- and millimeter wave regions. It also appears that the concepts developed here may eventually be extended toward both higher and lower frequencies, and to the description of radically different types of tunneling phenomena (Bardeen, 1980) as well. Accordingly, it seems appropriate to summarize the photon-assisted tunneling theory in a format that will hopefully provide improved access to a wider spectrum of the scientific community. Most of the new and technically important effects observed in quasiparticle mixers are apparently unique, and can only be fully described within this framework. So far, all attempts to draw useful correspondences with more familiar quantum detection systems have been unsuccessful. The physical origins of conversion gain and negative differential resistance, for example, are currently understood only in terms of the theory presented here. Future progress in understanding and utilizing these phenomena will likely be realized through insights achieved within this context.

The following sections will summarize the basic aspects of photon-assisted tunneling theory, and the quantum mixer analysis that has been developed from it. A formal derivation of these results and references to previous work may be found in the original literature (Tucker, 1979). Here the emphasis will be placed on presenting a conceptual basis for these ideas, and on practical use of the theory to characterize the performance of a mixer. This line of development is surprisingly straightforward. The physics of a single-particle tunnel junction can be understoon by examining the problem of two weakly coupled energy levels using simple time-dependent perturbation theory. The total junction response may then be determined by superposition for every pair of levels on opposite sides of the barrier. Use of the theory to describe practical quasiparticle mixers is also relatively uncomplicated. A single analytic model has been found to characterize the performance of most fundamental and harmonic mixers under typical operating conditions. The major assumptions used in constructing this model represent only slightly idealized versions of the conditions already demonstrated to be experimentally favorable. This "universal" quasiparticle mixer model will in any case be the only available guide for most applications, because generalizations of virtually any sort would require computer simulations of formidable complexity.

2. PHOTON-ASSISTED TUNNELING AND DIRECT DETECTION

The original observation of photon-assisted tunneling, schematically illustrated in Fig. 1a, was made using an SIS diode. This is actually a form of Josephson junction in which the capacitance effectively shunts the oscillating pair currents at frequencies $\omega_J = 2eV_0/\hbar$ near the gap voltage $V_0 = 2\Delta/e$. Under these conditions, the tunneling of single-electron quasiparticles will determine the shape of the dc I-V curve over a substantial range of bias voltage. The sharp onset of quasiparticle current at $V_0 = 2\Delta/e$ is a direct reflection of the energy gap in the single-particle excitation spectrum for identical superconductors on opposite sides of the barrier, as illustrated in Fig. 1b. At the bias voltage shown in the diagram, no dc current can flow because the absence of available single-particle states on the right prevents electrons on the left from tunneling. The absorption of a single microwave quantum $\hbar\omega$, however, can supply the extra energy required to tunnel into empty states above the energy gap, as shown. The bias voltage illustrated in Fig. 1b corresponds to being on the first step below $2\Delta/e$ in Fig. 1a. The sequence of these steps below the gap voltage in the presence of applied microwave radiation represents the series of bias thresholds for which the absorption of at least $n = 1, 2, 3, \ldots$ photons is required for single-particle tunneling. The heights of the steps are directly determined by the amplitude of the microwave field.

Two essential observations can be made with respect to the results illustrated in Fig. 1. The sharp induced step structure of Fig. 1a obviously reflects the very abrupt onset of quasiparticle current at the full gap voltage. This rise is perfectly sharp for an ideal SIS junction, and is due to the crossing of mildly divergent densities of states on both sides of the barrier as shown in Fig. 1b. In reality, of course, this current onset has some finite width determined by a variety of physical mechanisms. The existence of the induced photon steps in Fig. 1a will therefore be evident only for frequencies sufficiently high that $\hbar\omega/e$ is large compared to the voltage width of the dc current nonlinearity. At lower frequencies, an ac field will smoothly average the dc I-V curve, and the behavior will be essentially classical. The second important observation is that the process of photon-assisted tunneling, illustrated in Fig. 1b, is capable of generating one additional carrier flowing in the external circuit for each microwave photon absorbed in the junction. This implies that such a device has, at least in principle, the ability to detect individual quanta at very long wavelengths.

In the presence of an applied potential:

$$V(t) = V_0 + V_1 \cos \omega t \tag{1}$$

Tien and Gordon (1963) showed that the dc tunneling current will be given by:

$$I_{dc} = \sum_{n=-\infty}^{\infty} J_n^2(\alpha) I_0(V_0 + n\hbar\omega/e) \tag{2}$$

Here $I_0(V_0)$ represents the unmodulated dc I-V curve, and the argument of the Bessel functions is proportional to the strength of the ac field:

$$\alpha = \frac{eV_1}{\hbar\,\omega} \tag{3}$$

Notice that eq.(2) is just the sort of expression required to explain the step structure in Fig. 1a. In the presence of an ac field, any feature on the dc I-V curve that is sharp on a scale $\hbar\omega/e$ will be reflected at values of dc bias voltage displaced by integral multiples of the photon energy.

The impact of photon-assisted tunneling on device properties is most easily understood by examining the response to small ac signals. Expanding the Bessel functions in eq. (2) to lowest order in the ac potential yields the following result for the rectified current:

$$\Delta I_{dc} = \frac{1}{4} V_1^2 \left\{ \frac{I_0(V_0 + \hbar\omega/e) - 2I_0(V_0) + I_0(V_0 - \hbar\omega/e)}{(\hbar\omega/e)^2} \right\} \tag{4}$$

The expression in brackets on the right is a finite difference form which reflects the absorption or emission of a single photon in the tunneling process. It reduces to the second derivative of the dc I-V curve, d^2I_0/dV_0^2, when the photon energy $\hbar\omega/e$ is small compared to the voltage scale of the nonlinearity, reproducing the usual classical result. The dissipative component of current induced at the applied ac frequency is given by an analogous expression:

$$I_\omega = V_1 \left[\frac{I_0(V_0 + \hbar\omega/e) - I_0(V_0 - \hbar\omega/e)}{2\hbar\omega/e} \right] \tag{5}$$

Here the classical conductance of the junction, dI_0/dV_0, is replaced by a corresponding first-difference form involving single photon absorption or emission for weak ac signals. The expressions of eqs. (4) and (5) thus represent simple and intuitively appealing generalizations of the classical analysis to account for photon-assisted tunneling.

The current responsivity for a direct or video detector is defined as the change in dc current per unit ac signal power absorbed. Utilizing eqs. (4) and (5) gives:

$$\begin{aligned} R_i &= \Delta I_{dc}/\tfrac{1}{2}V_1 I_\omega \\ &= \frac{e}{\hbar\omega}\left[\frac{I_0(V_0 + \hbar\omega/e) - 2I_0(V_0) + I_0(V_0 - \hbar\omega/e)}{I_0(V_0 + \hbar\omega/e) - I_0(V_0 - \hbar\omega/e)}\right] \\ &= \frac{1}{2}\,\frac{d^2I_0/dV_0^2}{dI_0/dV_0}, \quad \text{Classical limit} \\ &= \frac{e}{h\omega}, \quad \text{Quantum limit} \end{aligned} \tag{6}$$

The standard classical result is, of course, recovered when the slope of the dc I-V curve changes slowly on a voltage scale $\hbar\omega/e$. This classical expression would predict that such a detector could be made arbitrarily sensitive by increasing the nonlinearity of the diode. Photon-assisted tunneling theory, however, shows that there is an absolute limit to the current responsivity for any given frequency. This quantum limit, $R_i = e/\hbar\omega$, represents the tunneling of one additional electron across the barrier for each signal photon absorbed. It is independent of the detailed shape of the dc I-V curve, so long as the photon energy is large compared to the voltage scale of the dc nonlinearity. An example in which this limit can be practically achieved is illustrated in Fig. 1b, and corresponds to dc biasing of an SIS junction on the first step below the gap voltage in Fig. 1a. In this case $I_0(V_0 + \hbar\omega/e) >> I_0(V_0)$ and

$I_0(V_0 - \hbar\omega/e)$, so that the bracketed quantity in eq. (6) approaches unity. The general result of eq. (6) characterizes direct detection in any single-particle tunnel junction, and describes a continuous transition between classical behavior and quantum response as the photon energy becomes comparable to the voltage scale of the dc I-V nonlinearity.

The sensitivity of a tunnel junction as a video detector will be limited by shot noise due to the bias current. For dc voltages $eV_0 \gtrsim 2kT$ large compared to thermal energies, the mean square noise current is given by the usual expression:

$$< I_n^2 > = 2eI_0(V_0)B \tag{7}$$

where B is the output bandwidth. The noise-equivalent power (NEP) may then be estimated as:

$$\begin{aligned} NEP &\simeq \frac{<I_n^2>^{\frac{1}{2}}}{\eta R_i} \\ &\simeq \frac{\hbar\omega B}{\eta}\left[\frac{I_0(V_0)}{eB}\right]^{\frac{1}{2}}, \text{ Quantum limt} \end{aligned} \tag{8}$$

where η is a factor describing the impedance matching of signal power into the diode. The quantum-limited value for NEP shown here has a simple physical interpretation. The quantity $N = I_0(V_0)/eB$ is the average number of bias current electrons tunneling through the barrier during a resolution time 1/B determined by the output bandwidth. Since these individual events are uncorrelated, the mean fluctuation in their number during this period is $N^{\frac{1}{2}}$. The quantum limited value for NEP in eq. (8) therefore represents the absorption of $N^{\frac{1}{2}}$ photons per resolution time in order to generate a signal current equal to the mean square noise. Equation (8) is, in fact, identical to the corresponding result for a photomultiplier, a photodiode, or a photoconductor. For these more familiar quantum detectors, the dc bias current in the above expression would be replaced by the dark current or by generation-recombination currents, respectively. The general form of eq. (8) is characteristic of all types of direct photon detectors in which the dominant source of noise is not due to the signal itself.

The dc bias current for an SIS junction video detector may be made very small, as indicated in Fig. 1a. Experiments at 36 GHz (Richards et al., 1980) have yielded responsivities approaching the quantum limit and extremely low NEP's in excellent quantitative agreement with eqs. (6) and (8). In theory, though not in practice, it would be possible to reduce the bias current so that

$N = I_0(V_0)/eB < 1$ by sufficiently lowering the temperature in a perfect SIS junction. In this case, the noise generated by the signal itself would no longer be negligible. The rectified current $\Delta I_{cd} = \eta R_i P_s$, proportional to the absorbed signal power, should, in principle, have been added to the bias current in computing the shot noise of eq. (7). If this were to become the dominant noise course, eq. (8) immediately yields an NEP $\sim \hbar\omega B$ for good coupling that equals the absolute limit imposed by the quantum nature of electromagnetic radiation. By analogy with other photon detectors a noisy photomultiplier, for example - it might be expected that this ultimate limit could be approached by utilizing heterodyne techniques. This result is indeed predicted by the full quantum mixer theory, and has been experimentally observed.

3. THE QUANTUM RESPONSE OF NONLINEAR TUNNEL JUNCTIONS

The currents induced in a single-particle tunnel junction can be computed in terms of just two functions: the voltage across the junction at all previous times $V(t)$, and the measured dc I-V curve $I_0(V)$. The voltage enters as a time-dependent phase factor, because it can be considered to modulate the energy of every electron level on one side of the tunnel barrier. The additional time dependence in the Schrodinger wave function for these levels due to applied ac fields may be written in the form:

$$\exp\left\{-\frac{ie}{\hbar}\int^{t} dt'[V(t')-V_0]\right\} = \int_{-\infty}^{\infty} d\omega' W(\omega') e^{-i\omega' t} \qquad (9)$$

Here the Fourier transform of this phase factor has been defined with the dc bias explicitly removed. This definition facilitates writing the following expression for the average tunneling current in a form suitable for analysis in the frequency domain:

$$\langle I(t)\rangle = Im \int_{\infty}^{\infty} d\omega' d\omega'' W(\omega') W^*(\omega'') e^{-i(\omega'-\omega'')t} \cdot j(V_0+\hbar\omega'/e) \qquad (10)$$

The physics of the junction is contained in the complex response function $j(V)$. From eq. (9) it may be seen that if the applied ac potential vanishes, then $W(\omega) = \delta(\omega)$. Using this special case, eq. (10) shows that the imaginary part of $j(V)$ must be just the dc I-V characteristic:

$$Im\ j(V) = I_0(V) \qquad (11)$$

The real part of the response function $j(V)$ characterizes the reactive components of the tunneling current. It is related to the imaginary part by a Kramers-Kronig transform that reflects the requirements of causality. This Kramers-Kronig transform of the dc I-V curve may be defined in the form:

$$Re\ \ j(V) \equiv I_{KK}(V) = P\int_{\infty}^{\infty} \frac{d\bar{V}}{\pi} \frac{[I_0(\bar{V}) - \bar{V}/R_N]}{\bar{V} - V} \tag{12}$$

Here it is assumed that the dc current becomes ohmic in the limit of large bias voltage, exhibiting a normal resistance R_N. The second term in the integrand of eq. (12) thus implies that only the nonlinear portion of $I_0(V)$ need be considered in evaluating the reactive part of the response. All physical quantities depend on differences between values of *Re* j(V), and not on its absolute magnitude. The particular definition chosen in eq. (12) turns out to be very convenient, however, both conceptually and for computational purposes as well. The reactive response indeed arises entirely from the dc nonlinearity, and this nonlinear region may be approximately bounded when computing the Kramers-Kronig transform $I_{KK}(V)$ in this form. It should be pointed out that the reactances calculated in terms of $I_{KK}(V)$ are nonclassical, and are present in addition to the ordinary junction capacitance. This effect is negligible so long as the photon energies $\hbar\omega/e$ are small compared to the voltage scale of the dc nonlinearity. The current can then be described as a classical modulation of the dc I-V characteristic. In the quantum regime, however, the time-dependent current becomes a non-local function of the applied voltage. The quantum reactances computed using $I_{KK}(V)$ must then be included in order to obtain a physically consistent description of the tunnel junction.

Equations (9)-(12) completely characterize the ac response of a single-particle tunnel junction in terms of its measured dc I-V characteristic. A derivation of these expressions, based on a somewhat simplified model, is presented in the following section. Although this microscopic analysis provides deeper insight into several aspects of photon-assisted tunneling, a working knowledge of the theory can be obtained by examining some predictions made using the results summarized here. The remainder of this section will describe the response to a sinusoidal ac potential, and subsequent sections will characterize the mixing properties of a quasiparticle tunnel junction.

The sinusoidal potential of eq. (1) produces a time-dependent phase factor which, according to eq. (9), may be written in the form:

$$\exp\left\{-\frac{ie}{\hbar}\int^{t} dt' V_1 \cos\omega t'\right\} = \exp\left\{-i\left[\frac{eV_1}{\hbar\omega}\right]\sin\omega t\right\} = \sum_{n=-\infty}^{\infty} J_n(eV_1/\hbar\omega)e^{-in\omega t} \tag{13}$$

The Fourier transform of this function is then:

$$W(\omega') = \sum_{n=-\infty}^{\infty} J_n(\alpha)\,\delta(\omega'-n\omega) \tag{14}$$

where $\alpha = eV_1/\hbar\omega$ as before. Substituting this result into eq. (10) for the induced current gives:

$$\begin{aligned}\langle I(t)\rangle &= Im \sum_{n,m=-\infty}^{\infty} J_n(\alpha) J_{n+m}(\alpha)\, e^{+im\omega t} j(V_0 + n\hbar\omega/e) \\ &= a_0 + \sum_{m=1}^{\infty} [2a_m \cos m\omega t + 2b_m \sin m\omega t]\end{aligned} \tag{15}$$

The average current thus contains components at all the various harmonic frequencies, with magnitudes:

$$\begin{aligned}2a_m &= \sum_{n=-\infty}^{\infty} J_n(\alpha)[J_{n+m}(\alpha) + J_{n-m}(\alpha)] I_0(V_0 + n\hbar\omega/e) \\ 2b_m &= \sum_{n=-\infty}^{\infty} J_n(\alpha)[J_{n+m}(\alpha) - J_{n-m}(\alpha)] I_{KK}(V_0 + n\hbar\omega/e)\end{aligned} \tag{16}$$

These rather complicated series expressions cannot in general be evaluated in closed form. The corresponding classical result, on the other hand, appears relatively simple by comparison. For a sinusoidal modulation of the dc current:

$$I^{CL}(t) = I_0(V_0 + V_1 \cos\omega t) = a_0^{CL} + \sum_{m=1}^{\infty} 2a_m^{CL} \cos m\omega t \tag{17}$$

where:

$$2a_m^{CL} = \frac{2}{\pi}\int_0^{\pi} d(\omega t) \cos m\omega t\; I_0(V_0 + V_1 \cos \omega t) \tag{18}$$

Notice first that the quantum expression of eq. (15) contains additional reactive terms, which are seen to depend on the Kramers-Kronig transform $I_{KK}(V)$ of the dc I-V characteristic. In addition, the simple Fourier transform of eq. (18) is replaced by a complicated Bessel series involving the dc current $I_0(V_0 + n\hbar\omega/e)$ evaluated at voltages displaced from the dc bias point by integral multiples of $\hbar\omega/e$. It is this feature of the quantum theory that accounts for the dissipative absorption or emission of particular numbers of photons in the tunneling process. This effect becomes an unimportant consideration at sufficiently low frequencies. When the photon energy is small on the scale of the dc I-V nonlinearity, the quantum-reactive terms may be shown to vanish, and the series expressions for the harmonic current components reduce to their classical values.

The predictions of the theory in the quantum regime are very different, however. Setting m=0 in eq. (16) immediately yields the Tien-Gordon (1963) result for the dc current given by eq. (2). Similarly, the dissipative current component induced at the applied frequency is found to be:

$$2a_1 = \sum_{n=-\infty}^{\infty} J_n(\alpha)[J_{n+1}(\alpha) + J_{n-1}(\alpha)]I_0(V_0 + n\hbar\omega/e) \tag{19}$$

Expanding this expression in the small signal limit $\alpha << 1$ then yields the finite difference form for the ac conductivity quoted in eq. (5). These relatively simple consequencies of the photon-assisted tunneling theory were utilized in the preceding section to demonstrate that nonlinear tunnel junctions will respond to individual quanta at high frequencies.

4. A MICROSCOPIC MODEL FOR QUASIPARTICLE TUNNEL JUNCTIONS

The basic dynamics of single-particle tunnel junctions, summarized in eqs. (9)-(12), will be derived in this section using a simplified model. This derivation is not essential to the use of these results in describing quasiparticle mixer performance, and might therefore be omitted by those not concerned with the underlying device physics. Certain theoretical nuances will be sacrificed here to achieve greater insight into the essential processes. The results, however, are identical to those obtained using more elaborate and general formulations of the problem; some comments will be offered as to why this should be so.

Consider two electron states ψ_i and ϕ_j on opposite sides of a tunnel junction. These are taken to be energy eigenstates of an unperturbed Hamiltonian in the absence of coupling through the barrier:

$$H_0 \psi_i = E_i \psi_i \tag{20}$$

$$H_0 \phi_j = E_j \phi_j$$

Any solution to the time-dependent Schrodinger equation:

$$i\hbar \frac{\partial\psi}{\partial} = H_0\psi \tag{21}$$

for this two-level system can then be written in the form:

$$\psi(t) = c_i \psi_i \, e^{-iE_i t/\hbar} + c_j \phi_j \, e^{-iE_j t/\hbar} \tag{22}$$

where $c_{i,j}$ are arbitrary complex constants, subject to appropriate normalization.

Now consider the effect of a small perturbation, H_1, which allows for tunneling between these two states:

$$\begin{aligned} H_1 \psi_i &= T_{ji} \phi_j \\ H_1 \phi_j &= T_{ij} \psi_i \end{aligned} \tag{23}$$

The tunneling matrix elements $T_{ji} = T^*_{ij}$ are taken to be sufficiently small that the transfer of electrons across the barrier may be treated to lowest order in the coupling. In general, any solution to the complete Schrodinger equation:

$$i\hbar \frac{\partial \psi}{\partial t} = (H_0 + H_1)\psi \tag{24}$$

can still be written in the form of eq. (22), but now the amplitudes $c_i(t)$ and $c_j(t)$ will be functions of time. Substituting eq. (22) into eq. (24) then yields the following equations of motion for these coefficients:

$$\begin{aligned} i\hbar \frac{dc_i}{dt} &= T_{ij} c_j(t) \, e^{+i(E_i-E_j)t/\hbar} \\ i\hbar \frac{dc_j}{dt} &= T_{ji} c_i(t) \, e^{-i(E_i-E_j)t/\hbar} \end{aligned} \tag{25}$$

Let us assume that an electron is initially in state ψ_i on the left hand side of the junction at t=0, and that the coupling is small enough so that the amplitude for being in state ϕ_j on the right can be computed to lowest order. Approximating $c_i(t) \simeq 1$ gives:

$$\begin{aligned} c_j(t) &= \frac{T_{ji}}{i\hbar} \int_0^t dt' e^{-i(E_i-E_j)t'/\hbar} \\ &= T_{ji} \frac{[e^{-i(E_i-E_j)t/\hbar} - 1]}{(E_i - E_j)} \end{aligned} \tag{26}$$

Note that this amplitude for tunneling through the barrier will be extremely small and rapidly oscillating unless the initial and final states are very close in energy. This is, of course, a direct consequence of the Heisenberg uncertainty principle.

The probability that the electron originally in state ψ_i on the left has tunneled into state ϕ_j on the right is given by the absolute square of eq. (26):

$$|c_j(t)|^2 = \frac{|T_{ij}|^2}{\hbar^2} \frac{\sin^2(\Delta\omega_{ij}t/2)}{(\Delta\omega_{ij}/2)^2} \tag{27}$$

where:

$$\Delta\omega_{ij} \equiv (E_i - E_j)/\hbar \tag{28}$$

There will actually be a continuous density of final energy states $\rho_f(E_j)$ on the right hand side of the junction into which this electron can tunnel. The total transition rate out of the initial state ψ_i may be calculated by summing the rate of change of the tunneling probability over all final states in the limit $t \to \infty$:

$$\begin{aligned} W_{i\to f} &= \lim_{t\to\infty} \int_{-\infty}^{\infty} dE_j \; \rho_f(E_j) \frac{d}{dt} |c_j(t)|^2 \\ &= \frac{2|T_{ij}|^2}{\hbar} \rho_f(E_i) \int_{-\infty}^{\infty} d(\Delta\omega_{ij}) \frac{\sin(\Delta\omega_{ij}t)}{\Delta\omega_{ij}} \\ &= \frac{2\pi}{\hbar} |T_{ij}|^2 \rho_f(E_i) \end{aligned} \tag{29}$$

The limit $t\to\infty$ is taken in order to eliminate transient effects associated with the origin of time, $t=0$ in this case, at which the coupling was mathematically turned on. In this limit, the function $\sin(\Delta\omega_{ij}t)/\Delta\omega_{ij}$ becomes so sharply peaked about the resonance condition $E_i=E_j$ that both the tunneling matrix element and the density of final states may be evaluated at the initial energy and extracted from the integral.

The final result of eq.(29) is known as Fermi's "Golden Rule", and is widely employed in quantum physics. Here it can be used to calculate the dc current tunneling through the junction. In the presence of a bias potential, all electrons on the left side of the barrier can be considered displaced in energy by eV_0. Conservation of energy then requires that there be transitions only between initial and final states related by $E_i + eV_0 = E_j$. The current flowing from left to right is then computed by summing the transition rate of eq. (29) over all initial and final energy levels:

$$I_{\ell\to r}(V_0) = \frac{2\pi e}{\hbar}\int_{-\infty}^{\infty} dE_i dE_j |T_{ij}|^2 \rho_L(E_i)\rho_R(E_j) \cdot f(E_i)[1 - f(E_j)]\delta(E_i - E_j + eV_0) \tag{30}$$

Here $\rho_{L,R}(E)$ are the densities of states in the left and right hand electrodes, respectively. The Fermi occupation factors:

$$f(E) = \frac{1}{e^{(E-\mu)/kT}+1} \tag{31}$$

account for the statistical probability that the initial state is occupied and the final state empty. By symmetry, the transition probability is the same going from right to left, so that the only change in evaluating the reverse current is in the statistical factors:

$$I_{r\to\ell}(V_0) = \frac{2\pi e}{\hbar}\int_{-\infty}^{\infty} dE_i dE_j |T_{ij}|^2 \rho_L(E_i)\rho_R(E_j) f(E_j)[1 - f(E_i)]\delta(E_i - E_j + eV_0) \tag{32}$$

The dc I-V characteristic is then given by the difference of these two components:

$$I_0(V_0) = \frac{2\pi e}{\hbar}\int_{-\infty}^{\infty} dE_i dE_j |T_{ij}|^2 \rho_L(E_i)\rho_R(E_j) [f(E_i) - f(E_j)]\delta(E_i - E_j + eV_0) \tag{33}$$

The response of the tunnel barrier to a time-dependent applied potential requires only minor modifications to the preceding calculation. If the energy of each state on the left side of the junction is modulated according to:

$$E_i(t) = E_i + eV(t) \tag{34}$$

then the factor $e^{-iE_i t/h}$ in eqs. (25) and (26) must be replaced by:

$$\exp\left\{-\frac{i}{h}\int^{t} dt'[E_i + eV(t')]\right\}$$
$$= e^{-i(E_i+eV_0)t/\hbar}\int_{-\infty}^{\infty} d\omega' W(\omega')e^{-i\omega' t} \qquad (35)$$

In this expression, the phase factor arising from the time-dependent part of the applied potential has been represented by its Fourier transform as defined in eq. (9). Substituting this form in place of $e^{-iE_i t/h}$ into eqs. (25) and (26) then gives to lowest order:

$$\frac{dc_j}{dt} = \frac{T_{ji}}{i\hbar} e^{-i(E_i-E_j+eV_0)t/\hbar}\int_{-\infty}^{\infty} d\omega' W(\omega')e^{-i\omega' t} \qquad (36)$$

and:

$$c_j(t) = \frac{T_{ji}}{\hbar}\int_{-\infty}^{\infty} d\omega' W(\omega')\left\{\frac{e^{-i[(E_i-E_j+eV_0)/\hbar + \omega']t} - 1}{[(E_i-E_j+eV_0)/\hbar + \omega']}\right\} \qquad (37)$$

The rate at which electrons tunnel from a filled initial state ψ_i on the left into an empty state ϕ_j on the right may then be written:

$$\frac{d}{dt}|c_j(t)|^2 = 2\,Re\left\{c_j(t)\frac{dc_j^*}{dt}\right\}$$
$$= Im\int_{-\infty}^{\infty} d\omega' d\omega'' W(\omega')W^*(\omega'')e^{-i(\omega'-\omega'')t}$$
$$\cdot \frac{2|T_{ij}|^2}{\hbar^2}\left\{\frac{e^{+i[(E_i-E_j+eV_0)/\hbar + \omega']t} - 1}{[(E_i-E_j+eV_0)/\hbar + \omega']}\right\} \qquad (38)$$

By symmetry, this is also the rate for electrons tunneling from an occupied state ϕ_j on the right into an unoccupied state ψ_i on the left. The average net current flowing through the barrier is therefore given by:

$$\langle I(t)\rangle = e\int_{-\infty}^{\infty} dE_i dE_j \rho_L(E_i)\rho_R(E_j)[f(e_i) - f(E_j)]$$
$$\cdot \lim_{t\to\infty}\frac{d}{dt}|c_j(t)|^2 \qquad (39)$$

The factor containing the energy dependences in eq. (38) may written in the form:

$$\frac{e^{+i[(E_i-E_j+eV_0)/\hbar + \omega']t} - 1}{[(E_i-E_j+eV_0)/\hbar + \omega']}$$

$$= \frac{-2\sin^2\frac{1}{2}[(E_i-E_j+eV_0)/\hbar + \omega']t + i\sin[(E_i-E_j+eV_0)/\hbar + \omega']t}{[(E_i-E_j+eV_0)/\hbar + \omega']}$$

$$\xrightarrow{t\to\infty} -P\left\{\frac{1}{[(E_i-E_j + eV_0)/h + \omega']}\right\} + i\pi\delta\left\{(E_i-E_j + eV_0)/h+\omega'\right\} \tag{40}$$

In the limit $t\to\infty$, the term $2\sin^2\left\{\frac{1}{2}\Delta\omega t\right\}$ will oscillate so rapidly that it may by replaced by its average value, unity, except at $\Delta\omega=0$ where it vanishes. When integrated over energy, the real component of eq. (40) will therefore become a principal value integral as indicated. The imaginary part is of the form $\sin(\Delta\omega t)/\Delta\omega$. Again, for $t\to\infty$ this factor will be so highly peaked about $\Delta\omega=0$ that it may be replaced by a delta function. Then combining eqs. (38)-(40) finally yields:

$$\begin{aligned}\langle I(t)\rangle = Im\int_{-\infty}^{\infty} d\omega' d\omega'' W(\omega')W^*(\omega'')e^{-i(\omega'-\omega'')t} \\ \cdot \frac{2e}{\hbar^2}\int_{-\infty}^{\infty} dE_i dE_j |T_{ij}|^2 \rho_L(E_i)\rho_R(E_j)[f(E_i) - f(E_j)] \\ \cdot\left\{-P\left[\frac{1}{[(E_i-E_j + eV_0)/\hbar + \omega']}\right. + i\pi\delta(E_i-E_j+eV_0)/\hbar+\omega'\right]\right\}\end{aligned} \tag{41}$$

This form is identical to the result quoted in eq. (10). The imaginary part of the response function $j(V_0 + h\omega'/e)$ may be immediately identified with the dc I-V characteristic of eq. (33):

$$Im\ j(V_0 + \hbar\omega'/e) = I_0(V_0 + \hbar\omega'/e) \tag{42}$$

This is seen to be in agreement with eq. (11).

The real part of the response function may be expressed using eq. (41) in the form:

$$Re\ j(V_0 + \hbar\omega'/e) = P\int_{-\infty}^{\infty}\frac{d\bar{V}}{\pi}\frac{1}{\bar{V} - (V_0 + \hbar\omega'/e)}$$

(Continued)

$$\cdot \frac{2\pi e}{\hbar} \int_{-\infty}^{\infty} eE_i dE_j |T_{ij}|^2 \rho_L(E_i)\rho_R(E_j)$$

$$\cdot [f(E_i) - f(E_j)]\, \delta(E_i - E_j + e\bar{V}) \tag{43}$$

$$= P\int_{-\infty}^{\infty} \frac{d\bar{V}}{\pi} \frac{I_0(\bar{V})}{\bar{V} - (V_0 + \hbar\omega'/e)}$$

This quantity is formally divergent, assuming that $I_0(V)$ approaches ohmic behavior at large bias voltages. The physical response of the junction, however, depends only on differences between this function evaluated at various arguments. As an example, the amplitudes for the reactive current components generated by a sinusoidal potential in eq. (15) may be rewritten using eq. (16) in the form:

$$2b_m = \sum_{n=-\infty}^{\infty} J_n(\alpha)J_{n+m}(\alpha)[I_{KK}(V_0 + n\hbar\omega/e) - I_{KK}(V_0 + (n+m)\hbar\omega/e)] \tag{44}$$

The function $I_{KK}(V)$ defined by eq. (12) is equivalent to the result of eq. (43) for the real part of the response function *Re* j(V), except that the divergence associated with the ohmic conductance at large bias has been removed. The particular choice for eliminating this divergence in eq. (12) is not unique, but it has both conceptual and computation advantages as previously discussed.

The results of eqs. (9)-(12) which characterize the dynamics of a single-particle tunnel junction have been derived in this section using a simplified model. More general formulations of this problem (Werthamer, 1966; Tucker, 1979) are capable of including many-body effects that cannot be described within the system of single-electron energy eigenstates assumed here. The end result is nevertheless the same. Simple manifestations of interaction between the single-particle excitations, such as lifetime broadening, affect the overall densities of states. These appear only in calculating the dc I-V characteristic, so that they are automatically included if the measured dc I-V curve is used in the analysis. For a superconductor, the single-particle excitations of the BCS ground state are linear combinations of electrons and holes. This introduces certain "coherence factors" into the linear response functions for various external probes (Schrieffer, 1964). In the case of single-particle tunneling, however, these "coherence factors" cancel out (Cohen, et al., 1962; Schrieffer, 1964). The dc current can then be calculated directly using the total quasiparticle density of states for a superconductor relative to its normal value:

$$\rho_s(E) = \rho_N(0) \frac{|E|}{(E^2 - \Delta^2)^{\frac{1}{2}}} \tag{45}$$

as illustrated in Fig. 1b.

One final observation on the microscopic analysis of a single-particle tunnel junction may be instructive. Examination of eq. (41) shows that the dissipative component of the response, characterized by *Im* j(V), involves only processes in which energy is conserved through the delta function. This part of j(V) can therefore be directly related to the dc I-V characteristic. The "quantum reactances" of the tunnel junction, however, are expressed in terms of *Re* j(V), and given by the principal value integral. Looking back to eq. (40), these reactive terms may be seen to arise from the rapid quantum-mechanical "sloshing" of electrons between states on opposite sides of the barrier whose energies cannot be matched through the absorption or emission of photons. This behavior is analogous to the well-known ac Josephson effect, in which the quantum "sloshing" of condensed Cooper pairs across a junction produces a macroscopic current at a frequency corresponding to the energy difference at the Fermi level $\omega_J = 2eV_0/\hbar$. Here there is a distribution of single-particle states on both sides of the junction, and their combined oscillatory currents cancel out for a dc bias. The ac response, however, does show these effects at frequencies sufficiently high that energies cannot be matched over significant ranges by the exchange of integral numbers of photons with the applied field. This occurs in the quantum regime, where the dc I-V curve is highly nonlinear on the scale of the photon energy.

5. THE "UNIVERSAL" QUASIPARTICLE MIXER MODEL

A relatively simple model for quasiparticle heterodyne mixers will be presented using the theory summarized in Section 3. A more general and complete development, along with a summary of classical mixer analysis, was the principal subject of a previous paper (Tucker, 1979). Only the concepts and results essential to a practical working knowledge of the theory will be discussed here.

The large signal local oscillator drive is taken to be sinusoidal:

$$V_{LO}(t) = V_{LO} \cos \omega t \tag{46}$$

All higher harmonics are assumed shorted by the junction capacitance. The average current induced by the local oscillator together with the bias voltage V_0 is then determined by equs. (15) and (16). The resulting dc current is given by eq. (2). Since the higher harmonics are shorted, the only other physically relevant components of the current are those at the applied frequency:

$$I_{LO}^{\omega}(t) = I_{LO}' \cos \omega t + I_{LO}'' \sin \omega t \tag{47}$$

where from eq. (16):

$$I_{LO}' = \sum_{n=-\infty}^{\infty} J_n(\alpha)[J_{n+1}(\alpha) + J_{n-1}(\alpha)]I_0(V_0 + n\hbar\omega/e)$$

$$I_{LO}'' = \sum_{n=-\infty}^{\infty} J_n(\alpha)[J_{n+1}(\alpha) - J_{n-1}(\alpha)]I_{KK}(V_0 + n\hbar\omega/e) \tag{48}$$

and $\alpha = eV_{LO}/\hbar\omega$ characterizes the amplitude of the local oscillator waveform.

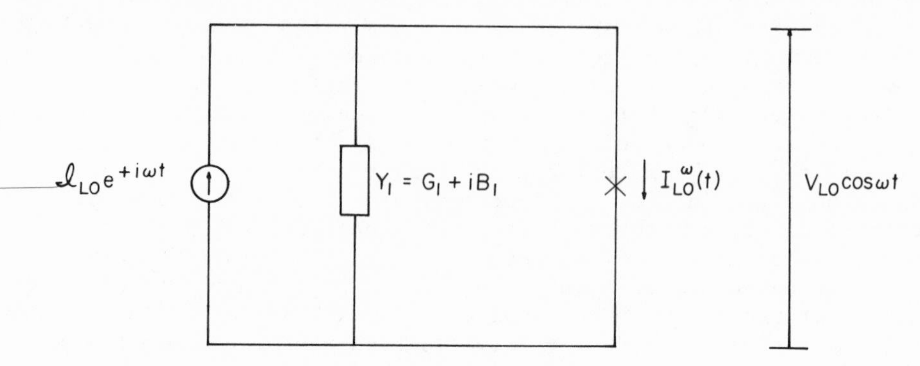

Fig. 2. Equivalent circuit representing the large-signal local oscillator drive a frequency ω for a quasiparticle mixer diode, assuming all higher harmonics to be effectively shorted through the junction capacitance.

The equivalent circuit at this frequency is shown in Fig. 2. The local oscillator is represented as a current generator with an effective source admittance $Y_1 = G_1 + iB_1$ determined by the input waveguide and mounting structure. The imaginary part B_1 will also contain a contribution ωC due to the ordinary junction capacitance. Then according to Fig. 2:

$$\begin{aligned} \mathcal{I}_{LO}(t) &= I_{LO}(t) + Re\left\{Y_1 V_{LO} e^{+i\omega t}\right\} \\ &= (I_{LO}' + G_1 V_{LO})\cos\omega t + (I_{LO}'' - B_1 V_{LO})\sin\omega t \end{aligned} \tag{49}$$

The local oscillator power incident on the mixer is represented by available power from the current source in the equivalent circuit:

$$P_{LO} = \frac{|\mathcal{I}_{LO}|^2}{8\,G_1} = \frac{1}{8G_1}\left[(I_{LO}' + G_1 V_{LO})^2 + (I_{LO}'' - B_1 V_{LO})^2\right] \tag{50}$$

This expression, together with eq. (48), gives the local oscillator power in terms of V_{LO} and the effective source admittance at the pump frequency. In most experiments, the value of P_{LO} is known quite accurately, and values for G_1 and B_1 can be inferred. A systematic method for experimentally determining this source admittance in SIS mixers is described by Shen (1981). The large signal problem, within the context of this model, then consists of an iterative solution of eq. (50) to determine the amplitude of the local oscillator drive V_{LO} in terms of the applied power and known source admittance.

The small-signal mixing properties of the tunnel junction may be calculated once the amplitude for the local oscillator waveform has been established. A schematic diagram for a heterodyne mixer is shown in Fig. 3. The strong pumping at frequency ω mixes the output ω_0 with all sidebands:

$$\omega_m = m\omega + \omega_0 \qquad m=0, \pm 1, \pm 2, \ldots \tag{51}$$

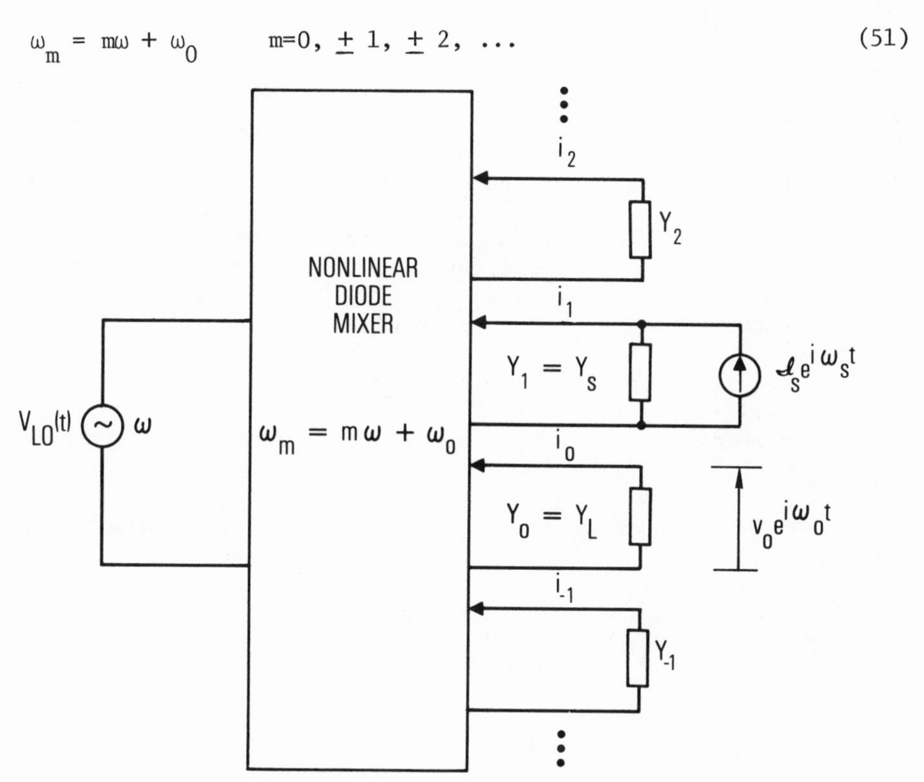

Fig. 3. Schematic diagram for a nonlinear diode heterodyne receiver. The various sideband frequency terminations are represented as admittances. This illustration shows a current source at the signal input frequency for a fundamental heterodyne receiver, together with the voltage component mixed down across the output frequency load.

The voltages and currents at the sideband frequencies may be represented in the form:

$$V_{SIG}(t) = Re\sum_{m=-\infty}^{\infty} v_m\, e^{+i\omega_m t}$$

$$i_{SIG}(t) = Re\sum_{m=-\infty}^{\infty} i_m\, e^{+i\omega_m t} \tag{52}$$

These voltage and current components will be linearly related for small signals by an admittance matrix of the form

$$i_m = \sum_{m'} Y_{mm'}\, v_{m'} \tag{53}$$

The values of the admittance matrix elements are determined by the strength of the local oscillator drive and by the nonlinear dc I-V characteristic of the junction. They are calculated by including the sideband frequency voltage components of eq. (52) along with the local oscillator waveform when determining the time-dependent phase factor in eq. (9). Only terms linear in the small signal potentials are retained in the Fourier transform $W(\omega')$ of this phase factor, and the result is then utilized in eq. (10) to determine the total current through the junction. Examining the average current induced at the various sideband frequencies then yields explicit expressions for the admittance matrix elements. This calculation is lengthy but straightforward. The results for the fundamental and harmonic mixer models to be discussed in this section are tabulated in Appendices A and B, respectively.

The effective termination at each sideband frequency in Fig. 3 is represented by a complex admittance Y_m. A set of current generators $\{I_m\}$ placed at every port ω_m of the mixer will then produce small signal current and voltage components across the junction satisfying:

$$\begin{aligned} I_m &= i_m + Y_m v_m \\ &= \sum_{m'} [Y_{mm'} + Y_m\, \delta_{m,m'}]v_{m'} \end{aligned} \tag{54}$$

Inverting these equations, the signal voltages produced by an arbitrary set of current generators are given by:

$$v_m = \sum_{m'} Z_{mm'}\, I_{m'} \tag{55}$$

where in matrix notation:

$$||Z_{mm'}|| = ||Y_{mm'} + Y_m \delta_{m,m'}||^{-1} \tag{56}$$

In particular, the output voltage may be written in the form:

$$V_0 = Z_{00} \sum_m \lambda_{0m} \mathcal{I}_m \tag{57}$$

where the quantity:

$$\lambda_{0m} = Z_{0m}/Z_{00} \tag{58}$$

does not depend on the output frequency load termination $Y_0 = Y_L$.

Once the matrix inversion of eq. (56) has been performed, the mixer conversion efficiency is easily obtained. The power available at the signal input in Fig. 3 is:

$$P_{IN} = |\mathcal{I}_S|^2/8\ G_S \tag{59}$$

The power down-converted and delivered into the load may be written:

$$\begin{aligned} P_{OUT} &= \frac{1}{2} G_L |v_0|^2 \\ &= \frac{1}{2} G_L |Z_{0s}|^2 |\mathcal{I}_S|^2 \end{aligned} \tag{60}$$

where s represents the index of the signal input port.

The conversion loss is then:

$$\begin{aligned} L_C &= P_{IN}/P_{OUT} \\ &= \frac{1}{4G_S G_L |Z_{0s}|^2} \end{aligned} \tag{61}$$

In these expressions, G_S and G_L represent the real parts of the source and load admittances Y_S and Y_L, respectively.

The noise properties of a heterodyne mixer require lengthy analysis, and only the results will be quoted here. The contribution to the mixer noise temperature generated by the combination of dc bias and local oscillator drive may be written in the form:

$$kT_M^{LO} = \frac{1}{4G_S|\lambda_{0s}|^2} \sum_{m,m'} \lambda_{0m}\lambda^*_{0m'} H_{mm'} \tag{62}$$

Here $H_{mm'}$ represents a current correlation matrix that relates fluctuations at the various sideband frequencies. These matrix elements are functions only of the local oscillator waveform and the dc I-V characteristic for a quasiparticle tunnel junction, and the results for the fundamental and harmonic mixer models are included in Appendices A and B.

The fundamental heterodyne mixer model includes only the signal, output, and image frequencies ω_1, ω_0, ω_{-1}, respectively. The local oscillator waveform is sinusoidal according to eq. (46), and all higher harmonics and their sidebands are assumed shorted by the junction capacitance. The output frequency is taken to be low compared to the pump $\omega_0 << \omega$, so that the approximation $\omega_1 \simeq \omega \simeq -\omega_{-1}$ may be used in computing the elements of the small signal admittance and current correlation matrices. Both of these will be 3x3 in this case, and explicit expressions for the various components are tabulated in Appendix A. The signal source admittance is represented by $Y_S = Y_1 = Y^*_{-1}$ in the limit of low output frequency, and the load admittance $Y_0 = G_L$ will be purely real. The following set of dimensionless parameters then prove convenient for expressing the predictions of this model:

$$L_0 = \frac{2G_{10}}{G_{01}} \qquad \eta = \frac{2G_{01}\,G_{10}}{G_{00}(G_{11} + G_{1-1})}$$

$$g_S = \frac{G_S}{G_{11} + G_{1-1}} \qquad g_L = \frac{G_L}{G_{00}}$$

$$\xi = \frac{G_{11} - G_{1-1}}{G_{11} + G_{1-1}} \qquad \beta = \frac{B_{10}}{G_{10}} \tag{63}$$

$$\gamma = \frac{B_{1-1}}{G_{11} + G_{1-1}} \qquad b_S = \frac{B_{11} + B_S}{G_{11} + G_{1-1}}$$

Inversion of the 3x3 augmented admittance matrix of eq. (56) leads to the following result for mixer conversion loss:

$$L_c = \frac{1}{4G_SG_L|Z_{01}|^2}$$

$$= \frac{L_0}{4\eta g_S g_L} \frac{[(\xi+g_S)(1+g_S) + (b_S^2-\gamma^2)]^2}{[(\xi+g_S)^2 + (b_S-\gamma)^2]} (g_L + g_L^0)^2 \tag{64}$$

In this expression, the output conductance of the mixer:

$$G_L^0 = \frac{1}{Z_{00}} - G_L \tag{65}$$

is represented by:

$$g_L^0 = \frac{G_L^0}{G_{00}} = 1 - \eta \frac{[(\xi+g_S) + \beta(b_S-\gamma)]}{[(\xi+g_S)(1+g_S) + (b_S^2-\gamma^2)]} \tag{66}$$

This quantity is simply the dynamic resistance or slope of the dc I-V curve in the presence of the local oscillator drive, since the limit of low output frequency has been taken in constructing the model.

In the original published version (Tucker, 1980), the reactive part of the signal input termination B_S was assumed to tune out the diagonal quantum reactances along with the ordinary junction capacitance, so that $b_S=0$. The generalization to $b_S \neq 0$ was used by Feldman (1981) in his discussion of this theory, and has been included in the results quoted here.

The contribution to the mixer temperature from local oscillator shot noise in this model may be computed in terms of the current correlation matrix elements listed in Appendix A and written in the form:

$$kT_M^{LO} = \frac{1}{4G_S|\lambda_{01}|^2} \Big\{ H_{00} + 2(\lambda_{01} + \lambda_{01}^*)H_{10} + (\lambda_{01}^2 + \lambda_{01}^{*2})H_{1-1} + 2|\lambda_{01}|^2 H_{11} \Big\} \tag{67}$$

where

$$\lambda_{01} = \frac{-G_{01}}{(G_{11}+G_{1-1})} \frac{[(\xi+g_S) - i(b_S-\gamma)]}{[(\xi+g_S)(1+g_S) + (b_S^2-\gamma^2)]} \tag{68}$$

Equations (63)-(68), together with the results of Appendix A, characterize the conversion efficiency and local oscillator component of noise temperature for a fundamental quasiparticle mixer within the context of this 3-port model. The output frequency has been assumed small compared to the local oscillator, $\omega_0 << \omega$, as it usually will be in practice. The major assumption made here is that current components at the higher harmonics 2ω, 3ω, ... and their sidebands are effectively shorted through the junction capacitance. This allows a simple self-consistent solution, given by eqs. (46)-(50), for the local oscillator waveform. If the higher harmonics are not rigorously shorted, the waveform of eq. (46) should contain components at all of these frequencies. A self-consistent solution to the large signal problem for the pumped nonlinear mixer then becomes much more difficult. This problem can be solved by the successive reflection technique of Kerr (1975), utilizing the time-domain formulation for the junction response given in Appendix C. The programming effort required, however, would prove very substantial. In addition, experimental results on SIS quasiparticle mixers indicate that a relatively large junction capacitance, in the range $\omega C R_N \simeq 5$, is beneficial in order to suppress harmonic conversion into unwanted sidebands (Kerr et al., 1981, Smith et al., 1981, McGrath et al., 1981). A capacitance of this magnitude may still be conveniently resonated out at the signal frequency without adversely affecting the bandwidth. Thus for both technical and computational reasons, the 3-port model summarized here represents the simplest available guide to understanding quasiparticle mixer performance in most practical applications.

Harmonic mixing in quasiparticle tunnel barriers has received little attention thus far. This is primarily due to the use of SIS junctions in the first observations of non-classical phenomena. These SIS diodes exhibit a stronger nonlinearity at the gap voltage than a super-Schottky or superconductor- insulator-normal metal (SIN) junction, and so photon-assisted tunneling effects may be seen at lower frequencies. Whenever the applied voltage in an SIS mixer is swept into the neighborhood of V=0, however, large noise associated with the Josephson pair current is observed (Richards et al., 1979; Dolan et al., 1979; Rudner and Claeson, 1979; Shen et al., 1980; Rudner et al., 1981 a,b). This is caused by the inability of the capacitance to continue shorting the Josephson currents at frequencies $\omega_J = 2\ eV/\hbar$ when the voltage becomes sufficiently small. An SIN diode, on the other hand, would have no Josephson pair current, and could therefore be operated as a harmonic mixer at zero dc bias. Such devices may eventually prove important in extending detection by photon-assisted tunneling into the frequency region beyond 100-200 GHz. At present, the fabrication technology for sufficiently small area and high current density junctions has been developed only for SIS diodes as a by-product of research on Josephson computer circuits (Greiner et al., 1980; Huang et al., 1980). Nevertheless, the success achieved with these devices as quasiparticle mixers may be expected to stimulate future efforts to fabricate SIN or super-Schottky diodes suitable for submillimeter wave detector and mixer applications.

The quasiparticle mixer model summarized in this section may easily adapted to the description of harmonic mixing under certain conditions. The dc I-V characteristic for an SIN or super-Schottky diode has approximate inversion symmetry, $I_0(-V) = -I_0(V)$, about zero bias. The sinusoidal local oscillator waveform of eq. (46) will therefore induce current components only at odd multiples of the pump frequency. Here it will be assumed that the capacitance effectively shorts the junction at 3ω and all higher harmonics. The large signal response to the local oscillator drive is then simply determined by eqs. (46)-(50) as before, except that here the dc bias voltage is $V_0 = 0$. The signal input for the harmonic mixer correspond to the port $\omega_2 = 2\omega + \omega_0$ in Fig. 3, and the effective source admittance $Y_S = Y_2$ now represents the embedding network at this frequency. The small signal admittance and current correlation matrices will both be 5x5 in this case. At zero dc bias, however, only the sidebands at $m=\pm 2$ are coupled to the output by symmetry. This effectively reduces the problem to a 3-port model of the same form previously encountered in the analysis of fundamental mixing. The matrix elements which determine the properties of the harmonic mixer are tabulated in Appendix B, again assuming the limit $\omega_0 \ll \omega$ of small output frequency. Equations (63)-(68) for conversion loss, output impedance, and noise temperature may then be taken over directly by making the substitution $1 \to 2$ in every subscripted quantity. These results will, in this case, characterize harmonic quasiparticle mixing in a symmetric tunnel junction at zero dc bias.

The quasiparticle mixer model summarized in this section is, in a sense, universal. It describes both fundamental and harmonic mixing in single-particle tunnel junctions under conditions that will be closely approximated in most practical applications. Any generalization will require much more elaborate calculations if it is to remain self-consistent, unless the effects under consideration can be treated as small perturbations. There is also some advantage in constructing an algebraically soluble model, and important insights have already been achieved by Feldman (1981) through analysis of its structure. Detailed predictions, however, require computer simulation. Some of the remarkable conclusions that come out of this theory are discussed in the following section.

6. CONVERSION GAIN AND NEGATIVE RESISTANCE

The most striking prediction of the quasiparticle mixer theory is conversion gain: simultaneous amplification along with frequency down-conversion of the incoming signal. This conversion gain is accompanied by large and even negative values of low frequency dynamic resistance in the pumped junction. The available gain actually becomes infinite for negative output impedances. These are new physical phenomena, predicted by the quasiparticle mixer model and now experimentally observed (Shen et al., 1980; Rudner et al., 1981 a,b; Kerr et al., 1981; Smith et al., 1981; McGrath et al., 1981). The other major prediction of the theory is that quantum limited mixer noise temperatures are possible under a broad range of conditions. This limit has been approached in experiments on individual SIS diodes (Shen et al., 1980; Smith et al., 1981; McGrath et al., 1981). The realization of conversion gain in this context is expected to yield practical micro- and millimeter wave receivers in the near future whose overall system temperatures approach the quantum noise limit.

A number of computer simulations, based on the model of the preceeding section, have been constructed for the purpose of comparing theory with experiment. Generally speaking, such efforts have been very successful. Conversion efficiencies of SIS quasiparticle mixers have been quantitatively interpreted, except in cases where the model assumptions are grossly violated, and the phenomena of conversion gain and negative resistance appear precisely as predicted. The analysis of a particular experiment normally begins by inputting the measured dc I-V characteristic $I_0(V)$. If this function is represented as piecewise-linear, the Kramers-Kronig transform of eq. (12) may be simply calculated. The detailed results obtained from the model in this way for individual experiments may be found in the literature cited above, and extensive model calculations are reported by Shen and Richards (1981). Here the predictions for an idealized SIS mixer will be discussed as an introduction to the gain phenomenon.

The original computer simulation of an SIS mixer (Tucker, 1980) utilized the quasiparticle current response function calculated by Werthamer (1966) for an ideal junction in the limit of zero temperature. The real and imaginary parts of j(V) for this case are illustrated in Fig. 4. The dc quasiparticle current vanishes out to the gap voltage $eV_0 = 2\Delta$, at which point there is a discontinuous onset of current leading to ohmic behavior at large bias. The Kramers-Kronig transform contains a mild logarithmic singularity at the gap, reflecting this infinitely steep current rise for the ideal junction.

Results of the quasiparticle mixer simulation for an ideal SIS diode will be summarized here under the following set of conditions. The signal frequency is taken to be $\hbar\omega = 0.40\Delta$, which corresponds to approximately 115 GHz for a typical Pb-alloy junction. The dc bias and amplitude of the local oscillator waveform are given by $eV_0 = 1.80\Delta$ and $eV_{LO} = 0.50\Delta$ respectively, and are held constant throughout the calculation. Finally, it was assumed that the diagonal quantum reactances $B_{11} = -B_{-1-1}$ are tuned out along the ordinary junction capacitance at the signal frequency, so that the parameter $b_s = 0$. This choise has virtually no impact on the results.

The performance of the diode in photon-assisted tunneling theory is determined by the values of the response function in Fig. 4 at integral multiples of the photon energy $(V_0 + n\hbar\omega/e)$ above and below the dc bias voltage. At a frequency $\hbar\omega = 0.40\Delta$, the gap edge in Fig. 4 lies half-way between $eV_0 = 1.80\Delta$ and $e(V_0 + \hbar\omega/e) = 2.20\Delta$; and the separation here is sufficient that the results should

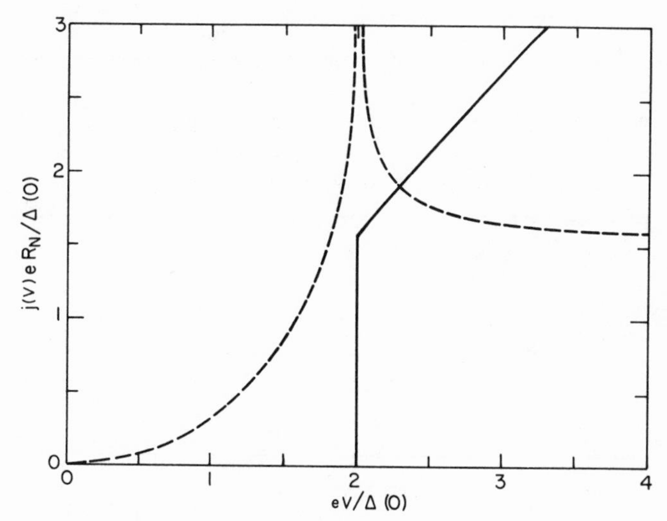

Fig. 4. Quasiparticle current response function for an ideal SIS junction between identical superconductors in the limit $kT \ll \Delta$ of low temperatures. The solid line represents the dc I-V characteristic $Im\ j(V) = I_0(V)$, and the dashed curve is its Kramer-Kronig transform $Re\ j(V) = I_{KK}(V)$.

be relatively insensitive to a small rounding of the dc I-V curve in a real diode. This SIS mixer is, in effect, dc biased in the middle of the first photon step below the gap as illustrated in Fig. 1, and the photon energy here is large compared to the voltage scale of the current onset.

The conversion loss expression of eq. (64) may be written in the form:

$$L_C = L_C^0 \frac{(R_L + R_L^0)^2}{4R_L^0 R_L} \tag{69}$$

Here L_C^0 represents the conversion loss evaluated for $g_L = g_L^0$. The low frequency dynamic resistance, or output impedance, of the pumped diode is given according to eq. (66) by:

$$R_L^0 = (G_L^0)^{-1} = (G_{00}g_L^0)^{-1} \tag{70}$$

The functional dependence of the conversion loss on the load resistance:

$$R_L = (G_L)^{-1} = (G_{00}g_L)^{-1} \tag{71}$$

is therefore seen to be given by a simple mismatch relation.

The computed values of the parameters L_C^0 and R_L^0 under the particular set of conditions described above are shown in Fig. 5a for a wide range of effective source impedance R_S. When the low

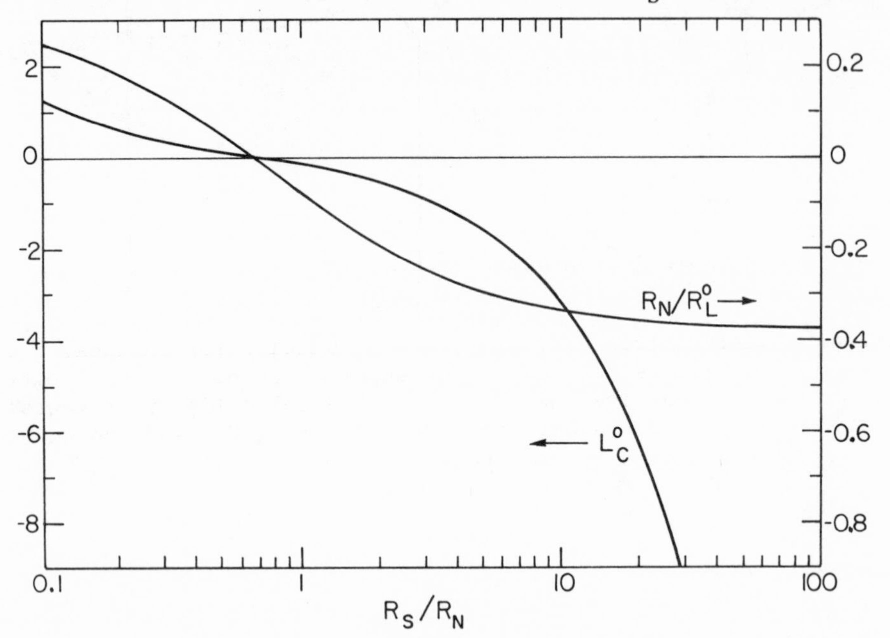

Fig. 5a. Computed results of the SIS heterodyne receiver simulation in the limit $kT \ll \Delta$ for $\hbar\omega = 0.4\Delta$, $eV_0 = 1.8\Delta$, and $eV_{LO} = 0.5\Delta$. (a) The parameters L_C^0 and R_L^0 appearing in eq. (69).

frequency dynamic resistance $R_L^0 > 0$ is positive, the value of L_C^0 represents the minimum conversion loss that can be achieved for optimal matching $R_L = R_L^0$ of the mixer to the load. As the source impedance in Fig. 5a is increased, the minimum conversion loss is seen to decrease below unity while the output resistance becomes large compared to the normal state value R_N. Conversion gain becomes possible for $L_C^0 < 1$, and the available gain continues to increase without limit as the slope $1/R_L^0$ on the first photon step below the gap voltage in Fig. 1a approaches the horizontal. At the point where R_L^0 becomes infinite, the available conversion gain becomes infinite as well, and it remains infinite as the dynamic resistance of the pumped junction goes negative. This may be seen in eq. (69), where the conversion loss can be reduced to zero as the load impedance approaches $R_L = -R_L^0$ for negative values. A smooth transition between regions of conversion loss, gain, and infinite available gain is thus predicted to occur as the dynamic resistance becomes large, infinite, and then negative.

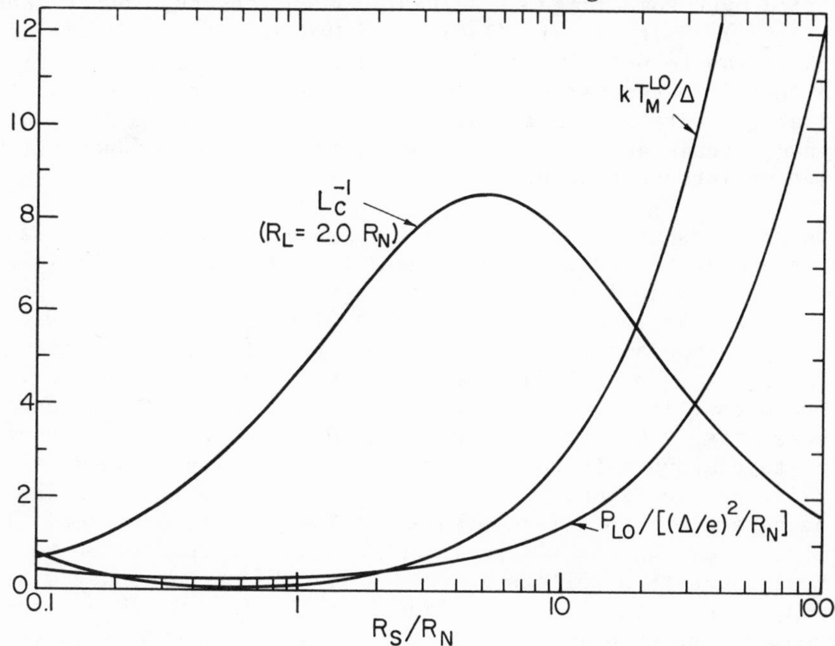

Fig. 5b. Computed results of the SIS heterodyne receiver simulation in the limit $kT \ll \Delta$ for $\hbar\omega = 0.4\Delta$, $eV_0 = 1.8\Delta$, and $eV_{LO} = 0.5\Delta$. (b) The conversion efficiency L_C^{-1} for $R_L = 2.0\ R_N$, the local oscillator contribution T_M^{LO} to the mixer noise temperature, and the total incident power P_{LO}, all as functions of the effective source impedance R_S seen by the junction at the signal frequency.

Figure 5b shows the computed conversion gain L_C^{-1} for this example assuming a constant load impedance $R_L = 2.0\ R_N$. In this case, $R_L < |R_L^0|$ over the entire range of negative resistance, so that the conversion efficiency will be finite and continuous everywhere. The total applied local oscillator power and mixer noise temperature, computed using eqs. (50) and (67) respectively, are also indicated in Fig. 5b. Large values of conversion gain are seen to be predicted in regions where the local oscillator contribution to the mixer noise temperature is less than or comparable to the quantum limited value $kT_M \simeq \hbar\omega = 0.40\Delta$ at this frequency.

The simulation discussed here was carried out holding the dc bias voltage V_0 and the amplitude of the local oscillator waveform V_{LO} constant, while varying the source impedance R_S seen by the diode at the signal frequency. Experiments are usually performed in the reverse manner. The source impedance is adjusted using the mixer mount for optimum performance at a single operating point, and is then held constant as the dc bias and local oscillator power are varied over some range of values. Recent observations by Kerr et al. (1981), Smith et al. (1981), and McGrath et al. (1981) of the transition to negative differential resistance and infinite available gain nevertheless demonstrate precisely the behavior described above. The available gain was found to increase along with the dynamic resistance, and to become infinite at the onset of the negative resistance region.

In these experiments, no singular behavior in the mixer signal was observed at the transition to negative resistance. The measured output power could, in fact, be characterized by a constant current generator, proportional to the incoming signal, in parallel with a large output impedance R_L^0. This may be appreciated by examining the expression given in eq. (64) for the conversion loss. The region of large output impedance, either positive or negative, corresponds to values of $g_L^0 \simeq 0$. Small changes in local oscillator power or bias voltage about this point will have a major impact on $R_L^0 = (G_{00}g_L^0)^{-1}$, but would not be expected to significantly alter the remaining parameters in this expression. The equivalent circuit interpretation of the mixer's performance near the transition to negative resistance then follows immediately. The output power will, according to eq. (64), be proportional to the load impedance and independent of the dynamic resistance R_L^0 when this quantity is very large.

7. COMMENTS ON SERIES ARRAYS

The work of the group at Chalmers University in Sweden (Rudner et al., 1981 a,b) has been very influential in establishing the potential of series arrays for quasiparticle mixer applications. Their results demonstrate that a series of SIS junctions can, under suitable conditions, function as a single mixer element with conversion efficiencies comparable to those achieved with individual diodes. The voltage scale of the dc I-V characteristic for an N element array of identical junctions is, of course, expanded by a factor N. The steep current rise at the gap voltage will thus appear at $eV_0 = 2N\Delta$. The widths of the induced photon-assisted tunneling steps, illustrated in Fig. 1a, will also be increased to $N\hbar\omega/e$. An accurate quantitative interpretation of the conversion efficiencies observed in these experiments was, in fact, obtained using the quasiparticle mixer model modified by replacing $\hbar\omega/e$ everywhere in the theory with $N\hbar\omega/e$.

The use of series arrays has been shown to have several potential advantages, particularly for high frequency applications. The nominal roll-off frequency for an SIS diode, determined by $\omega CR_N \simeq 1$, is proportional to the critical current density of the junction and independent of its area. For operation above about 100 GHz, values of $j_c \sim 10^4$ A cm^{-2} are required (Harris and Hamilton, 1978). Such high current densities, however, imply that the area of a single diode must be of submicron dimensions in order to realize a normal state resistance level $R_N \sim 50\Omega$ required for efficient impedance matching. This fabrication constraint can be relaxed by using a series array of larger area junctions to achieve the same total impedance. Larger areas also facilitate the magnetic biasing normally utilized with SIS junctions to help suppress the effects of Josephson pair currents. The magnetic field strengths needed to insert a flux quantum into each junction can, in this way, be kept sufficiently small that the quasiparticle portion of the dc I-V characteristic is not adversely affected.

There is another potential advantage of series arrays that is perhaps less evident. Although single junction devices can approach quantum limited sensitivity, their dynamic range is very small. Substituting an N element array of equivalent total impedance is expected to increase the dynamic range of a quasiparticle mixer by a factor of N^2, according to the following argument (Feldman and Tucker, 1980). Large conversion efficiency is intimately related to the detailed shape of the step structure induced on the dc I-V curve by the local oscillator, as discussed in the preceding section. The optimum applied local oscillator power will correspond to a particular value of the parameter $\alpha = eV_{LO}/\hbar\omega$ for a single junction mixer. The same step structure for N identical junctions in series will then require an amplitude NV_{LO} across the array. Since the total resistance of the array must be comparable to that of the single junction for proper impedance matching, this scaling requirement for the voltage implies that the optimized local oscillator power will increase as N^2. The dynamic range over which the mixer responds linearly to the incoming signal will represent a small fraction of this applied local oscillator power, and so it too should scale as N^2 for a series array.

The conversion efficiency on the nth photon step below the gap voltage has been noted to peak for values of $\alpha = eV_{LO}/N\hbar\omega$ near the first maximum of $J_n(\alpha)$ in all SIS mixer experiments to date, independent of the number of junctions and the frequency involved (Rudner et al., 1981 b). The optimized local oscillator power should then be roughly proportional to $\omega^2 N^2/R_N$, and this relation does appear to explain the observed behavior. The ratio is approximately 100, for example, between the Chalmers experiments on 6-element arrays at 73 GHz (Rudner et al., 1981 a,b) and Berkeley measurements on comparable single junction mixers at 36 GHz (Shen et al., 1980). The observed local oscillator powers were close to 1μW and 10nW, respectively, and the optimized power for the different arrays at 73 GHz was found to be inversely proportional to their normal state resistances. Experimental data taken under quite different conditions thus supports the basic validity of the argument summarized here.

The dynamic range and required local oscillator power for a quasiparticle mixer are therefore expected to increase as N^2 when a single junction is replaced by an identical N element series array of equal total impedance, and this effect may eventually find important technical applications. The conversion efficiency of the mixer will remain unchanged by such a substitution. This may be appreciated by noting that the admittance matrix elements tabulated in Appendices A and B are invariant under the following transformation:

$$\begin{aligned} &I_0(V) \to N\, I_0(V/N) \\ &V_0,\ V_{LO} \to N\, V_0,\ N\, V_{LO} \qquad (72) \\ &h\omega/e \to Nh\omega/e \end{aligned}$$

For a series array of N identical junctions, the applied voltage will be equally divided across the individual elements. The area of each must therefore be scaled by the factor N with respect to the single junction in order to preserve the same total impedance. Both the current and the voltage scales are thus expanded by N for the equivalent series array. If the measured array dc I-V characteristic is used as an input to the quasiparticle mixer model, the only modification required to compute the admittance matrix is the substitution $\hbar\omega/e \to N\hbar\omega/e$ as noted by Rudner, et al. (1981 a,b). The conversion efficiency for the array will be indentical to that of the equivalent single junction mixer characterized by the transformation of eq. (72), where the values for dc bias and local oscillator drive are also appropriately scaled, while the local oscillator power and dynamic range are increased by a factor N^2 (Feldman and Rudner, 1981; Smith and Richards, 1981; Woody, 1981).

The substitution of $\hbar\omega/e \to N\hbar\omega/e$ into the analysis for an N element array is simply a consequence of the voltage division between identical elements in series. The scaling of the conversion properties discussed above may be formally derived from photon-assisted tunneling theory on the basis of this assumption. These arguments are quite lengthy, however, and will not be presented here. Instead, a few comments are offered that will hopefully illustrate the physical nature of the response in series arrays.

A most important observation is that the appearance of $N\hbar\omega/e$ in the theory for an array does not imply that photons are somehow being simulataneously absorbed in all N junctions. Replacing $\hbar\omega/e \to N\hbar\omega/e$ in equ. (6) yields a limiting value for the current responsivity of an N element direct detector:

$$R_i = \frac{e}{N\hbar\omega}, \text{ Quantum limit} \tag{73}$$

The proper physical interpretation of this result is that the absorption of an individual photon results in the tunneling of one additional electron across a single junction of the array, and that this event represents a current pulse:

$$\delta i(t) = \frac{e}{N}\,\delta(t-t_0) \tag{74}$$

through the entire structure. The presence of non-zero capacitance insures that charges do not build up unevenly along the array, and that the absorption of photons in indivudal junctions will thus be statistically independent.

The noise properties of series arrays are directly related to this interpretation. The current pulse of eq. (74) may be Fourier transformed according to:

$$\delta i(t) = \int_{-\infty}^{\infty} d\omega\ \delta i(\omega) e^{i\omega t} \tag{75}$$

where:

$$\delta i(\omega) = \frac{e}{2\pi N}\, e^{-i\omega t_0} \tag{76}$$

Filtered to a bandwidth B about frequency ω, the time-dependent current due to the pulse at t_0 becomes:

$$\delta i_\omega(t) = \int_{\omega-\pi B}^{\omega+\pi B} d\omega' [\delta i(\omega') e^{i\omega' t} + \delta i(-\omega') e^{-i\omega' t}] \tag{77}$$

For a constant dc potential applied across the array, the total contribution to the square of the current in this frequency band is:

$$\int_{-\infty}^{\infty} dt[\delta i_\omega(t)]^2 = 8\pi^2 B|\delta i(\omega)|^2 \tag{78}$$

Assuming the dc bias voltage to be sufficiently large that forward tunneling pulses predominate, the average number of these events per unit time is given by $I_0(V_0)$ divided by the current e/N per pulse:

$$\bar{N} = N\, I_{dc}(V_0)/e \tag{79}$$

The total mean square noise current for the array may then be determined by combining eqs. (76), (78), and (79) to yield:

$$\langle I_n^2 \rangle = 8\pi^2 B |\delta i(\omega)|^2 \overline{N}$$
$$= \frac{2eI_0(V_0)B}{N} \tag{80}$$

The shot noise produced by a given value of dc current is therefore seen to be reduced by a factor N for the series array.

In comparing a series array with a single junction of equal total impedance, however, it should be recognized that the scaling relations of eq. (72) imply that the dc current must be increased by a factor N for comparable operation. The shot noise of eq. (80) will therefore wind up being the same for the array as for an equivalent single junction when both are biased at the same impedance level. This result may be shown to hold also for the noise properties of a mixer with applied local oscillator drive (Feldman and Rudner, 1981; Smith and Richards, 1981; Woody, 1981). That is, the current correlation matrix elements given in Appendices A and B will have the same values for an ideal N element series array as for the equivalent single junction mixer provided that the dc bias and local oscillator voltages are scaled as in eq. (72). Although an explicit demonstration in terms of the full theory is too complicated for consideration here, the foregoing analysis of the shot noise under a constant dc voltage bias should make this result quite plausible. Notice, in fact, that the current correlation matrix elements in Appendices A and B are invariant under the transformation of eq. (72) if the final scaling relation is rewritten:

$$e \rightarrow e/N \tag{81}$$

This is a more physically suggestive form in any case, since eq. (74) shows that the current impulse for an electron tunneling through a single element is divided by a factor N when referred to the terminals of the entire array.

Summarizing the conclusions discussed in this section, a series array of N identical quasiparticle tunnel junctions is expected to have the same conversion efficiency and mixer noise temperature as a single junction of equal total impedance when placed in the same receiver. The optimized local oscillator power and dynamic range, however, will be increased by a factor N^2.

Finally, the ultimate quantum noise limited sensitivity of a heterodyne receiver $T_M \simeq \hbar\omega/k$, should be the same for an ideal array as for the single junction (Feldman and Rudner, 1981). This assertion is, of course, independent of the semiclassical noise theory embodied in the current correlation matrix elements, since quantum fluctuations have not been included. Nevertheless, such a conclusion appears inevitable within the context of the preceding discussion. The absorption of an individual photon in one of the junctions can generate a current pulse, as in eq. (74), through the entire array. Even though the magnitude of that pulse is reduced by a factor N referred to the complete structure, heterodyne techniques should be capable of detecting at the level of such individual events as already demonstrated in single junction mixers (Shen et al., 1980). The discussion presented below eq. (8) may also be modified, using the results of eqs. (73) and (80), to show that a hypothetically ideal SIS array video detector would have a noise equivalent power NEP $\simeq \hbar\omega B$ equal to the quantum limited value.

All experimental measurements on arrays thus far support the conclusions presented here, except those on mixer noise temperature. Noise temperatures for 6-element arrays at 73 GHz were estimated at roughly $T_M \simeq 20K \pm 90K$ (Rudner et al., 1981 b). A 14-junction array operated at 115 GHz yielded $T_M \simeq 70K \pm 40K$ (Kerr et al., 1981). Although the standard deviations quoted here are all quite large, substantial increases in the mixer noise temperatures of arrays compared with the best single junctions are clearly observed. This effect may be due to the influence of small inhomogeneities among individual junctions within the arrays or to inductance in the connections between them. The mixer noise arising from the combined dc bias and local oscillator drive depends in a complicated way on interference effects between fluctuations that are generated at the various sidebands and mixed down to the output frequency. Reducing this component of the noise to values comparable to or below the quantum limit is the result of delicate cancellations in relative phase, and such effects may very well be more sensitive to array imperfections than the conversion efficiency and other properties. If this is indeed the case, single junction mixers will continue to provide the lowest possible noise temperatures; but future applications requiring expanded dynamic range are likely to utilize series arrays with only small sacrifices in sensitivity.

APPENDIX A.

The model for fundamental mixing in a quasiparticle tunnel junction described in Section 5 considers only the signal, output, and image sidebands ω_1, ω_0, ω_{-1}, respectively, in the limit $\omega_0 << \omega$ of low output frequency. The local oscillator waveform in eq. (46) is taken to be sinusoidal, with normalized amplitude:

$$\alpha = eV_{LO}/\hbar\omega \tag{A1}$$

All higher harmonics and their sideband frequencies are assumed shorted through the junction capacitance. The 3x3 small signal admittance and current correlation matrices for the pumped diode are evaluated in the original literature (Tucker, 1979, Tucker, 1980) with the following results.

The admittance matrix elements which determine the mixing:

$$Y_{mm'} = G_{mm'} + iB_{mm'} \tag{A2}$$

are given by:

$$G_{00} = \sum_{n=-\infty}^{\infty} J_n^2(\alpha) \frac{d}{dV_0} I_0(V_0 + n\hbar\omega/e)$$

$$G_{10} = G_{-10} = \frac{1}{2} \sum_{n=-\infty}^{\infty} J_n(\alpha)[J_{n-1}(\alpha) + J_{n+1}(\alpha)] \frac{d}{dV_0} I_0(V_0 + n\hbar\omega/e)$$

$$G_{01} = G_{0-1} = \frac{e}{\hbar\omega} \sum_{n=-\infty}^{\infty} J_n(\alpha)[J_{n-1}(\alpha) - J_{n+1}(\alpha)] I_0(V_0 + n\hbar\omega/e)$$

$$G_{11} = G_{-1-1} = \frac{e}{2\hbar\omega} \sum_{n=-\infty}^{\infty} [J_{n-1}^2(\alpha) - J_{n+1}^2(\alpha)] I_0(V_0 + n\hbar\omega/e)$$

$$G_{1-1} = G_{-11} = \frac{e}{2\hbar\omega} \sum_{n=-\infty}^{\infty} J_n(\alpha)[J_{n-2}(\alpha) - J_{n+2}(\alpha)] I_0(V_0 + n\hbar\omega/e) \tag{A3}$$

and

$$B_{00} = B_{01} = B_{0-1} = 0$$

$$B_{10} = -B_{-10} = \frac{1}{2} \sum_{n=-\infty}^{\infty} J_n(\alpha)[J_{n-1}(\alpha) - J_{n+1}(\alpha)] \frac{d}{dV_0} I_{KK}(V_0 + n\hbar\omega/e)$$

$$B_{11} = -B_{-1-1} = \frac{e}{2\hbar\omega} \sum_{n=-\infty}^{\infty} [J_{n-1}^2(\alpha) - 2J_n^2(\alpha) + J_{n+1}^2(\alpha)] I_{KK}(V_0 + n\hbar\omega/e)$$

$$B_{1-1} = -B_{-11} = \frac{e}{2\hbar\omega} \sum_{n=-\infty}^{\infty} [J_{n-2}(\alpha)J_n(\alpha) - 2J_{n-1}(\alpha)J_{n+1}(\alpha) + J_n(\alpha)J_{n+2}(\alpha)] \cdot I_{KK}(V_0 + n\hbar\omega/e) \tag{A4}$$

The current correlation matrix elements characterizing the noise properties of the tunnel junction in this model are:

$$H_{00} = 2e \sum_{n=-\infty} J_n^2(\alpha)\coth[(eV_0 + n\hbar\omega)/2kT]\, I_0\, (V_0 + n\hbar\omega/e)$$

$$H_{10} = H_{-10} = H_{01} = H_{0-1}$$

$$= e \sum_{n=-\infty} J_n(\alpha)[J_{n-1}(\alpha) + J_{n+1}(\alpha)]\coth[(eV_0 + n\hbar\omega)/2kT]$$

$$\cdot\, I_0\, (V_0 + nh\omega/e)$$

$$H_{11} = H_{-1-1} = e \sum_{n=-\infty}^{\infty} [J_{n-1}^2(\alpha) + J_{n+1}^2(\alpha)]\coth[(eV_0 + n\hbar\omega)/2kT]$$

$$\cdot\, I_0(V_0 + n\hbar\omega/e)$$

$$H_{1-1} = H_{-11} = 2e \sum_{n=-\infty}^{\infty} J_{n-1}(\alpha)J_{n+1}(\alpha)\coth[(eV_0 + n\hbar\omega)/2kT]$$

$$\cdot\, I_0(V_0 + n\hbar\omega/e) \tag{A5}$$

In general, these expressions for the current correlation matrix are valid for output frequencies $\hbar\omega_0 << 2kT$ which are small compared to thermal energies as well as the local oscillator frequency.

APPENDIX B.

The model of Section 5 can also be used to characterize harmonic mixing in a single-particle tunnel junction at zero dc bias, assuming the I-V characteristic has inversion symmetry $I_0(-V) = - I_0(V)$ about this point. Only the even numbered sideband frequencies will be coupled to the output under these conditions, so that the mixing may again be described in terms of signal, output, and image frequencies ω_2, ω_0, ω_{-2}. The output frequency is taken to be small compared to both the local oscillator $\omega_0 << \omega$ and to thermal energies $\hbar\omega_0 << 2kT$, as before. Here the source admittance $Y_S = G_S + iB_S$ describes the effective diode termination at the signal frequency $\omega_2 \simeq 2\omega$, while all higher harmonics and their sidebands are assumed shorted by the junction capacitance. Equations (63)-(68) then characterize the conversion efficiency, output impedance, and noise temperature for such a harmonic mixer when the substitution $1\rightarrow 2$ is made in every subscripted quantity.

The complete small signal admittance and current correlation matrices are both 5x5 for this case. Under the above assumptions, however, there will be no coupling of the signal and output to the sidebands $\omega_1 \simeq \omega \simeq -\omega_{-1}$ near the pump frequency. The admittance matrix elements which determine the harmonic mixing properties in this model are then given by:

$$G_{00} = \sum_{n=-\infty}^{\infty} J_n^2(\alpha) \frac{d}{dV_0} I_0(V_0 + n\hbar\omega/e) \Big|_{V_0=0}$$

$$G_{20} = G_{-20} = \frac{1}{2} \sum_{n=-\infty}^{\infty} J_n(\alpha)[J_{n-2}(\alpha) + J_{n+2}(\alpha)] \frac{d}{dV_0} I_0(V_0 + n\hbar\omega/e) \Big|_{V_0=0}$$

$$G_{02} = G_{0-2} = \frac{e}{2\hbar\omega} \sum_{n=-\infty}^{\infty} J_n(\alpha)[J_{n-2}(\alpha) - J_{n+2}(\alpha)] I_0(n\hbar\omega/e)$$

$$G_{22} = G_{-2-2} = \frac{e}{4\hbar\omega} \sum_{n=-\infty}^{\infty} [J_{n-2}^2(\alpha) - J_{n+2}(\alpha)] I_0(n\hbar\omega/e)$$

$$G_{2-2} = G_{-22} = \frac{e}{4\hbar\omega} \sum_{n=-\infty}^{\infty} J_n(\alpha)[J_{n-4}(\alpha) - J_{n+4}(\alpha)] I_0(n\hbar\omega/e) \tag{B1}$$

and:

$$B_{00} = B_{02} = B_{0-2} = 0$$

$$B_{20} = -B_{-20} = \frac{1}{2} \sum_{n=-\infty}^{\infty} J_n(\alpha)[J_{n-2}(\alpha) - J_{n+2}(\alpha)] \frac{d}{dV_0} I_{KK}(V_0 + n\hbar\omega/e) \quad V_0=0$$

$$B_{22} = -B_{-2-2} = \frac{e}{4\hbar\omega} \sum_{n=-\infty}^{\infty} [J_{n-2}^2(\alpha) - 2J_n^2(\alpha) + J_{n+2}^2(\alpha)] I_{KK}(n\hbar\omega/e)$$

$$B_{2-2} = -B_{-22} = \frac{e}{4\hbar\omega} \sum_{n=-\infty}^{\infty} [J_{n-4}(\alpha)J_n(\alpha) - 2J_{n-2}(\alpha)J_{n+2}(\alpha)$$

$$+ J_n(\alpha)J_{n+4}(\alpha)] I_{KK}(n\hbar\omega/e)$$

The required components of the current correlation matrix for this model are:

$$H_{00} = 2e \sum_{n=-\infty}^{\infty} J_n^2(\alpha)\coth(n\hbar\omega/2kT)I_0(n\hbar\omega/e)$$

$$H_{20} = H_{-20} = H_{02} = H_{0-2}$$

$$= e \sum_{n=-\infty}^{\infty} J_n(\alpha)[J_{n-2}(\alpha) + J_{n+2}(\alpha)]$$

$$\cdot \coth(n\hbar\omega/2kT)I_0(n\hbar\omega/e)$$

$$H_{22} = H_{-2-2} = e \sum_{n=-\infty}^{\infty} [J_{n-2}^2(\alpha) + J_{n+2}^2(\alpha)]$$

$$\cdot \coth(n\hbar\omega/2kT)I_0(n\hbar\omega/e)$$

$$H_{2-2} = H_{-22} = 2e \sum_{n=-\infty}^{\infty} J_{n-2}(\alpha)J_{n+2}(\alpha)$$

$$\cdot \coth(n\hbar\omega/2kT)I_0(n\hbar\omega/e) \tag{B3}$$

APPENDIX C.

The prediction of photon-assisted tunneling theory for the average quasiparticle current may be expressed in the time domain, as an alternative to the frequency representation summarized in eqs. (9)-(12). The result may be written in the form:

$$\langle I(t)\rangle = \frac{V(t)}{R_N} + Im\left\{U^*(t)\int_{-\infty}^{t} dt'\bar{\chi}(t-t')U(t')\right\} \tag{C1}$$

Here the time-dependent phase factor induced by the applied potential is given by:

$$U(t) = \exp\left\{-\frac{ie}{\hbar}\int^{t} dt'V(t')\right\} \tag{C2}$$

and the response function characterizing the nonlinear behavior of the junction is found to be:

$$\bar{\chi}(t) = \frac{2}{\pi}\int_0^\infty d\omega\left[I_0(\hbar\omega/e) - \frac{\hbar\omega}{eR_N}\right]\sin\omega t \tag{C3}$$

The dc I-V characteristic is assumed to be antisymmetric, $I_0(-V) = -I_0(V)$, and to become ohmic with resistance R_N at large bias.

This formulation may be obtained utilizing the results quoted in Section 2 of a previous work (Tucker, 1979). In the notation of that paper, eqs. (2.9), (2.10), and (2.14) may be combined to give:

$$\langle I(t)\rangle = Im\left\{U^*(t)\int_{-\infty}^{t} dt'\,\frac{4i}{e}\,\chi''_{+-}(t-t')U(t')\right\} \tag{C4}$$

where according to eq. (2.17):

$$\chi''_{+-}(\omega) = \frac{e}{4\pi}\,I_0(\hbar\omega/e) \tag{C5}$$

Computing the Fourier transform directly using the total dc I-V characteristic would, however, encounter problems associated with the ohmic divergence at large argument. In eq. (C3), this difficulty is circumvented by subtracting out the ohmic contribution from the response function:

$$\chi(t) = \frac{4i}{e}\,\chi''_{+-}(t)$$

$$= \frac{i}{\pi}\int_{-\infty}^{\infty} d\omega\; I_0(\hbar\omega/e)e^{-i\omega t} \tag{C6}$$

It then remains only to demonstrate that the average current through an ohmic tunnel junction is simply $V(t)/R_N$ within the context of this theory.

For an ohmic junction, the time integration in eq. (C4) may be written:

$$\int_{-\infty}^{t} dt'\,\frac{4i}{e}\,\chi''_{+-}(t-t')U(t')$$

$$= \frac{i}{\pi}\int_{-\infty}^{\infty} d\omega\,\frac{\hbar\omega}{eR_N}\int_{-\infty}^{t} dt'U(t')e^{-i\omega(t-t')} \tag{C7}$$

An integration by parts then leads to:

$$\int_{-\infty}^{t} dt' U(t') e^{-i\omega(t-t')} = \frac{1}{i\omega}\left[U(t) + \frac{ie}{\hbar}\int_{-\infty}^{t} dt' V(t') U(t') e^{-i\omega(t-t')}\right] \tag{C8}$$

Substituting this expression back into eq. (C7) gives:

$$\int_{-\infty}^{t} dt' \frac{4i}{e} \chi''_{+-}(t-t') U(t') = \frac{\hbar}{\pi e R_N} U(t) \left\{ \int_{-\infty}^{\infty} d\omega + \frac{i\pi e}{\hbar} V(t) \right\} \tag{C9}$$

The formal divergence thus winds up in the real part of the factor contained within brackets here. When this result is inserted into the current expression eq. (C4), the divergence is elimated and the response for an ohmic tunnel junction is found to be:

$$\langle I(t) \rangle = \frac{V(t)}{R_N} \tag{C10}$$

as anticipated.

ACKNOWLEDGEMENTS

A portion of the work presented here was performed at the NASA Goddard Institute for Space Studies, New York, under a National Research Council senior associateship. I wish to thank P. Thaddeus and A. R. Kerr for their generous hospitality during this period. The comments on noise in series arrays evolved from a discussion with A. Davidson, M. J. Feldman, A. D. Smith, and D. P. Woody. I have also benefited from technical discussions with T. A. DeTemple, K. E. Irwin, W. R. McGrath, P. L. Richards, S. E. Schwarz, P. H. Siegel, and from extensive conversations with A. R. Kerr and M. J. Feldman.

REFERENCES AND BIBLIOGRAPHY

Bardeen, J. (1980). Phys. Rev. Lett. 45, 1978.

Cohen, M.H., Falicov, L.M., and Phillips, J.C. (1962). Phys. Rev. Lett. 8, 316.

Dayem, A.H., and Martin, R.J. (1962). Phys. Rev. Lett. 8, 246.

Dolan, G.J., Phillips, T.G., and Woody, D.P. (1979). Appl. Phys. Lett. 34, 347.

Dolan, G.J., Linke, R.A., Sollner, T.C.L.G., Woody, D.P., and Phillips, T.G. (1981). IEEE Trans. Microwave Theory Tech. MTT-29, 87.

Feldman, M.J. (1981). J. Appl. Phys., to be published.

Feldman, M.J., and Rudner, S. (1981). This volume.

Feldman, M.J., and Tucker, J.R. (1980). Private Communication.

Greiner, J.H., Kircher, C.J., Klepner, S., Lahiri, S., Warnecke, A., Basavaiah, S., Yen, E., Baker, J.M., Brosious, P.R., Huang, H.-C. W., Murakami, M., and Ames, I. (1980). IBM J. Res. Develop. 24, 195.

Harris, R.E., and Hamilton, C.A. (1978). In Future Trends in Superconductive Electronics, edited by B.S. Deaver, Jr., et al., AIP Conf. Proc. 44, New York, pp. 448-458.

Huang, H.-C. W., Basavaiah, S., Kircher, C.J., Harris, E.P., Murakami, M., Klepner, S.P., and Greiner, J.H. (1980). IEEE Trans. Electron Devices ED-27, 1979.

Kerr, A.R. (1975). IEEE Trans. Microwave Theory Tech. MTT-23, 828.

Kerr, A.R., Pan, S.-K., Feldman, M.J., and Davidson, A. (1981). In Proceedings of the 16th International Conference on Low Temperature Physics LT-16, Physica 108B, 1369.

McColl, M., Millea, M.F., and Silver, A.H. (1973). Appl. Phys. Lett. 23, 263.

McColl, M., Millea, M.F., Silver, A.H., Bottjer, M.F., Pedersen, R.J., and Vernon, F.L. (1977). IEEE Trans. Magn. MAG-13, 221.

McGrath, W.R., Richards, P.L., Smith, A.D., van Kempen, H., Batchelor, R.A., Prober, D., and Santhanam, P. (1981). Appl. Phys. Lett., to be published.

Phillips, T.G., Woody, D.P., Dolan, G.J., Miller, R.E., and Linke, R.A. (1981). IEEE Trans. Magn. MAG-17, 684.

Richards, P.L., and Shen, T.-M. (1980). IEEE Trans. Electron Devices ED-27, 1909.

Richards, P.L., Shen, T.-M., Harris, R.E., and Lloyd, F.L. (1979). Appl. Phys. Lett. 34, 345.

Richards, P.L., Shen, T.-M., Harris, R.E., and Lloyd, F.L. (1980). Appl. Phys. Lett. 36, 480.

Rudner, S., and Claeson, T. (1979). Appl. Phys. Lett., 34, 711.

Rudner, S., Feldman, M.J., Kollberg, E., and Claeson, T. (1981a). IEEE Trans. Magn. MAG-17, 690.

Rudner, S., Feldman, M.J., Kollberg, E., and Claeson, T. (1981b). J. Appl. Phys. to be published.

Schrieffer, J.R. (1964). Theory of Superconductivity, Benjamin, New York.

Shen, T.-M. (1981). IEEE J. Quantum Electron. QE-17, 1151

Shen, T.-M., and Richards, P.L. (1981). IEEE Trans. Magn. Mag-17, 677.

Shen, T.-M., Richards, P.L., Harris, R.E., and Lloyd, F.L. (1980). Appl. Phys. Lett. 36, 777.

Silver, A.H., Pedersen, R.J., McColl, M., Dickman, R.L., and Wilson, W.J. (1981). IEEE Trans. Magn. MAG-17, 698.

Smith, A.D., and Richards, P.L. (1981). Manuscript in draft form.

Smith, A.D., McGrath, W.R., Richards, P.L., van Kempen, H., Prober, D., and Santhanam, P. (1981). In Proceedings of the 16th International Conference on Low Temperature Physics LT-16, Physica 108B, 1367.

Tien, P.K., and Gordon, J.P. (1963). Phys. Rev. 129, 647.

Tucker, J.R. (1975). In Proceedings of the 14th International Conference on Low Temperature Physics LT-14, edited by M. Krusius and M. Vuorio, North Holland, Amsterdam, vol. 4, pp. 180-184.

Tucker, J.R. (1979). IEEE J. Quantum Electron. QE-15, 1234.

Tucker, J.R. (1980). Appl. Phys. Lett. 36, 477.

Tucker, J.R., and Millea, M.F. (1978). Appl. Phys. Lett. 33, 611.

Tucker, J.R., and Millea, M.F. (1979). IEEE Trans. Magn. MAG-15, 288.

Vernon, F.L., Millea, M.F., Bottjer, M.F., Silver, A.H., Pedersen, R.J., and McColl, M. (1977). IEEE Trans. Microwave Theory Tech. MTT-25, 286.

Werthamer, N.R. (1966). Phys. Rev. 147, 255.

Woody, D.P. (1981). Manuscript in draft form.

MIXING WITH SIS ARRAYS

M.J. Feldman*

NASA Goddard Institute for Space Studies
2880 Broadway
New York, N.Y. 10025, U.S.A.

and

S. Rudner

Chalmers Institute of Technology
S-412 96 Gothenburg, Sweden

1. INTRODUCTION

The various chapters in this volume attest to the rapid development of the superconductor-insulator-superconductor (SIS) quasiparticle mixer. At every stage in this development, the results from series-arrayed junctions have kept pace with the single-junction mixers. The earliest SIS mixer results demonstrated single-sideband (SSB) noise temperatures as low as 14 K (Richards et al., 1979), operation at frequencies as high as 115 GHz (Dolan et al., 1979), and SSB conversion as good as 5.8 dB loss (Rudner and Claeson, 1979). Even then, were all these attributes combined, the SIS mixer would have been superior to the best conventional devices. Rudner and Claeson (1979) used arrays of 40 SIS elements at 10 GHz; their results showed no trace of quantum mixing effects.

Applying the quantum theory of mixing, Tucker (1980) predicted that the SIS mixer can supercede the classical proscription of conversion gain, and can have a noise temperature so low as to be effectively limited only by quantum noise. Both predictions were verified by Shen et al. (1980) at 36 GHz. Rudner et al. (1981a)

*NAS-NRC Research Fellow

soon demonstrated the non-classical conversion using arrays at 75 GHz. In both cases the SSB conversion was slightly better than 3 dB loss. Hartfuss and Gundlach (1981) have recently had similar results (with a single-junction mixer at 70 GHz). Rudner et al. (1980) also found satisfactory performance for long arrays.

Most recently, infinite available conversion gain was observed using a single-junction mixer at 36 GHz (Smith et al., 1981) and an array mixer at 115 GHz (Kerr et al, 1981). This means that arbitrarily high gain would be achieved with an appropriately matched IF load. The only other published SIS array mixer results (Richards and Shen, 1980) contains little detailed information.

The original impetus for using arrays of SIS junctions for mixing was a practical consideration. To assure that the junction capacitance would not shunt out the nonlinearity, all of the earlier expements were designed with the ωRC product close to unity. Since R is determined by impedance matching considerations, the junction area must be very small to limit the capacitance, especially at high frequency. For instance, Dolan et al. (1979, 1981) chose a junction of area 0.4 μm^2 at 115 GHz. Alternatively, the same overall ωRC product may be achieved by using larger junctions in a series array. This is discussed in Section 1.1. The arrays of larger junctions are less susceptible to electrical transients which may destroy small single junctions, and they can be made with less sophisticated equipment and technique. These are crucial factors for many laboratories.

Perhaps for this reason, the great majority (∿80%) of the samples reported as SIS mixers have been arrays. Much of this work appears in Rudner et al. (1981b), which we shall review rather closely in Sections 2. and 3. This large data set has provided the first detailed quantitative verification of the quantum theory of mixing, as well as a variety of insights into the design of practical SIS mixers.

Recent results have shown that an SIS mixer will give superior performance with its ωRC product substantially greater than unity (as discussed in Section 5.), if care is taken to tune out the capacitance. This somewhat mitigates the original reason for using arrays, but other advantages of arrays have become apparent. One crucial consideration is the limited dynamic range of single-junction SIS mixers. The dynamic range increases as the square of the number of junctions arrayed (Section 3.1.). Also, any junction used in an SIS mixer should have a linear dimension large enough that Josephson effect noise, discussed in Section 4.2., may be suppressed by an applied magnetic field. This is a problem for single-junction mixers at high frequency.

The difficulties in using arrays are apparent. Each array has some heterogeneity in its junctions' characteristics. Since the same currents flow through all of the junctions, the voltages do not divide evenly among them. This can interfere with the mixer's performance, and makes theoretical comparisons more difficult. In Section 1.1. we conclude that these problems are acceptably small in practice. Another consideration is that the series inductance along an array tends to be large, for a given overall device impedance. Section 2.1. reviews how this problem has been circumvented in past experiments; it will be more serious at higher frequencies.

Section 4.1. discusses the noise temperature of SIS mixers. We conclude that noise, and in particular quantum noise, is in no way worse for array than for single-junction mixers.

1.1. Arrays

The simplest assumption which can be made in treating a series-array mixer is that it behaves exactly like a single-junction mixer in which all relevant voltages and impedances are multiplied by N, the number of junctions in the array. When we compare the experimental array results with the theory, we shall see that this assumption is often quite successful. The implication is that the array is a simple sum of its elements, with no complexity introduced because $N > 1$.

What requirements are necessary that an array behave in this simple manner? Let us examine Fig. 1. Figure 1a represents an array of SIS junctions, connected by small series inductances, each element consisting of a nonlinear resistance, a nonlinear reactance, and a capacitance.

The first assumption we shall make is that the current is in-phase all along the array, or equivalently, that the array can be represented as a two-terminal device. (We do not yet consider how the current is shared among the various branches of each individual element.) If the currents are in-phase all along the array, then the ordering of the array elements is not important, Thus Fig. 1a is equivalent to Fig. 1b. This first assumption is approximately true in an experimental situation if (1) the array is short compared to the effective wavelength at the frequencies of interest, or (2) the array is in shunt across a transmission line such that the electric field has the same value at each element. We shall present experimental results for both of these configurations.

The other assumption we shall make is that all of the elements of the array are identical. If this is true, then the two example current loops drawn in Fig. 1b carry equal currents. This means that the total current carried by the horizontal connecting bars between the array elements is zero. These bars can thus be eliminated so that Fig. 1b becomes Fig. 1c. The impedances are then added in series to give Fig. 1d.

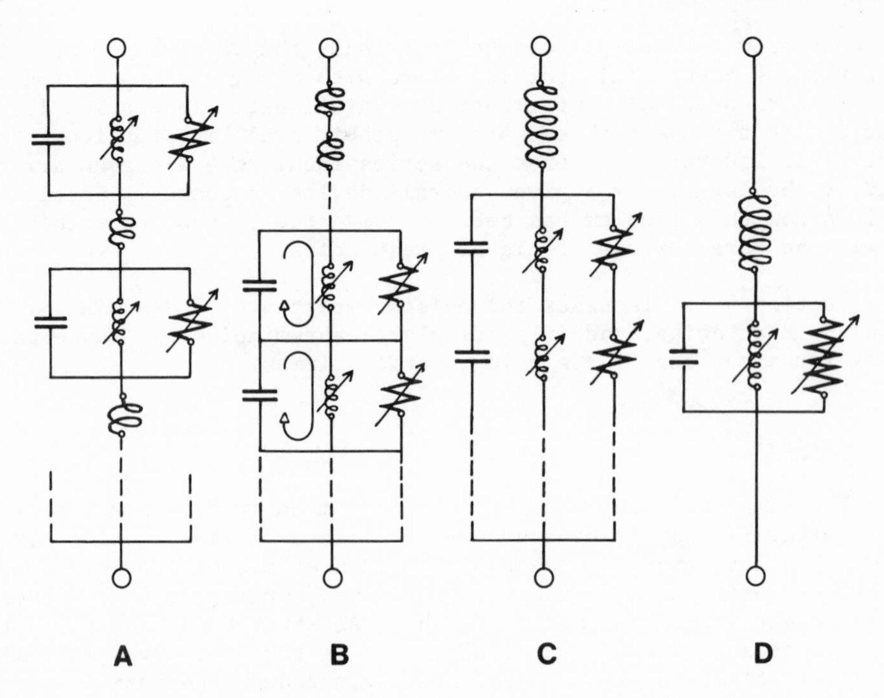

Fig. 1. Construction showing that, given certain assumptions, the equivalent circuit for an array of SIS junctions reduces to that of a single element, with all impedances multiplied by N.

Figure 1d is identical in form to a sigle element of Fig. 1a. Thus a series array behaves exactly like any one of its elements, but with all impedances multiplied by N, providing that (1) the current is in-phase all along the array, and (2) all of the elements of the array are identical. Note that the array RC product is independent of N.

This argument may seem too obvious to warrant such a detailed discussion. Nevertheless, it is important for a number of reasons. Firstly, the two requirements for "simple" array behavior are necessary design criteria in the situations we consider. We shall discuss experimental results for which these requirements are violated: for arrays not short compared to a wavelength, and for arrays with large heterogeneity. In both cases the SIS mixer performance deteriorates. Secondly, an exact solution for the equations of an array which does not behave in this simple manner is extremely difficult, even in the case N = 2; the experiments cannot easily be compared to the theory.

A third reason is that this simple approach has occasioned some controversy. Shen and Richards (1981) have stated an opposing view, that each element of an array must be treated individually, considering the other junctions as part of the external circuitry, and obtaining the overall device properties by superposition. But superposition will not work in general because the equations are nonlinear. After the problem has been linearized to calculate the small-signal response, this approach is valid but impractical: the embedding impedances of all of the other junctions depend upon the behavior of the junction in question. If the series impedance of the other junctions is simply added to the source impedance, incorrect results are predicted. For instance, each junction in an array would see an extremely large source impedance, which would imply that an SIS array mixer can display considerable conversion over a very narrow range of LO power (Shen and Richards, 1981), contrary to experiment.

How closely must the two requirements be fulfilled for the junctions in an array to behave in a simple additive manner? Certainly they are never perfectly satisfied, so the use of arrays will always introduce some ambiguity into the interpretation of experimental results. Our approach is pragmatic. We straightforwardly apply the theory with all impedances multiplied by N and see whether theory and experiment agree. The agreement is in fact found to be excellent within a certain range of parameter space. Note that for another array device, the unbiased Josephson junction parametric amplifier, Wahlsten et al. (1978) followed the same approach and found excellent agreement between theory and experiment in almost all respects. We conclude that SIS arrays can indeed be made sufficiently homogeneous.

2. EXPERIMENTAL

2.1 Design

A most important criterion for the design of SIS array mixers is that the effect of the array's series inductance be minimized. Rudner et al. (1980, 1981a, 1981b) have chosen to minimize the inductance by sandwiching the array between two ground planes. Their scheme is illustrated in Fig. 2. Figure 2a shows the glass substrate with an evaporated pattern; the pattern includes the junction array (Fig. 2b), an antenna, and the filtered IF and DC bias connection. The substrate was mounted in full-height waveguide as shown in Fig. 2c. Because an array is used the junctions could be relatively large (in this case 5 μm square) and were made with relatively unsophisticated equipment, by evaporation through bimetallic stencil masks. This procedure was very versatile; a large number of diverse samples could be made and tested. The ground planes on either side of the array (Fig. 2c) reduced its series inductive impedance and also formed a stripline with the array. The 6-junction arrays (Fig. 2b) were only $\lambda/30$ in length, making the phase change insignificant over the array, but 36-junction arrays were also tested which were about half a wavelength long. A series of measurements of passive reflec-

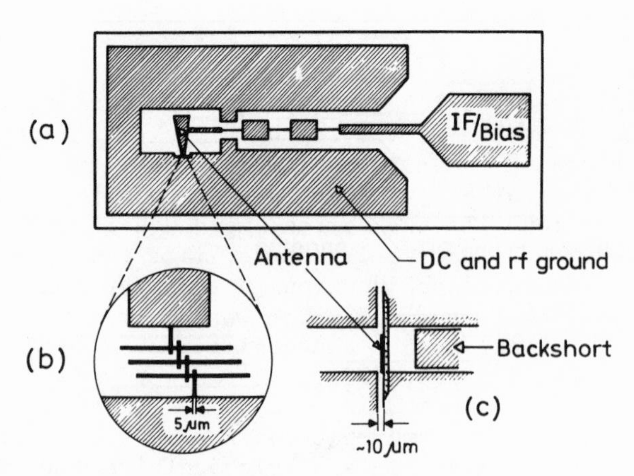

Fig. 2. The experimental design used by Rudner et al. (1981b). The glass substrate (a) with an evaporated pattern, shown to scale, includes (b) a 6-junction array in a small recess. The substrate is also shown (c) mounted in full-height waveguide with the array sandwiched between two ground planes.

tion loss was made in this configuration for non-superconducting arrays with various resistances. This indicated that the antenna transformed the waveguide impedance to $\sim$50 Ω with a mismatch loss of $\sim$0.5 dB.

Kerr et al. (1981) followed a different approach in dealing with the array series inductance. Their array was mounted in shunt across a reduced-height waveguide with an adjustable backshort, thus ensuring in-phase currents along the array. The mounting circuit was designed in scale-model experiments to compensate the series inductance and parallel capacitance, and to permit a range of desired source impedance. Recent calculations indicate that these goals were in fact achieved.

2.2. Conversion Results

The most promising results with an SIS array mixer were achieved by Kerr et al. (1981) at 115 GHz and are shown in Fig. 3. The negative differential resistance region on the pumped DC I-V curve, Fig. 3b, indicates that this mixer has infinite "available" conversion

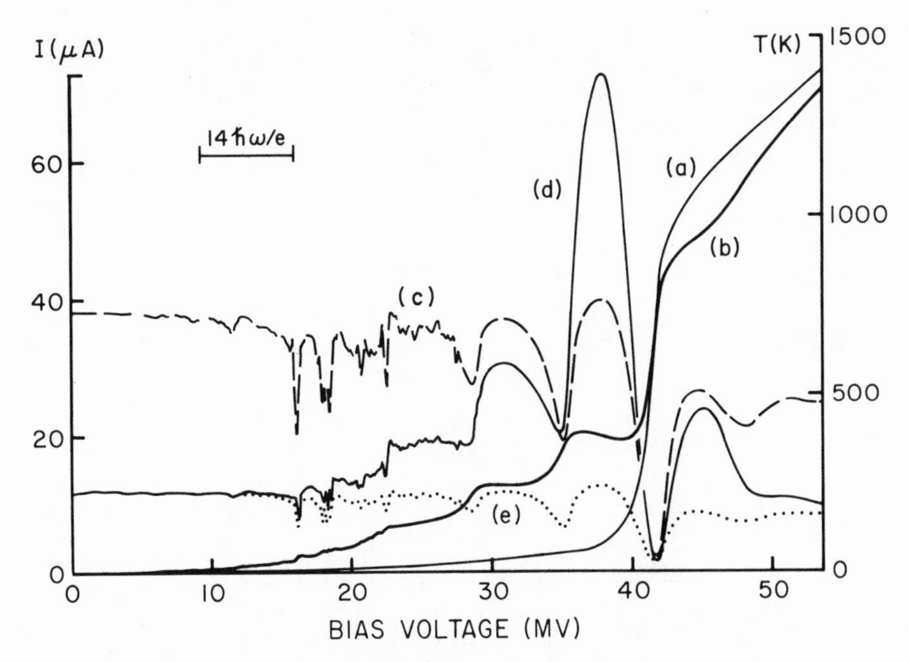

Fig. 3. Experimental results for a 14-junction SIS array mixer at 115 GHz. (a) The unpumped DC I-V curve. (b) The DC I-V curve with LO power P_{LO} = 0.375 μW, showing a region of negative differential resistance. Curves (c)-(e) show the IF radiometer output with (c) a noise source applied to the mixer's IF port, (d) a monochromatic source applied to the mixer's signal port, and (e) no signal or IF power applied; all with P_{LO} = 0.375 μW. The array normal resistance (R_A) measured from (a) is 600 Ω. From Kerr et al. (1981).

gain. This means that essentially infinite gain would be achieved into an appropriately matched IF load, limited only by the dynamic range of the mixer. In fact, the R_A = 600 Ω array is very badly matched into the 50 Ω load. This completely accounts for the relatively poor realized conversion of 11.5 dB loss (see Feldman, 1981).

The structure seen in the curves of Fig. 3 below about 28 mV is caused by the Josephson effect, as evidenced by its acute sensitivity to changes in the applied magnetic field. Beyond this voltage all of the structure in Fig. 3 occurs on the scale of 14 ħω/e, the "photon voltage" for this 14-junction array. There is no trace of any structure on the scale of ħω/e. The same is true of essentially all array results.

In the work of Rudner et al. (1981b) a large number of arrays of various composition and impedance were compared as SIS mixers at

75 GHz. We shall quite thoroughly review these experiments, because this large data set provides a clear test of the quantum theory of mixing, as well as a variety of insights into the design of practical SIS mixers.

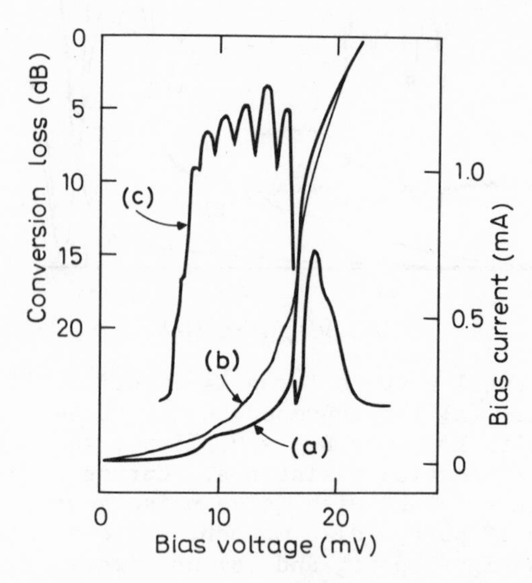

Fig. 4. DC I-V curves for a 6-junction Pb(In) array at 73.5 GHz (a) without P_{LO} and (b) with P_{LO} = 1.62 μW, chosen to optimize the conversion loss. (c) SSB conversion loss again with P_{LO} = 1.62 μW. R_A = 11.9 Ω. From Rudner et al. (1981a,b).

Some typical results for 6-junction arrays of Pb-O-Pb with small amounts of In added to the base electrode [these arrays are called "Pb(In)"] are shown in Figs. 4 and 5. Except at small voltage, all structure in the curves occurs on the scale of $6\hbar\omega/e$. The best SSB conversion measured was 2.0 ± 0.9 dB loss, an early verification that quantum mixing violates the classical prohibition of mixer gain. The

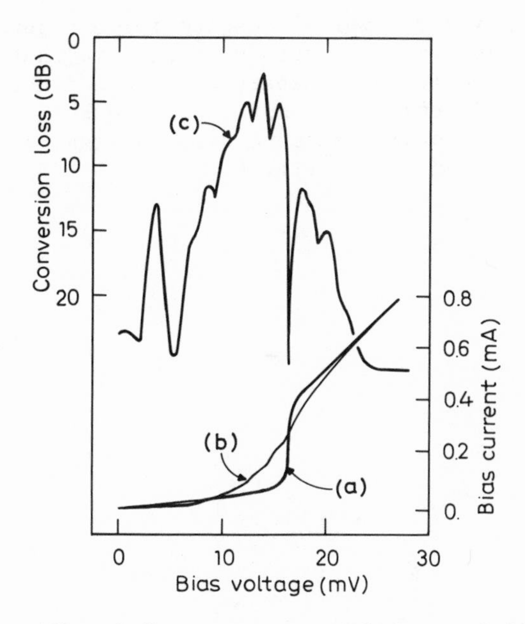

Fig. 5. Same as Fig. 4 for an array with P_{LO} = 0.97 μW and R_A = 26.3 Ω. Unpublished result from Rudner et al. (1981a,b).

conversion was very stable: the LO power could be varied over 2 dB with less than 0.5 dB change in conversion.

Note that photon-assisted tunneling structure is considerably less evident above the energy gap. This seems to be a general feature of SIS array mixers. The explanation is clear. Any dispersion in the junction critical current (I_J) is mirrored by a dispersion in the junction resistances, because the product $I_J R$ is fixed. These reinforce one another in this region to create a large dispersion in

DC voltage across the individual junctions, for a given DC current. Therefore any structure which would appear on a single-junction I-V curve tends to be averaged away.

Figure 6 shows some typical results for a 6-junction array of pure Pb-O-Pb. There was good reason to believe that the pure Pb arrays had considerably more dispersion in the junctions' leakage current below the energy gap, compared to the Pb(In) arrays. Therefore the distinct quantum structure seen in Figs. 4 and 5 in this region is quite washed out in Fig. 6.

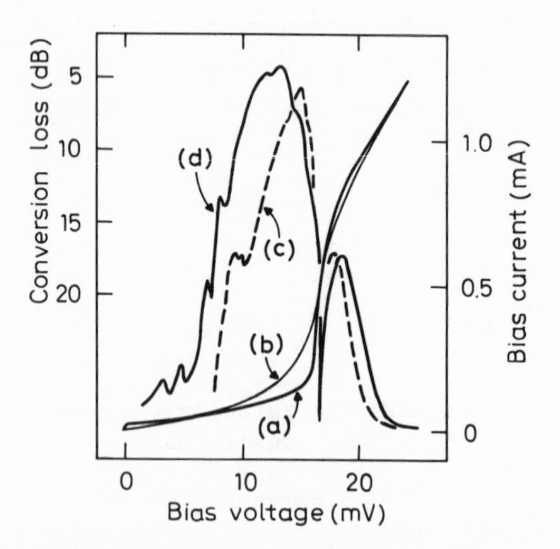

Fig. 6. DC I-V curves for a 6-junction pure Pb array at 73.5 GHz (a) without P_{LO}, (b) with P_{LO} = 0.75 μW, (c) SSB conversion loss again with P_{LO} = 0.75 μW, and (d) SSB conversion loss with optimized P_{LO} = 1.54 μW. R_A = 15.6 Ω. From Rudner et al. (1981b).

A number of 36-junction arrays of pure Pb-O-Pb were also tested (see also Rudner et al., 1980). These arrays generally showed no quantum structure; one sample showed structure of periodicity only 0.75 of the expected $36\hbar\omega/e$. This is not too surprising. The most significant fact about the 36-junction experiments is that the arrays were about one-half wavelength long. There is thus a large spread in the LO voltage (and hence the DC voltage as well) across the individual junctions.

The better conversion results for the three kinds of arrays tested are shown in Fig. 7, as a function of the array resistance R_A. Each point represents the maximum conversion achieved with a given array. The total number of samples tested was about twice that shown in Fig. 7. The clear trends seen in Fig. 7 are not simply due to impedance mismatch. Rather (Section 3.), uncompensated reactances

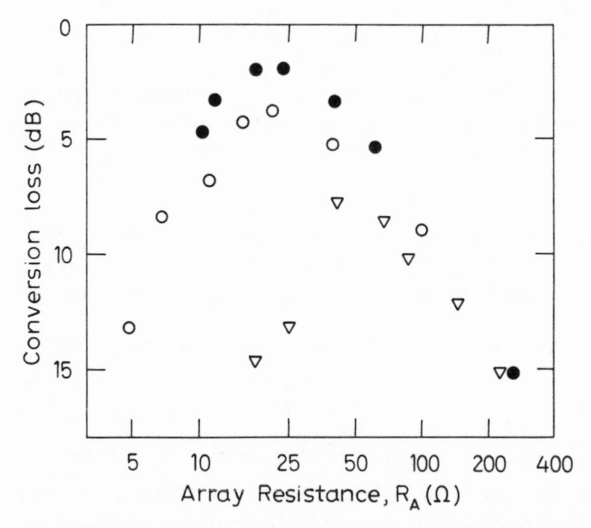

Fig. 7. Maximum conversion for various arrays achieved by Rudner et al. (1981b). Three kinds of arrays were tested: 6-junction Pb(In) arrays (●), 6-junction pure Pb arrays (○), and long 36-junction pure Pb arrays (▽).

are important at large R_A, and the deteriorating sharpness of the junction nonlinearity is important at small R_A (i.e. at high current density).

3. THEORETICAL CONFIRMATION

The experimental results of Rudner et al. (1981b) deserve close attention because they provide the only detailed quantitative verification of the quantum theory of mixing. The results from other experiments which have been considered in light of the theory have shown general qualitative agreement or, at best quantitative agreement at one point and disagreement elsewhere (Shen et al., 1980). The theory, developed by Tucker (1979, 1980), is thoroughly discussed in another chapter of this volume.

In applying the quantum theory of mixing to their experimental results Rudner et al. (1981b) made five distinct assumptions: They assumed (1) that the array was short compared to a wavelength and (2) that all of the junctions in the array were identical. As discussed in Section 1.1. these two assumptions mean that the voltage $N\hbar\omega/e$ could be substituted throughout for $\hbar\omega/e$ in the theory. They assumed (3) that the array capacitance shorted out all of the harmonics, but (4) that the array capacitance and series inductance are resonated out at the fundamental. Finally, the theory is only applicable if (5) the currents are entirely due to single-particle tunneling, with no pair or resistive leakage current, no multiparticle tunneling current, etc. Each of the five assumptions listed was violated for some range of the experimental parameters. But when all of the assumptions were well enough satisfied, the theory and experiments agreed quite well.

The nonlinear quantum reactance which appears in the theory was ignored. Feldman (1981) has shown that this is reasonable. The source and load impedances were each measured to be 50 Ω.

Using the experimental unpumped DC I-V curve for each array, they applied the Tucker theory, with no free parameters, to calculate the conversion loss as a function of both the DC voltage (V_{DC}) and the RF voltage (V_{LO}) across the junctions. The <u>maximum</u> conversion values, maximized with respect to both V_{DC} and V_{LO}, were compared for each conversion peak in the theoretical and experimental data. The maximum conversion values should be relatively insensitive to a small amount of dispersion in V_{DC} and V_{LO} among the individual junctions.

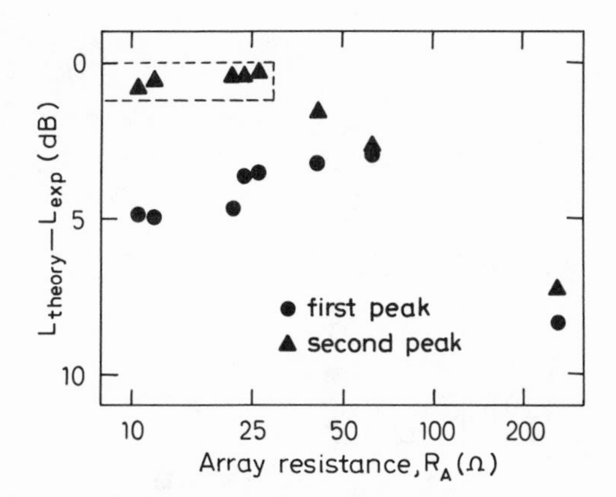

Fig. 8. Discrepancy between the optimum theoretical and experimental conversion loss for each Pb(In) array tested by Rudner et al. (1981a,b), for both the first and second peaks below the energy gap voltage.

The results of this comparison for the Pb(In) arrays are shown in Fig. 8. If the theory and experiments perfectly agreed, all of the points would fall along the 0 dB line. All of the Pb(In) samples tested were included in Fig. 8, except for two arrays which had been previously noted to be extremely heterogeneous in their Josephson critical currents. The simple theory should not give good results for these heterogeneous arrays, and in fact it did not. In addition, there was less complete data for the third conversion peak of the Pb(In) arrays and for the 6-junction pure Pb arrays, all of which fell quite precisely along the smooth trends seen in Fig. 8.

The most important region of Fig. 8 is the second peak data for $R_A \lesssim 25\ \Omega$, which includes the maximum conversion which was seen in those experiments. That region, within the dashed lines in Fig. 8 and extending to somewhat lower R_A, is shown with a magnified scale in Fig. 9. In addition to the five data points seen in Fig. 8, for the second conversion peak of the Pb(In) arrays, additional data has been added in Fig. 9: for the third conversion peak of the Pb(In) arrays, and for the maximum conversion with each pure Pb array, whether in the second or third peak. Three points are lower limits, because the third peak conversion had not been fully maximized in the experimental record for these arrays.

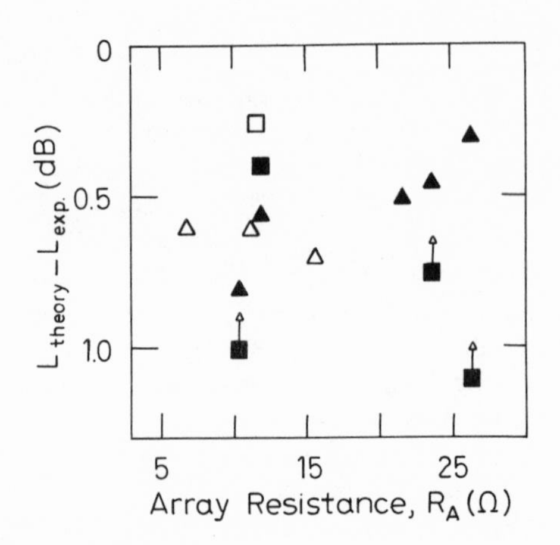

Fig. 9. The dashed area in Fig. 8, and its extension to smaller R_A, shown with a magnified scale. In addition to the five points (▲) seen in Fig. 8, other theoretical-experimental discrepancy data has been added: for the third conversion peak (■) of the Pb(In) arrays, and for the maximum conversion with each pure Pb array, whether in the second (△) or third (□) peak. This figure constitutes the first quantitative verification of the quantum theory of mixing. Unpublished results from Rudner et al. (1981b).

Each of the thirteen points in Fig. 9 represents a completely independent experiment, matched by the quantum theory with no adjustable parameters. Figure 9 includes all available data in this region ($R_A \lesssim 25$ Ω), except for a few arrays which had been previously noted to be especially heterogeneous. The scatter of the points is much less than the formal experimental uncertainty of ±0.9 dB. The mean of the points is quite close to the independently measured antenna mismatch loss of ∿0.5 dB. The message of Fig. 9 is clear: In at least one region of parameter space, the quantum theory of mixing is quantitatively verified.

The five assumptions made by Rudner et al. (1981b) in applying the theory are listed at the beginning of this section. Each assumption was violated in some other region of experimental parameter space (which explains the various features of Fig. 8):

1. The 36-junction Pb arrays were about one-half wavelength long. The simple theory predicted a larger conversion, by 4.8 ± 1.0 dB, than was actually observed, for each array.

2. A few arrays were noted to be excessively heterogeneous. The simple theory predicted the conversion of these arrays incorrectly. In addition, the normal array heterogeneity had a number of observable effects, some already mentioned. The theoretically computed conversion peaks are generally very broad, with very deep and very pointed cusps between them. A small amount of averaging caused by array heterogeneity tends to fill the cusps and make the peaks less broad, as in Fig. 4c, Fig. 5c, and more severely in Fig. 6c, where the cusps have almost disappeared.

3. The large discrepancy between theory and experiment seen in Fig. 8 for the first peak conversion at low R_A, up to 5 dB, has been observed in other SIS experiments. Shen et al. (1980) interpreted this effect as being due to significant harmonic conversion near the gap voltage. Figure 8 supports this: The discrepancy in the first peak conversion smoothly disappears (relative to the second peak data) as the array R_AC product increases and the capacitance is more effective in suppressing the harmonics.

4. The array series inductance does not appear to be a problem in these experiments. If it were, the discrepancies plotted in Fig. 8 would increase for smaller values of R_A. But in two cases, 6-junction arrays were accidently mounted wholly within the waveguide rather than in the ground-plane protected recess. These arrays thus had higher inductance than the others, and they had optimum conversion about 4 dB worse than the trend of the others. The array capacitance does appear to be a problem, if only at larger values of R_A. The rolloff of the optimum experimental conversion compared to theory seen at large R_A in Fig. 8 implies that in this experimental configuration the capacitance cannot be sufficiently compensated for large R_AC products (note that $\omega R_AC \approx 1.5$ for $R_A \approx 25\ \Omega$).

5. The structure seen at one-half of the energy gap voltage in the DC I-V curves of the Pb(In) arrays (e.g. Figs. 4a and 5a) is not due to single-particle tunneling and gave effects not accounted for in the simple theory. Rudner et al. (1981b) discuss this at length.

Why did the Pb arrays give a few dB less conversion than the Pb(In) arrays (Fig. 7)? Their DC I-V curves were equally successful in predicting the maximum conversion, using the quantum theory. Therefore the reduced performance of the Pb arrays was entirely due to the inferior quality of their DC I-V curves. Inferior quality is also the reason for the deterioration of the conversion for the lowest R_A arrays (Fig. 7), which was more severe than simple impedance mismatch could produce. The DC I-V curve of Fig. 4a is distinctly less ideal than that of Fig. 5a.

One further question must be asked. Why were Kerr et al. (1981) able to observe negative differential resistance on their pumped DC I-V curve? Their unpumped curve, our Fig. 3a, is certainly no sharper than those of Rudner et al. (1981b) for $\omega R_A C > 3$, for which negative differential resistance is also predicted on the first peak, but not observed. The answer is certainly that Kerr et al. (1981) were more successful in tuning out the array capacitance, even though they had $\omega R_A C \approx 6$.

3.1. Required LO Power

A simple intuitive argument gives the required LO power for SIS quantum mixing. All of the conversion peaks are strongly oscillatory functions of P_{LO}. Although this dependence is quite complicated, Rudner et al. (1981b) noted that, to a very rough approximation, the conversion on peak number n below the energy gap voltage varies like the Bessel function J_n of argument $\alpha = eV_{LO}/N\hbar\omega$. This means in particular that α, and hence V_{LO}, is reasonably well determined for maximum conversion. Also to rough approximation, the LO input impedance of an SIS mixer is on the order of its normal state resistance R_A. Therefore we can estimate that the required LO power for SIS quantum mixing is $P_{LO} \approx V_{LO}^2/2R_A$, where $V_{LO} = N\hbar\omega\alpha/e$ and α is chosen to maximize $J_n(\alpha)$, where n is the quantum peak number employed. This formula is remarkably successful: it predicts the required P_{LO} for all published SIS quantum mixers. It correctly estimates the required P_{LO} to within 2 dB, covering a range in published experimental P_{LO} from 1 nW to 30 μW.

Since this formula seems generally applicable, we conclude that the required LO power for SIS mixing increases as the square of the number of junctions arrayed, for a given R_A. This conclusion also follows from a much simpler version of the same argument. Operating near the energy gap voltage, the total voltage across an array mixer is N times as large as for a single-junction mixer. To give the same overall impedance, the junctions must be chosen so that the array mixer's currents are N times as large as well. Thus the required powers, LO and DC, increase as N^2.

The mixer's dynamic range should increase accordingly, as N^2. Therefore an array should be used if a moderate to large dynamic range is required. Shen et al. (1980) found that their SIS mixer began to saturate at a signal power 0.003 of P_{LO}. If this is generally true, a single-junction SIS mixer operating on the first peak at 115 GHz would be limited to a signal power of order 10 pW. This is inconveniently small for most applications.

Also, note that P_{LO} increases as the square of the operating frequency, everything else held constant.

4. NOISE IN ARRAYS

4.1. Noise Temperature

It is extremely important to determine whether the noise temperature T_M of an SIS mixer increases with the number of junctions arrayed. In particular, it is known (Shen et al., 1980) that a single-junction SIS mixer can have T_M as low as $\hbar\omega/k$, the "quantum limited" noise temperature. Is the lower limit of T_M for an array mixer $N\hbar\omega/k$? Were this the case, array mixers would be undesirable for low-noise receivers. We maintain the contrary: the noise temperature of an array mixer is not necessarily worse than that of a single-junction mixer.

Unfortunately, the experimental results are ambiguous on this point; the noise temperatures of SIS array mixers have not been measured as precisely as the single-junction mixers. Rudner and Claeson (1979) found T_M = 10-40 K in the classical regime, whereas $N\hbar\omega/k$ = 17 K. Rudner et al. (1981b) found $T_M = 20 \pm 90$ K for all of their arrays. This is barely significant for the 36-junction arrays, where $N\hbar\omega/k$ = 130 K. Kerr et al. (1981), on the other hand, found for N = 14 and 10, T_M = 70 K and 50 K respectively (uncertainty = ±40 K), whereas $N\hbar\omega/k$ = 77 K and 55 K respectively. All of the noise temperatures quoted here are SSB values. Since the quantum noise power $\hbar\omega B$ appears over the entire bandwidth it should be compared to the double-sideband T_M. Then the experimental array T_M's are generally somewhat $< N\hbar\omega/k$. It is also worth mentioning the calculations of Kuzmin et al. (1981) for various Josephson effect array devices. The limiting noise temperatures of the externally pumped devices they considered were independent of N.

The intuitive arguments one might advance to suggest that the noise temperature, in particular the quantum-limited noise temperature, increases with N do not bear close scrutiny. We list a few such arguments here:

1. At least one signal photon must be absorbed by each junction of an array to produce an IF current; thus the minimum detectable signal power is N times as large for an array. But, were this true, T_M would increase much faster than proportionally to N, because all of the signal photons would have to be present simultaneously. In fact, a single signal photon will produce a current. It is impossible to determine which of the junctions in the array absorbed the photon without uncoupling the junctions. That would be a quantum-mechanically distinct measurement.

2. The minimum detectable power for an SIS array operating as a video detector increases as N, because the quantum-limited responsivity $\sim 1/N$ (from Tucker, 1979, Eqs. 5.4 and 5.13). By analogy, the noise temperature in the mixer mode should behave the same. But such an analogy is not tenable; a phase-insensitive detector does not have quantum noise, as discussed in Section 4.1.1. In specific, Tucker (1980) has shown that an SIS mixer could have T_M far below $\hbar\omega/k$ if only shot noise is considered as for the video mode. The source of the quantum noise lies elsewhere.

3. One may argue that shot noise in an SIS array mixer increases with N. Compared to a single junction of the same overall impedance, an array consists of junctions of relative impedance 1/N and thus with Josephson critical current that is N times as large. To operate at the same fraction of I_J requires N times the current and hence N times the shot noise for each junction. Thus the array shot noise must increase more rapidly than N. The outline of this argument is true, but the last sentence is demonstrably false. The demonstration is quite general:

Consider (Fig. 10a) a series array of identical elements, each with impedance Z. The array current I generates shot noise in each element with mean square amplitude $\langle i_n^2 \rangle = 2eIB$. Figure 10a can be represented by the Thévenin equivalent circuit Fig. 10b, where $\langle v_n^2 \rangle = 2eIB|Z|^2$. The noise sources are added incoherently to produce Fig. 10c, with $\langle v_n^2 \rangle = 2eIBN|Z|^2$. This can now be transformed to the Thévenin equivalent circuit of Fig. 10d, giving $\langle I_n^2 \rangle = 2eIB/N$. Note that the available shot noise power from the array is thus the same as that from any one of its constituent elements.

We wish to compare the array shown in Fig. 10 to an "equivalent" single junction, not shown, which might substitute for the array in any device application. This single junction must be designed to have the same overall impedance as the entire array, NZ, and so to establish a comparable operating point the single junction must be driven by 1/N times the voltage and 1/N times the current required for the array. Thus the equivalent single junction has current I/N. This generates a shot noise of mean square amplitude $\langle I_n^2 \rangle = 2e(I/N)B$ for the single junction. But now we see that this is identical to the $\langle I_n^2 \rangle$ found for the entire array in Fig. 10d: The shot noise current

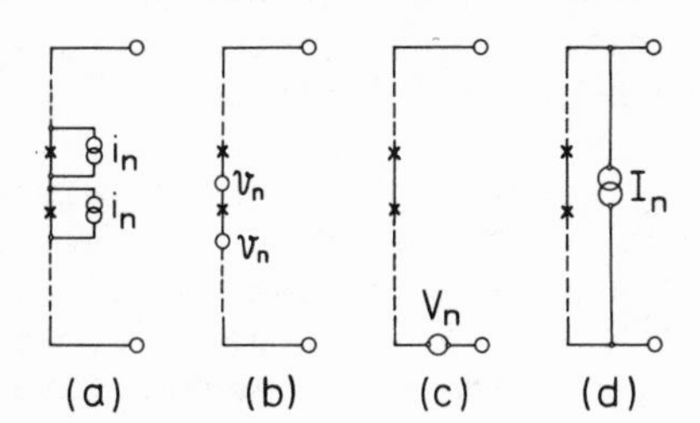

Fig. 10. Construction showing that the shot noise current for an array is identical to that for a single junction, now shown, with the same overall impedance as the array.

for an array is identical to that for a single junction of the same overall impedance. It is clear that an even simpler construction suffices for Johnson noise. Thus the noise properties of an array are no worse than those of a single junction with the same overall impedance given these simple equivalent circuits.

4.1.1. Quantum Noise

It is generally agreed that an SIS mixer cannot have a noise temperature less than $\hbar\omega/k$, the quantum limited noise temperature. But there is no published discussion of quantum noise for SIS mixers, so it is perhaps appropriate to consider this topic at length. In this section we discuss quantum noise in linear amplifiers. Then, in the next section, we apply these ideas to array mixers.

The word "quantum" is not used in the same sense as in the "quantum" theory of mixing, where the quantum-mechanical properties of the charge carriers is important. Rather, quantum noise arises because the radiation field consists of discrete photons. Thus a "classical" mixer will also have quantum noise.

Sometimes the quantum-limited noise temperature is written $\hbar\omega/k\ell n2$. This is correct for the standard definition of noise temperature: T_M is the temperature of the input termination of a noise-free equivalent of a device, which would result in the same output noise power (P_n) as the actual device connected to a noise-free input termination. The factors $\ell n2$ results from using the full Planck black body radiation formula for the P_n from the cold input termination, instead of the Rayleigh-Jeans limit: $P_n = kTB$. But most experiments measure T_M by comparison with an attenuated hot radiator, which is in the Rayleigh-Jeans limit. The noise temperature deduced from such a measurement will have the lower limit $\hbar\omega/k$.

Quantum noise in linear amplifiers was discussed extensively in the early 1960's, when it had become clear that both maser amplifiers and parametric amplifiers had limiting noise temperatures of about $\hbar\omega/k$. We shall review some of the general ideas that were developed. For a more detailed discussion, see Siegman (1964).

Some general properties of radiation sensitive devices can be deduced from the uncertainty principles. If one measures the energy E of a system and the precise time t at which the system possesses this energy, the uncertainties in these quantities are related by $\Delta E \Delta t \geqslant \hbar/2$. A less familiar form of the uncertainty principle results if E is the energy in a signal wave packet at some frequency ω. Then E is related to the number of quanta in the oscillation by $E = n\hbar\omega$, and the phase of the signal is $\phi = \omega t$. This leads to the equation

$$\Delta n \, \Delta\phi \geqslant 1/2$$

relating the uncertainties in the simultaneous measurement of the quantities n and ϕ. In other words, one cannot measure both the amplitude and the phase of a sinusoidal signal precisely.

Most radiation sensitive devices fall into either of two distinct categories. An amplifier attempts to reproduce its input radiation field, usually with larger amplitude, while maintaining the phase information. A detector responds to the power of its input and is insensitive to the input phase. An ideal detector can count n with no uncertainty ($\Delta n \to 0$) on the condition that ϕ is randomized ($\Delta\phi \to \infty$). A physical realization of such a quantum detector is the X-ray counter. One may also imagine other types of radiation sensitive devices with varying degrees of phase sensitivity (Serber and Townes, 1960).

The arrival time of the photons at a quantum detector will of course fluctuate. This fluctuation can be considered inherent in the input signal. In this sense an ideal detector is noiseless; it adds no noise to the fluctuations already present in its input. Can we imagine a noiseless linear amplifier, which perfectly reproduces its input, including fluctuations, with larger amplitude? A compelling argument given by Heffner (1962) shows that this is impossible. A noiseless linear amplifier would have the property that n_i photons received at its input would produce n_o output photons, where $n_o = Gn_i$ and G is the (integral) amplifier gain. Further, the output phase ϕ_o would be equal to the input phase ϕ_i plus some constant phase shift θ. Both G and θ can be precisely measured at large signal levels. An ideal detector attached to the output would enable us to measure n_o and ϕ_o with an uncertainty $\Delta n_o \, \Delta\phi_o = 1/2$. If the amplifier is noiseless, then we have performed a measurement of the input signal quantities with uncertainties $\Delta\phi_i = \Delta\phi_o$ and $\Delta n_i = \Delta n_o/G$. But this violates the uncertainty relation $\Delta n_i \, \Delta\phi_i \geqslant 1/2$ if $G > 1$. The conclusion is that a noiseless linear amplifier is impossible to construct. Carrying this argument further, Heffner (1962) shows that an ideal high-gain linear amplifier, one that minimizes the product $\Delta n \, \Delta\phi$ to 1/2, its uncertainty principle limit, has a noise temperature $\hbar\omega/k$, precisely the value derived from the detailed equations of both the maser and the parametric amplifier.

Where does the quantum noise come from? Is it generated in the amplifier or is it a property of the radiation field? Heffner treats the amplifier as a black box. The fluctuations seem to be an inherent attribute of the signal itself, as a result of the uncertainty principle. The noise can be treated as stimulated emission caused by the zero-point vacuum fluctuations which inevitably accompany the signal. (Note that a detector must absorb photons to produce an output, so it is not affected by zero-point fluctuations.) On the other hand, quantum noise can be derived from specific physical processes, such as shot noise or spontaneous emission, which appear to add noise to the signal as it is amplified. Thus the quantum noise output is present even without an input signal.

The answer of course is that in the quantum theory this question is meaningless. It is impossible to separate the measuring apparatus from the signal to be measured.

4.1.2. Quantum Noise in Array Mixers

In the last section we have reviewed some of the concepts which lead to the general result: the noise temperature of a high-gain linear amplifier is limited to the value $\hbar\omega/k$. This discussion did not specify that the input and output frequencies be the same. Linear frequency transformations are not ruled out. All that is required is that the output amplitude and phase are linearly related to the input quantities. A change in photon frequency is immaterial since the limiting uncertainty is dependent solely on the number of photons arriving, and not upon their energy. In this sense the gain is interpreted as photon number gain rather than the power gain.

An interesting example considered by Louisell et al. (1961) is the parametric up-converter, which has power gain by virtue of an increase in frequency but has unity photon gain. An ideal parametric up-converter has a limiting noise temperature and phase uncertainty of zero. Noise-free up-conversion does not violate the uncertainty principle because $\Delta n \, \Delta\phi$ remains the same. One may say that this device has neither spontaneous nor stimulated emission; it must absorb a low-frequency photon to produce a high-frequency photon.

Oliver (1961) and Hause and Townes (1962) found a limiting noise temperature of $\hbar\omega/k$ for the photoelectric heterodyne mixer. In light of the above discussion, this result must be valid for all heterodyne mixers. Even a classical mixer, which can have no power gain, has a very large photon gain by virtue of the frequency conversion, if the IF is low. And the phase information of the signal is preserved in the IF, presuming the LO phase is precisely known. Therefore the mixer is a high-gain linear amplifier in the sense discussed in the last section. As such, its limiting noise temperature is $\hbar\omega/k$.

To substantiate this, imagine an apparatus which consists of a mixer followed by an ideal parametric up-converter, chosen so that the overall input and output frequencies are identical. Taken as a whole the apparatus is a single-frequency high-gain linear amplifier and certainly must have a limiting noise temperature, referred to its input, of $\hbar\omega/k$. As noted, the parametric up-converter may be supposed to contribute zero noise. The entire noise must come from the mixer. The mixer thus has a limiting noise temperature of $\hbar\omega/k$, referred to *its* input. Clearly, ω is the frequency of the input signal.

After this general discussion, we may finally ask whether the quantum noise temperature of an array mixer is $N\hbar\omega/k$. This can be answered from two different viewpoints. First, let us consider the quantum noise to be inherent in the signal radiation field. We have seen that Heffner (1962) derives the quantum noise temperature $\hbar\omega/k$ without considering any specific properties of a particular device. The noise temperature clearly cannot depend upon whether the device consists of a single junction or an array.

On the other hand, consider quantum noise to arise within the mixer itself. To calculate the output noise, the quantum noise must be represented by one or many noise generators, in some equivalent circuit. Haus and Mullen (1964) have derived some equivalent circuits for quantum noise. Now the quantum noise equivalent circuit can be treated exactly like the equivalent circuits for shot noise or thermal noise. When this is done, we have already seen in the discussion of Fig. 10 in Section 4.1 that the noise properties of an array are no worse than of a single junction.

Thus we come to the same conclusion for either approach: the quantum-limited noise temperature of an array mixer, as of a single-junction mixer, is $\hbar\omega/k$.

4.2. Josephson Effect Noise

There is an additional source of noise which can be a problem for SIS mixers. This is quite evident in Fig. 11. Figure 11 shows typical chart recorder plots of the output noise power in the IF frequency band as a function of the DC bias voltage, for an SIS array mixer with applied LO power but with no applied signal. For larger

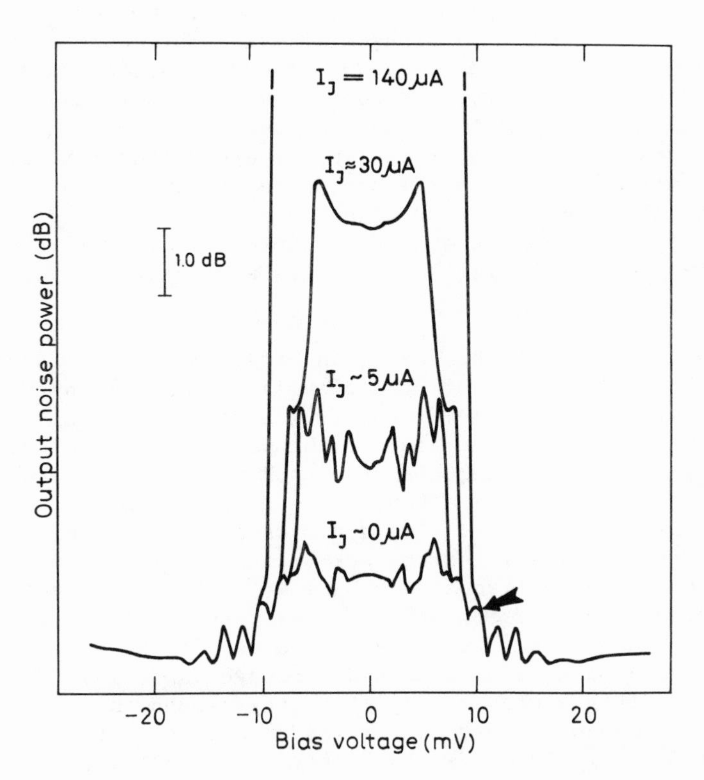

Fig. 11. The output noise power in the IF band from a typical array used by Rudner et al. (1981a,b), with constant LO but no signal power applied. The unpumped magnetic-field depressed Josephson critical current of the array (I_J) is listed, beginning with the zero-field value. The arrow indicates the definition of the noise threshold, V_N, for one curve.

values of V_{DC} the mixer is relatively quiet; the oscillatory noise seen in the figure is due to the converted RF input noise and to noise in the IF output circuit reflected by the output impedance of the mixer which is an oscillatory function of the bias voltage. But below a certain threshold bias voltage V_N the output IF noise in Fig. 11 suddenly becomes extremely large. When a magnetic field is applied to suppress the DC Josephson critical current this large noise output is reduced in two ways: the threshold bias voltage V_N is decreased and the magnitude of the noise is diminished. In Fig. 11 the mixer output noise temperature is thus reduced from more than 2000 K without magnetic field to about 65 K on the lowest trace. This strong dependence upon the Josephson critical current implies that this instability arises from the Josephson effect. The extremely sharp onset and large magnitude of the Josephson effect noise indicate that it is caused by some instability, and therefore can and should be completely avoided, whereas the other sources of noise we have mentioned are always present to some extent in any experimental mixer and can only be minimized. Although we are concerned only with the SIS mixer, note that the Josephson effect noise in Fig. 11 occurs even at zero DC bias and therefore should be a problem for various unbiased externally-pumped high-frequency devices employing Josephson junctions.

The explanation of the Josephson effect noise was first suggested by Dolan et al. (1981) and clearly established by Rudner et al. (1981a,b). When the I-V curve of a Josephson tunnel junction is slowly swept with a current source, the voltage hysteretically switches to zero from a certain drop-back voltage V_d on the quasiparticle branch. Thus it is impossible to maintain a stable bias at finite DC voltage below V_d. The dynamics of the switching are not understood in detail. The drop-back voltage is linearly related to the junction's plasma frequency: $V_d = kV_{pl}$, where $V_{pl} = (\hbar I_J/2eC)^{1/2}$ is the maximum plasma frequency voltage. Rudner et al. (1981a,b) found k = 3.0 and Shen (1981) k = 3.5 for their respective junctions. Consider an SIS mixer biased at a DC voltage $V_{DC} > V_d$. The applied LO causes a voltage swing of amplitude V_{LO} around this bias point. If V_{LO} is large enough so that the instantaneous voltage across a junction ever falls below V_d, the junction will then switch to its zero-voltage state very rapidly. The current must then increase to exceed the critical current before the junction can switch back to its quasiparticle branch. This hysteresis loop will be traversed on every cycle of the electromagnetic field.

This hysteretic switching due to the drop-back phenomenon is the cause of the Josephson effect noise in Fig. 11. It is clear that the noise threshold V_N occurs at a DC voltage such that $V_{DC} - V_d \approx V_{LO}$. A magnetic field decreases I_J and therefore decreases V_d and hence V_N, as observed. It is reasonable to suppose that the magnitude of the Josephson effect noise diminishes when I_J is reduced because then less time must be spent in traversing the hysteresis loop. (The output noise is reduced in approximate proportion to $\sqrt{I_J}$ as the magnetic field is increased.)

If an SIS mixer is biased securely above V_N, there is no noise from the mechanism we have discussed. The unstable region is never entered. There may, however, be a contribution to the mixer's noise temperature from noise implicit in the Josephson currents themselves, which exist in an SIS mixer at any DC voltage. Such noise would increase, gradually rather than abruptly, as the DC voltage is lowered, because the junction capacitance is a less effective shunt for the lower frequency Josephson oscillations. There is at present no experimental evidence for such a noise temperature contribution.

4.2.1. High Frequency Limitation

For low-noise operation an SIS mixer must avoid the Josephson effect noise region. Below about 100 GHz this is not a problem. But for high frequency operation the Josephson effect noise imposes severe design limitations on SIS mixers. This is because as the operating frequency is increased the optimum DC voltage bias point moves away from the energy gap towards lower voltage, while the optimum V_{LO} increases proportional to frequency (see Section 3.1.) to continue to sample the nonlinearity at the energy gap. Both of these tendencies decrease the quantity $V_{DC} - V_{LO}$. As we have seen, when this quantity becomes equal to or smaller than V_d, the Josephson effect noise occurs. This is already a problem for Dolan et al. (1981) at 115 GHz; because of the Josephson effect noise they cannot utilize their second conversion peak, which they predict to be 3 dB larger than their observed first peak conversion. A simpleminded extrapolation from published work (Shen et al., 1980; Dolan et al., 1981; Rudner et al., 1981b) indicates that at about 200 GHz the first conversion peak also moves into the Josephson effect noise region, making low-noise SIS mixing impossible.

One way to avoid this problem is to design an SIS mixer with decreased V_d. But the quantity $I_J R$ is a constant for a given superconductor, and the ωRC product is generally determined by other considerations. Then unless a magnetic field is used to decrease I_J, the drop-back voltage V_d, which is proportional to $(I_J/C)^{1/2}$, is fixed. Therefore a low-noise high-frequency SIS mixer must use a magnetic field to suppress I_J. But the very small junctions used in single-junction SIS mixers require very large fields to achieve this suppression, and these fields are large enough to begin to destroy the background superconductivity and so degrade SIS mixer performance. Phillips et al. (1981) show a very clear example of this at 230 GHz. Thus we can say that the rough upper frequency limit of a single-junction SIS mixer is 200 GHz.

An array mixer does not have this limitation. For the same overall impedances the arrayed junctions are larger in size and a magnetic field can be, and in general is, used to suppress the Josephson effect. Array SIS mixers should be satisfactory up to about 300 GHz and perhaps to twice that frequency. This is a very strong reason to choose arrays for high-frequency SIS mixers.

The Josephson effect noise limitation on high frequency SIS mixing is not absolute. There are a number of possible ways a single-junction SIS mixer may be designed to circumvent it. For instance, the discussion in Section 5. indicates that ωRC products considerably greater than unity are advantageous. This permits the use of larger junctions. One might also use a small-area tunnel junction which is relatively long and narrow, which allows a more moderate magnetic field for suppressing I_J. Another possibility is to operate an SIS mixer at a point of high differential DC resistance much closer to the energy gap than $\hbar\omega/2e$ (for instance, at $V_{DC} \approx$ 40 mV rather than 38 mV in Fig. 3). At high frequencies this would give a considerable increase in V_{DC}.

It is also possible that the Josephson effect noise may be avoided completely. Consider curve (e) of Fig. 3. This is an experimental plot of the output IF noise power for an SIS array mixer with applied LO power but with no applied signal: the same quantity as plotted in Fig. 11 for another experiment. And yet Fig. 3e does not exhibit a large Josephson effect noise such as is seen in Fig. 11. At intermediate DC bias voltage Fig. 3e does show the effects of impedance irregularities due to the remnants of the Josephson steps, also evident in curves (b) and (d), but the noise output here is no larger than at higher voltages. The reason for this is presumably that the array in Fig. 3 is effectively RF voltage biased: its impedance is much larger than the source impedance at all frequencies. Therefore the instantaneous voltage follows a stable path even below V_d and does not experience the hysteretic switching to zero voltage. If this explanation is correct, an SIS mixer may in general avoid the Josephson effect noise if it is driven from a relatively small RF source resistance. This condition is known to be advantageous for other reasons as well. Feldman (1981) shows that if infinite differential resistance on the pumped DC I-V curve is attained (a very small source resistance will prevent this), an SIS mixer should have a relatively small source resistance along with a relatively large IF load resistance to achieve high, stable, conversion.

4.2.2. SIN Mixer

An obvious way to eliminate Josephson effect noise and any other deleterious Josephson phenomena is to replace one of the superconducting tunnel electrodes with a normal metal. The superconductor-insulator-normal metal (SIN) tunnel junction has no Josephson currents. The SIN mixer has two disadvantages. Firstly, the normal electrode introduces a series resistance which can make the reactive tuning less efficient; this resistance can be made quite small and probably will not be a serious problem. More significantly, SIN junctions are considerably less nonlinear than SIS junctions. In fact, an ideal SIS junction has a perfectly sharp quasiparticle nonlinearity at all temperatures below its critical temperature, while an ideal SIN nonlinearity is perfectly sharp only at $T = 0$. Nevertheless, at high enough frequencies the SIN nonlinearity is sharper than the photon energy expressed in voltage units, and SIN junctions should be suitable for quantum mixing. There is good reason to believe that the SIN mixer will be competitive with the SIS mixer above perhaps 300 GHz.

Rudner et al. (1981b) have performed the first measurements of an SIN array mixer. The samples and the experimental apparatus were exactly the same as for their SIS measurements, except that the upper electrode of the junctions was made of Ag instead of Pb. The best results for a six-junction Pb(In)-O-Ag array at 73.5 GHz are shown in Fig. 12. There is no trace of quantum structure in the conversion curve, which is not surprising at this frequency. Still, the maximum conversion result $L = 9.1$ dB is encouraging.

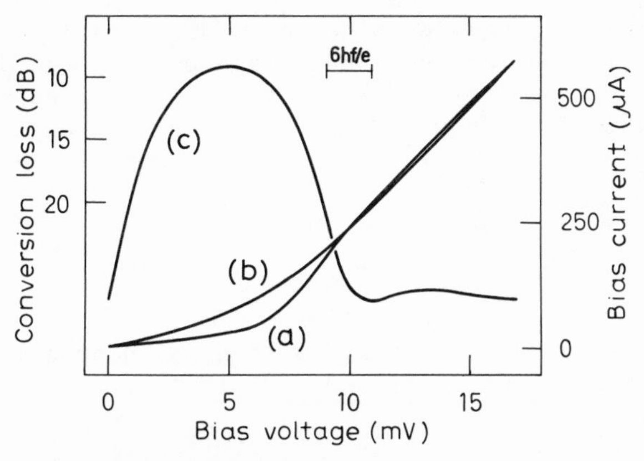

Fig. 12. Experimental results for an SIN mixer at 73.5 GHz. The nonlinear element was a 6-junction Pb(In)-O-Ag array with $R_A = 23.5\ \Omega$. DC I-V curves (a) without P_{LO} and (b) with $P_{LO} = 0.9\ \mu W$, and (c) SSB conversion loss again with $P_{LO} = 0.9\ \mu W$ are shown. Unpublished results from Rudner et al. (1981b).

5. CONCLUSION

It has generally been believed that an SIS mixer should be designed with its ωRC product $\lesssim 1$. This is *not* the same ωRC product that is usually considered for semiconductor devices, for which R refers to the parasitic series resistance external to the nonlinear element. For SIS mixing, R is the characteristic resistance of the nonlinear element itself. The distinction is crucial. The capacitance of an SIS junction can in principle be perfectly resonated; a semiconductor's capacitance cannot (without exacerbating the parasitic loss). Therefore an argument by analogy to semiconductor devices is misleading.

The experimental evidence to date indicates that *an SIS mixer in the quantum regime should have* $\omega RC \gg 1$. Dolan et al. (1979,1981) used a small ωRC product and found comparatively poor conversion. A number of experiments had $\omega RC \sim 1.5$; in each case a small conversion loss resulted. Both Smith et al. (1981) and Kerr et al. (1981) had ωRC close to 10 and found infinite available gain. The available conversion thus appears to improve with larger ωRC. This is presumably because the larger relative capacitance tends to minimize harmonic conversion effects. The use of a larger ωRC product of course requires a more sophisticated tuning circuit to prevent the capacitance from shunting out the nonlinearity at the fundamental frequency.

This result eases the requirement of small junction size, making it less likely that arrays will be used in any specific SIS mixer application. But there are two countervailing considerations. First of all, the dynamic range of single-junction mixers may be marginal for a practical device. Secondly, when SIS mixers are employed at higher frequency, the same problems of small junction size will recur. Our conclusion is that arrays should be used when the SIS mixer emerges from the laboratory as a practical device, especially at higher frequencies. We have shown that this will entail no loss of ultimate sensitivity.

ACKNOWLEDGMENTS

The authors acknowledge many substantial discussions with A.R. Kerr, J.R. Tucker, T. Claeson, and C.H. Townes.

REFERENCES

Dolan, G.J., Phillips, T.G., and Woody, D.P. (1979). *Appl. Phys. Lett.* *34*, 347.

Dolan, G.J., Linke, R.A., Sollner, T.C.L.G., Woody, D.P. and Phillips, T.G. (1981). IEEE Trans. Microwave Theory Tech. MTT-29, 87.

Feldman, M.J. (1981). Submitted to J. Appl. Phys.

Hartfuss, H.J., and Gundlach, K.H. (1981). Submitted to Int. J. Infrared and Millimeter Waves.

Haus, H.A., and Mullen, J.A. (1962). In Quantum Electronics, edited by P. Grivet and N. Bloembergen. Columbia University Press, New York, pp. 71-93.

Haus, H.A., and Townes, C.H. (1962). Proc. IRE 50, 1544.

Heffner, H. (1962). Proc. IRE 50, 1604.

Kerr, A.R., Pan, S.-K., Feldman, M.J., and Davidson, A. (1981). Physica 108B, 1369.

Kuzmin, L.S., Likharev, K.K., and Migulin, V.V. (1981). IEEE Trans. Magnetics MAG-17, 822.

Louisell, W.H., Yariv, A., and Siegman, A.E. (1961). Phys. Rev. 124, 1646.

Oliver, B.M. (1961). Proc. IRE 49, 1960.

Phillips, T.G., Woody, D.P., Dolan, G.J., Miller, R.E., and Linke, R.A. (1981). IEEE Trans. Magnetics MAG-17, 684.

Richards, P.L., and Shen, T.-M. (1980). IEEE Trans. Electron Devices ED-27, 1909.

Richards, P.L., Shen, T.-M., Harris, R.E., and Lloyd, F.L. (1979). Appl. Phys. Lett. 34, 345.

Rudner, S., and Claeson, T. (1979). Appl. Phys. Lett. 34, 713.

Rudner, S., Feldman, M.J., Kollberg, E., and Claeson, T. (1980). SQUID '80, edited by H.-D. Hahlbohm and H. Lübbig. W. de Gruyter, Berlin, pp. 901-906.

Rudner, S., Feldman, M.J., Kollberg, E., and Claeson, T. (1981a). IEEE Trans. Magnetics MAG-17, 690.

Rudner, S., Feldman, M.J., Kollberg, E., and Claeson, T. (1981b). J. Appl. Phys., to be published.

Serber, R., and Townes, C.H. (1960). In Quantum Electronics, edited by C.H. Townes. Columbia University Press, New York, pp. 233-255.

Shen, T.-M. (1981). IEEE J. Quantum Electron. QE-17, 1151.

Shen, T.-M., and Richards, P.L. (1981). IEEE Trans. Magnetics MAG-17, 677.

Shen, T.-M., Richards, P.L., Harris, R.E., and Lloyd, F.L. (1980). Appl. Phys. Lett. 36, 777.

Siegman, A.E. (1964). Microwave Solid-State Masers, McGraw-Hill, New York, pp. 412-431.

Smith, A.D., McGrath, W.R., Richards, P.L., van Kempen, H., Prober, D., and Santhanam, P. (1981). Physica 108B, 1367.

Tucker, J.R. (1979). IEEE J. Quantum Electron. QE-15, 1234.

Tucker, J.R. (1980). Appl. Phys. Lett. 36, 477.

Wahlsten, S., Rudner, S., and Claeson, T. (1978). J. Appl. Phys. 49, 4248.

FAST DETECTOR OF DISCONTINUOUS METAL FILM

FOR MILLIMETER THROUGH OPTICAL FREQUENCY RANGE

Sogo Okamura, Yoichi Shindo and Hiroshi Nakazato

Department of Electronic Engineering
Tokyo Denki University
Chiyoda-ku, Tokyo, 101, Japan

and

Yoichi Okabe and Sumiko Tamiya

Department of Electrical and
Electronic Engineering
University of Tokyo
Bunkyo-ku, Tokyo, 113, Japan

1. INTRODUCTION

In the microwave through millimeter-wave frequency range, semiconductor devices are very powerful as detectors and mixers. Especially gallium arsenide has high electron mobility and its Schottky barrier diode is one of the best devices in that frequency region. However, in the higher frequency region of submillimeter waves through the optical frequency region, the time sonstant determined from the contact capacitance and the spreading resistance can not be neglected for those semiconductor devices.

For this reason, metal-to-metal point-contacted diodes with tunneling effect or Josephson effect are frequently used in the higher frequency region, especially for precise frequency measurement of lasers in submillimeter wave and infrared ray regions. Along with low resistivity of metals, these two effects are considered to be very fast phenomena of electron transfers from one metal to the other.

Electron tunneling time through the barriers of metal-insulator-metal diodes is estimated to be of the order of 10^{-16} second and transition time of Josephson junctions is also of the order of

10^{-12} second, so the frequency response of such devices is mostly limited by the RC time constant. The time constant is mainly determined by the thickness of the insulator as shown in Fig. 1. Therefore the devices can be used as fast detectors for audio to optical frequency range, when the thickness of the insulator is very thin.

They, however, are not so stable for mechanical and electric shocks and their life-times are as short as the order of a day. To overcome this defect, metal-insulator-metal structures are fabricated by means of evaporation and lithography techniques, but it is very necessary to obtain a small contact area for short RC time constant.

In order to realize a small contact area, the idea was proposed that island structure in the very thin evaporated metal layer can

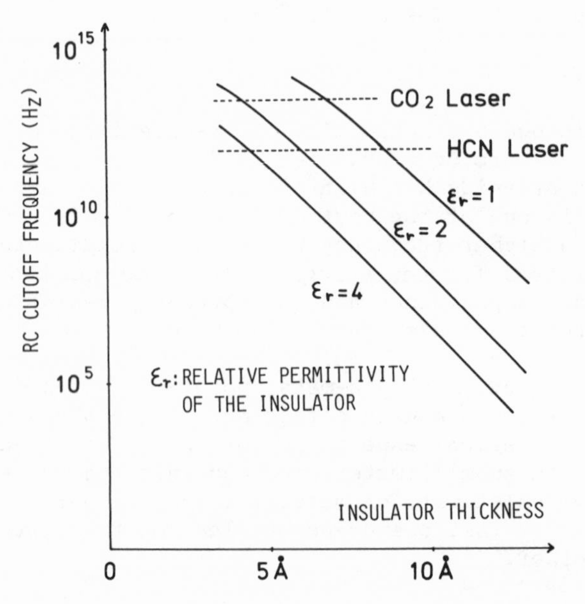

Figure 1. RC time constant of the metal-insulator-metal diode without bias voltage where work functions of metals are supposed 5.32 eV.

compose series and parallel connections of many metal-insulator-metal contacts, and the idea was checked by detection and mixing experiments of millimeter waves, HCN laser ray or CO_2 laser ray. Moreover, such a structure is a two-dimensional one, which might lead to new concepts of millimeter wave and optical circuit schemes.

2. METAL-TO-METAL POINT-CONTACTED DIODE

The metal-to-metal point-contacted diode has a structure where a metal whisker of several to several tens of micrometers in diameter, which is sharpened by electro-chemical etching, is contacted to a polished metal surface on a metal post. The nonlinearity of the electric conduction is explained by fast transition of electrons through one bulk metal to the other with the tunneling effect. As the transition time is estimated at the order of 10^{-16} second (Hartman, 1962), the diode has the potential to respond to frequencies up to the optical range (Evenson et al., 1970a,b; 1972; 1974; Abram et al., 1970; Ijichi et al., 1973; Okamura et al., 1974; Sakuma et al.,1974; Jenning et al., 1975; Farris et al., 1973).

An example of the structure of the diode is shown in Fig. 2 (Ijichi et al., 1973; Okamura et al., 1974). In the diode, the

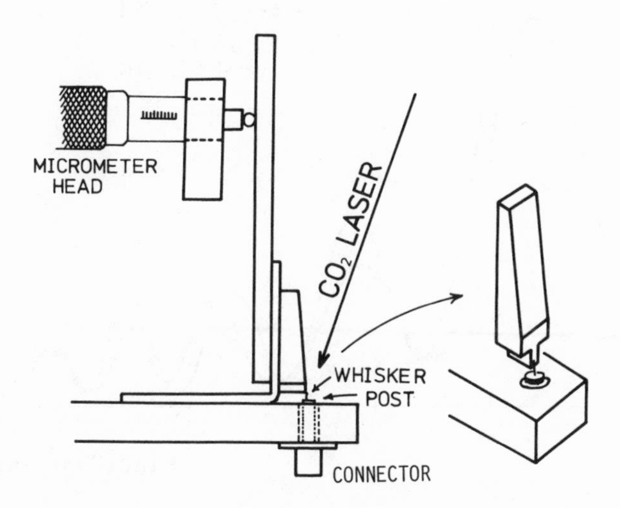

Figure 2. Structure of metal-to-metal point-contacted diode.

tungsten whisker also operates as a long-wave antenna for laser rays (Ijichi et al., 1973; Okamura et al., 1974; Matarrese et al., 1970; Twu et al., 1975). Figure 3 shows the experimental detector output as a function of the incident angle of the ray along the whisker. Theoretical values are also plotted by a solid line, which shows the directivity factor D(θ) given by the following equation(Ijichi et al., 1973; Okamura et al., 1974):

$$D_\theta = -jk_0 \int_0^L -e^{-(\alpha+jk_0)z} \sin\theta \; e^{jk_0 z\cos\theta} \, dz$$

$$= \frac{-jk_0 \sin\theta}{\alpha+jk_0(1-\cos\theta)} \left[e^{-L(\alpha+jk_0[1-\cos\theta])} -1 \right] \qquad (1)$$

where k_0 denotes the propagation constant ω/c, is the attenuation constant of the line, and L, the antenna length.

Figure 4 shows typical dc characteristics of the diode along with the detection-output dependence on vias voltage in Fig. 5.

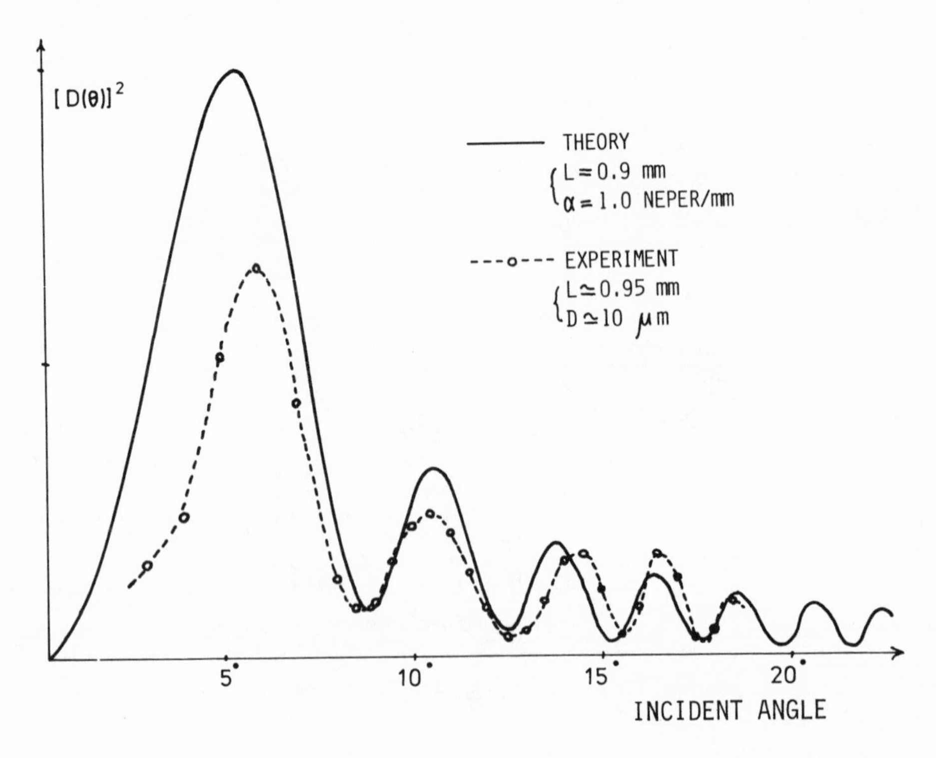

Figure 3. CO_2 laser detection power vs. incident angle of laser ray.

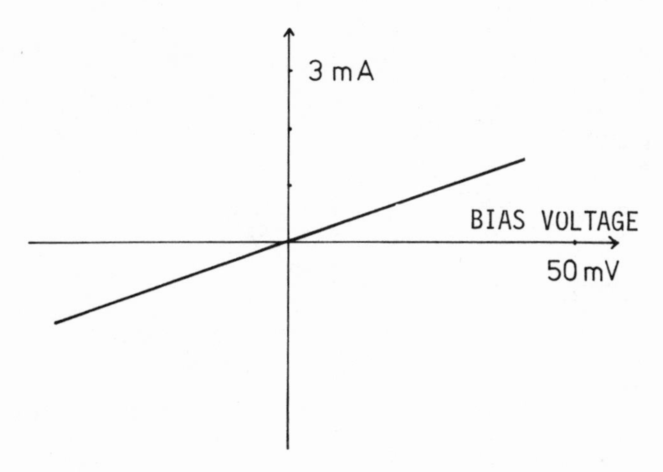

Figure 4. V-I characteristics of metal-to-metal point-contacted diode.

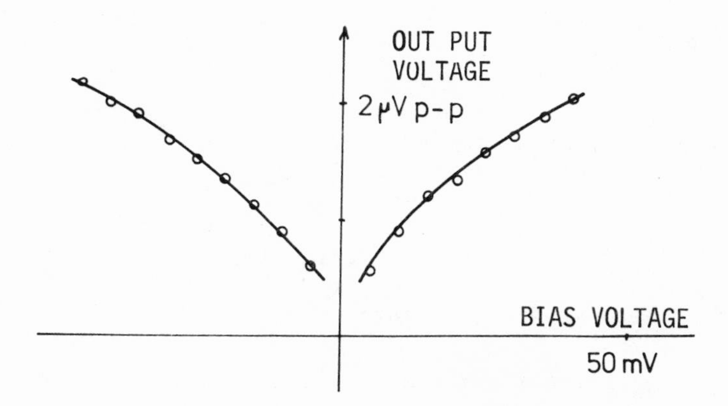

Figure 5. Detector output of metal-to-metal point-contacted diode vs. bias voltage.

The theoretical voltage-to-current relation of a metal-insulator-metal structure, where two metals of work functions, ϕ_1 and ϕ_2, are contacted through the insulator of thickness, s, is given by the following equation (Simmons, 1963 a, b):

$$J = J_0[\bar{\phi}\ e^{-A\bar{\phi}^{\frac{1}{2}}} - (\phi + eV)e^{-A(\phi + eV)^{\frac{1}{2}}}] \tag{2}$$

where

$$J_0 = \frac{e}{2\pi h(\beta\Delta s)^2} \ ,$$

$$\bar{\phi} = a - b(eV) \ ,$$

$$a = \phi_2 - b(\Delta\phi) - \frac{1.15\lambda s}{\Delta s} \ \ln\left[\frac{s_2(s-s_1)}{s_1(s-s_2)}\right],$$

$$b(x) = \frac{s_1+s_2}{2s} \ ,$$

$$A = \frac{4\pi\beta\Delta s}{h} (2m)^{\frac{1}{2}} \ ,$$

$$\beta \simeq 1.0 \ ,$$

$$\Delta s = |s_1 - s_2| \ ,$$

$$\Delta\phi = \phi_1 - \phi_2,$$

$$\lambda = \frac{e^2 \ \ln 2}{8\pi\varepsilon s} \ ,$$

m, effective mass of electron,

h, Plank's constant,

ε, dielectric constant of the insulator,

and s_1 and s_2 are two roots of

$$\phi(x) = \phi_2 - (\Delta\phi + eV)\left(\frac{x}{s}\right) - \frac{1.15\lambda s^2}{x(s-x)} = 0 \ . \tag{3}$$

The calculated results explain the measured ones in Figs. 4 and 5. However, the values of ϕ_1 and ϕ_2 estimated from the experimental results are smaller than the work functions of the metals.

The diode can also operate for the CO_2 laser ray (Ijichi et al., 1973; Okamura et al., 1974). Table 1 shows the frequency differences of p-branches, which are determined by mixing experiments using a 50 GHz local oscillator and 30 MHz receiver, and by the relation of

$$f_m = |f_{p(n)} - f_{p(n+2)}| \pm 30 \quad [MH_z]. \tag{4}$$

As shown above, metal-to-metal point-contacted diodes have high performances in the infrared and far-infrared frequency regions, but are not so stable for mechanical and electric shocks, and their life-times are as small as of the order of a day.

TABLE 1

FREQUENCY DIFFERENCES OF P-BRANCHES OF CO_2 LASER

Transition line of CO_2 Laser		Frequency difference (GHz)			
Branch No.	Frequency (GHz)	Theoretical value	Measured value		
P(16)	28431.9				
P(18)	28378.2	53.7	52.80		
P(20)	28324.5	53.7	53.50	53.51	53.52
P(22)	28271.1	53.4	54.25	54.27	
P(24)	28215.3	55.8	55.17	55.18	

3. METAL-INSULATOR-METAL DIODE OF SUPERPOSED STRUCTURE

In order to stabilize point-contacted diodes, the metal-insulator-metal structure is studied with vacuum evaporation and lithography techniques. The contact area, however, should be less than 10^{-10}cm^2 in order to obtain a small RC time constant. Presently, at the University of California, Berkeley (Wang et al., 1977) and the National Bureau of Standards, Boulder (Nahman),diodes with an area of the order of 10^{-10}cm^2 are fabricated. The upper 1 μm width metal strip was evaporated onto the other metal film of 100 μm thickness, whose surface was oxidized, and its detection characteristics are now checked for waves of 10.6, 3.39, 1.15 and 0.6328 μm wave-length.

4. DISCONTINUOUS THIN FILM DETECTOR

In the early stage of evaporation of metal in a vacuum, the metal composes a discontinuous island structure instead of a continuous thin film (Nishiura et al., 1975; Neugebauer, 1967). By further deposition, each island grows, absorbing evaporated metal atoms. Some islands merge, some islands contact to others, and gradually the structure becomes a continuous film.

Even when islands are isolated, the thin film has some conductivity. This is explained by the tunneling effect of electrons through the insulator (Nishiura et al., 1975; Hill, 1969). For a thin film where small islands are densely arranged, the direct tunneling effect is a major factor. When larger islands are separated by a further distance, the conduction with impurity levels in substrate material, or the effect of thermal excitation and thermal emission of electrons in the island, appears. These conduction mechanisms are summarized in Fig. 6, which was proposed by Neugebauer(1967). There was no explanation for the region A* , but it is considered that the main conduction mechanism is direct tunneling as for the region A.

Among these, the structure in the region A is considered as a set of many metal-insulator-metal contacts which are connected serially and parallelly. If an electromagnetic wave is applied to the film, the detected signals which are generated from the nonlinearity of each metal-insulator-metal contact would be summed up. Moreover, there is another advantage that this two-dimensional film device can be directly coupled to the solid circuit.

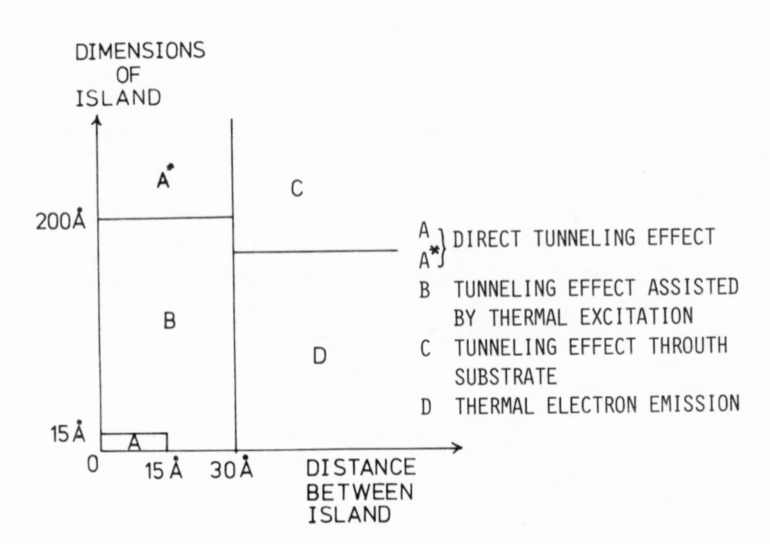

Figure 6. Electric conduction mechanism of island structure thin metal films.

The thin metal film of the island structure is mechanically stable and realizes stable small-area metal-insulator-metal contacts, and therefore becomes a fast detector in the far-infrared to infrared ray region (Okamura et al., 1976; Ijichi et al., 1977).

5. DISCONTINUOUS THIN FILM OF GOLD

By the evaporation of gold, nickel or aluminum, discontinuous thin film detectors were fabricated as shown in Fig. 7, and their detection properties were tested with 35 GHz millimeter waves and CO_2 laser ray (Okamura et al., 1976; Ijichi et al., 1977). Good performance was obtained for the gold evaporated film of 5 nm thick and the surface resistivity of several tens of kilo-ohms.

Metal was evaporated on a glass substrate at room temperature in vacuum of 10^{-3} to 10^{-4} Pa. The thickness of the film was controlled by observing the film resistance, but the resistance changed rapidly around the optimum point; then it was necessary to pay attention. Evaporation speed was set such that total evaporation time was around several minutes.

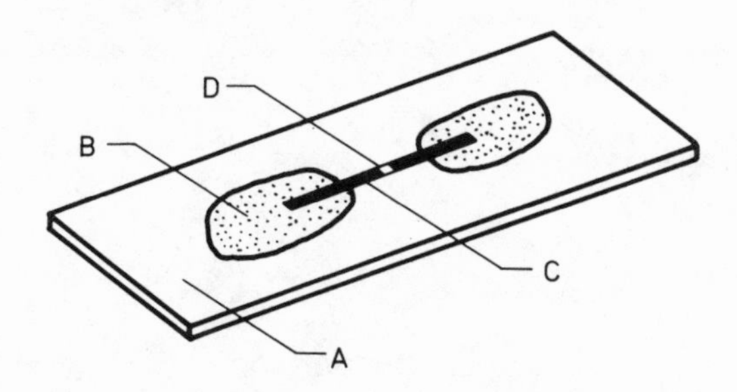

Figure 7. Schematic view of discontinuous gold film detector.

Thin film devices with nickel or aluminum did not show good detection performance for any resistance. The reason was considered that the separation of islands was increased by the oxidation of metals. Even with gold films, the detectors have poor sensitivity and very short life (about 80 hours). Rapid change of resistivity happens after the evaporation, but normally the resistance becomes stable within a week.

Figure 8 shows a photograph of the gold film observed by an electron microscope, and the film is composed of separated gold islands. The dc characteristics are shown in Fig. 9(a) along with the detection property of a CO_2 laser ray in Fig. 9(b). The ray was chopped by 375 Hz frequency, and the signal was observed by a locking amplifier. Results of measurements show that the phase of the detected signal is just opposite for the films of the resistivity of several kilo-ohms and several megohms, and for the films of intermediate resistance, both types co-exist.

Figure 10 indicates the detection property of 35 GHz millimeter waves modulated by 10 MHz hf waves. Sensitivity is not so good and is -35 dB compared with that of a semiconductor device of 1N23 without bias. It is observed that there are some detectors whose detection property deteriorates 70 to 80 hours after the fabrication when the millimeter wave oscillator is modulated by high frequency waves, but does not deteriorate when modulated by low frequency waves under 1 kHz. This fact and the former result of polarity change suggest

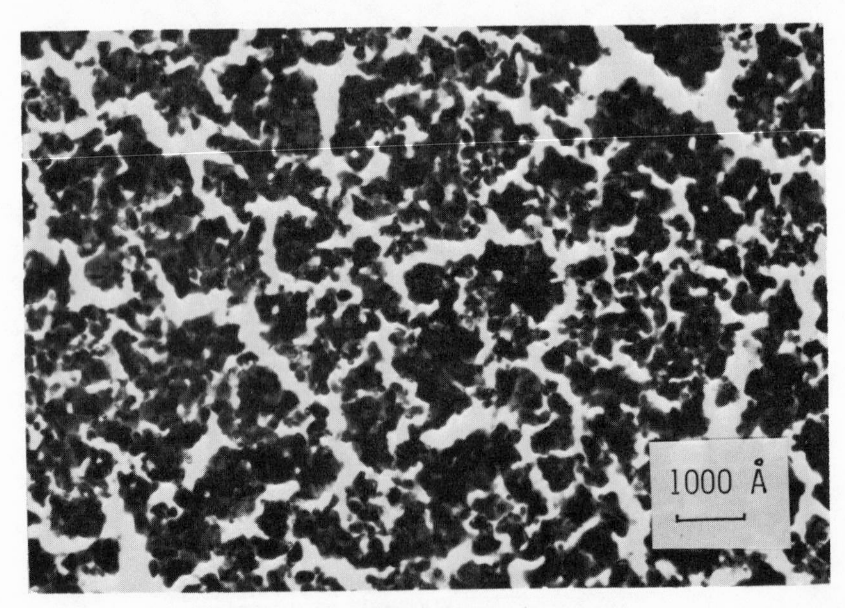

Figure 8. Photograph of surface of discontinuous gold film observed by electron microscope.

to us that there is a two-detection mechanism of high speed and low speed. The latter one is supposed to be the thermal mechanism; and the former one is the other mechanism such as tunneling effect.

The picture in Fig. 8 is a stable film of several days after evaporation, and it shows only the thermal property. The observation of the film just after the evaporation, on the other hand, indicates that there exist many small islands among larger islands. Therefore, for tunneling effect, small-islands structure is very important.

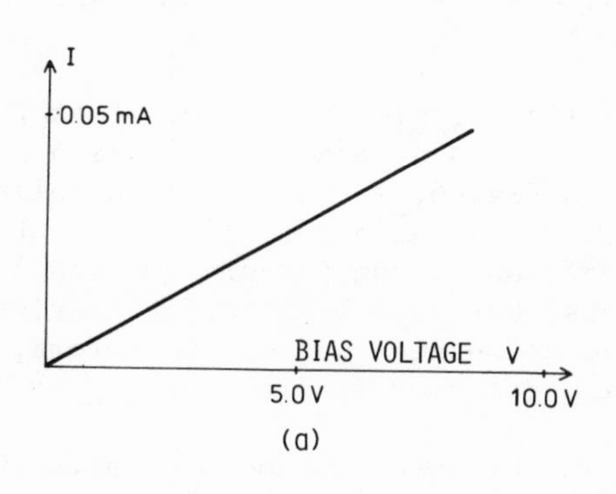

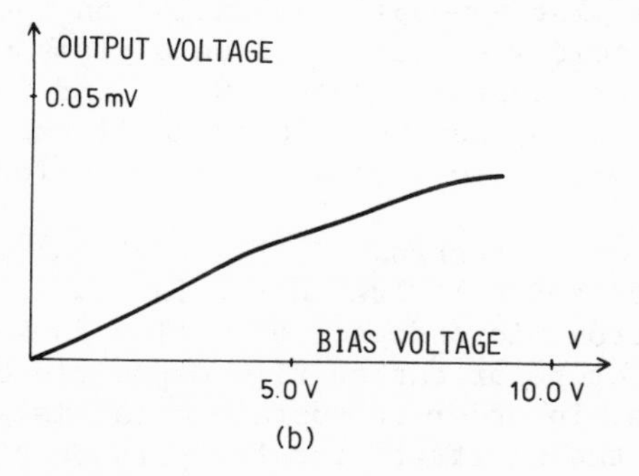

Figure 9. (a) V-I characteristics of discontinuous gold film, and (b) detection property of CO_2 laser.

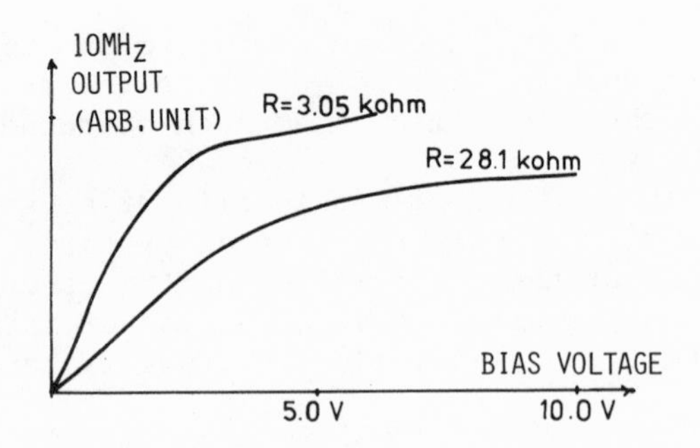

Figure 10. Detection property of 35 GHz millimeter wave modulated by 10 MHz hf wave.

6. DISCOUNTINUOUS THIN FILM OF PLATINUM AND PLATINUM-PALLADIUM ALLOY.

In order to realize stable thin film detectors of low noise and high sensitivity, it is necessary to fabricate thin films in the region A or A^* in Fig. 6, where direct tunneling is dominant conduction mechanism. Discontinuous films of platinum or platinum-palladium alloy are found to consist of very small islands separated with very narrow gaps, but have high surface resistivity of more than several hundred kilo-ohms square. Therefore, the impedance matching of this film detector is difficult.

Therefore, it was proposed (Okamura et al., 1976; Ijichi et al., 1977) to superpose platinum-palladium alloy on the discontinuous gold film. First gold was evaporated on a glass substrate until the resistance became approximately 100 kΩ, and in the next stage platinum-palladium alloy was superposed until the total resistance became several kilo-ohms to several tens of kiloohms.

Figure 11 shows photographs of films of (a) platinum, (b) platinum-palladium alloy and (c) platinum-palladium alloy on gold, observed by an electron microscope. Many tiny dots in these photographs show a structure of carbon film deposited over the discontinuous metal film, in order to sustain metal islands when the film is separated from the substrate for the purpose of observation by the electron microscope. Among these, the picture (c) shows that small platinum-palladium islands exist among larger gold islands.

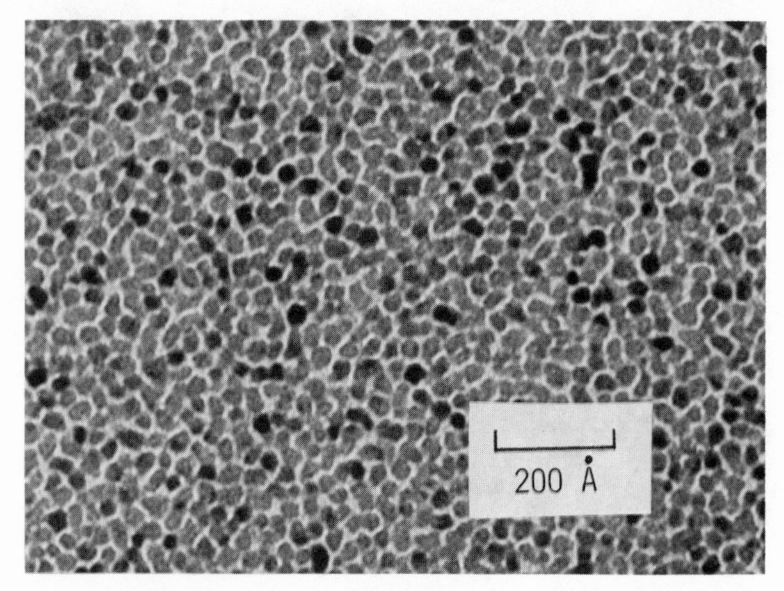

Figure 11 (a). Photograph of discontinuous film surface of platinum.

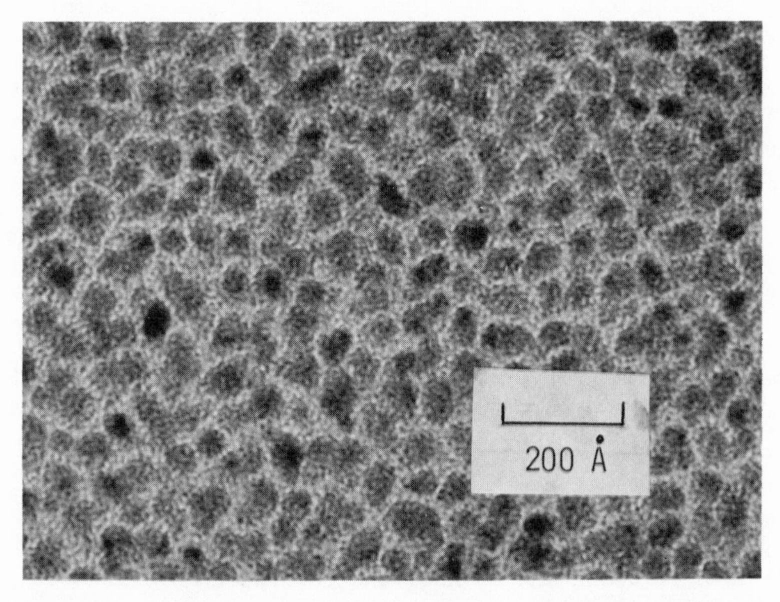

Figure 11 (b). Photograph of discontinuous film surface of platinum palladium alloy.

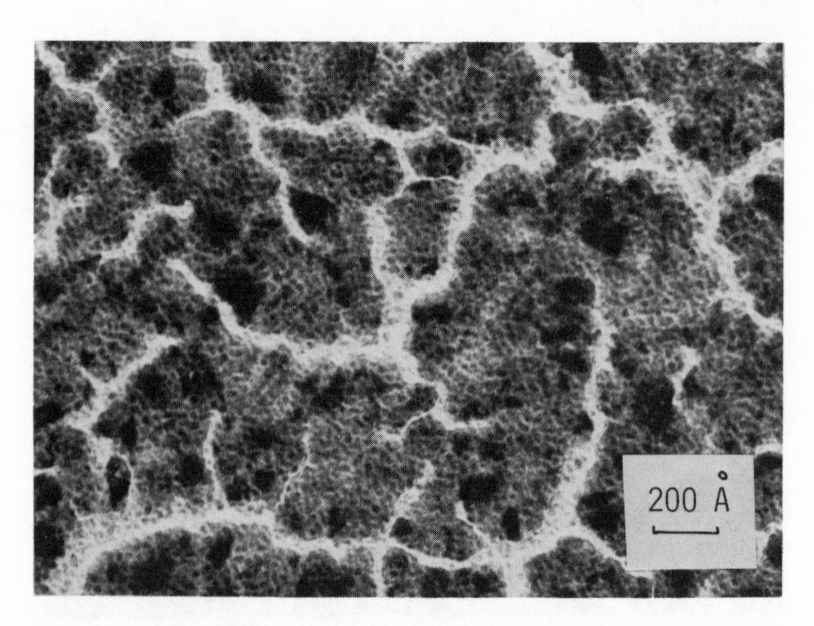

Figure 11(c). Photograph of discontinuous film surface of platinum-palladium alloy on gold observed by electron microscope.

Electron tunneling through these small islands produces stable electric conduction, and stability and the life of the film are greatly improved. The surface resistivity of the film is much lower than that of the platinum-palladium alloy film, but this resistivity is still high-impedance compared with the external circuit; then a reentrant cavity was used. The discontinuous metal film is deposited on a thin glass rod and is inserted in a reentrant cavity as shown in Fig. 12. The reentrant cavity is good for millimeter wave detection, but does not work satisfactorily for detecting CO_2 laser signals, because of the smallness of the coupling hole. The detector mount in Fig. 13 was then made, which could be used for both millimeter wave and CO_2 laser detection. Although this detector mount has poor impedance matching in millimeter wave frequencies, it is very useful for comparing the detection properties of various kinds of film detectors and also comparing the sensitivities for various kinds of input signals such as millimeter wave and CO_2 laser.

Voltage-to-current characteristics are shown in Fig. 14, and indicates small nonlinearity. The detection experiments were performed using millimeter waves of 35 GHz and 50 GHz which were modulated by 10 MHz, 5 MHz and 10 KHz, and CO_2 laser modulated by 1.6 kHz. The mixing experiments were also made using two millimeter waves around 50 GHz, whose frequencies were 30 MHz apart, and a beat

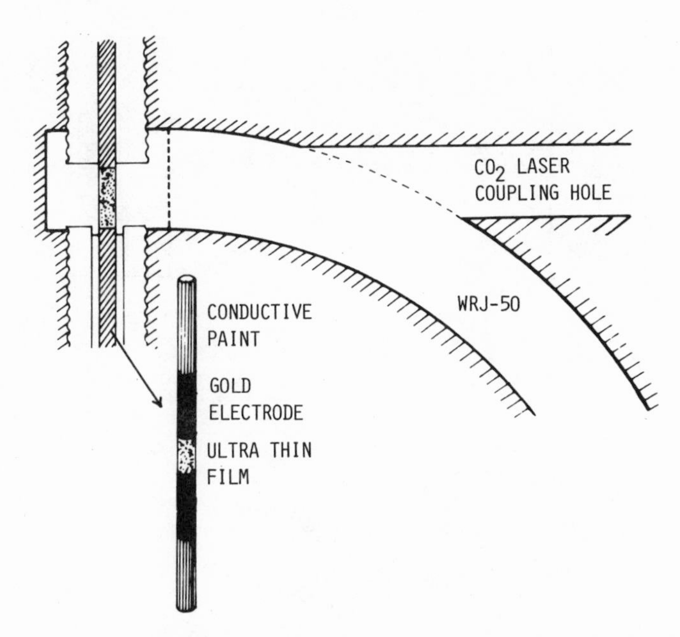

Figure 12. Reentrant cavity and thin metal film detector evaporated on thin glass cylinder.

of 30 MHz is noted. Figure 15 shows the schematic diagram of the measuring circuit.

The detection properties are plotted in Fig. 16. Figures 17 and 18 show sensitivity and resistance changes of thin films. In the sensitivity change, the variation of 20% caused by the remounting is included; then these figures indicate the stability of more than a month. The relation between the resistance of the detector and the detector output is shown in Fig. 19. The thin films become continuous at low resistance region, and impedance matching between the films and the circuit becomes difficult at high resistance region. Optimum sensitivity is obtained between ten kilo-ohms and a hundred kilo-ohms.

Figure 20 shows the summarized results of the experiments with the mount in Fig. 13. Input power of 7.3 mW for 54 GHz millimeter

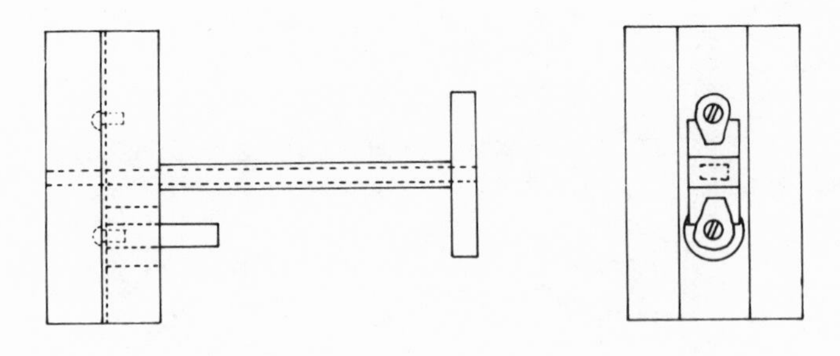

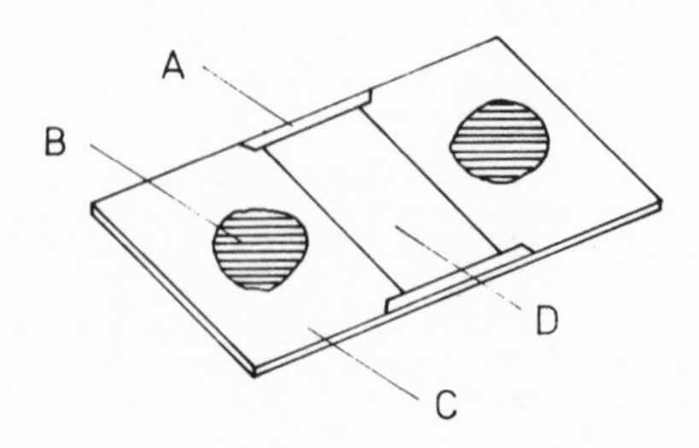

A SUBSTRATE (COVER GLASS)

B CONDUCTIVE PAINT

C GOLD ELECTRODE

D ULTRA-THIN METAL FILM

Figure 13. Detector and its mount for comparing various detection properties and noise measurement.

wave and 52 mW for CO_2 laser in the detection experiment, and input power of 33 and 6.6 mW in the mixing experiment of millimeter waves are applied to the same film detector. Output power is normalized to that of the mixing experiment, and plotted against bias voltage as shown in the figure. All these results show good coincidence and prove that the detection is caused by the nonlinearity of the film, that is, by the nonlinearity of the tunneling effect between metal islands. The calculated results are also shown in the same figure using the V-I characteristics and assuming the peak rf voltage of 0.967 V and 0.66 V. Estimated sensitivities are more than hundred times better than experimental ones, which may be explained by the very poor impedance matching of the detector. Obtained maximum responsivity of the detector was about 0.1 V/W.

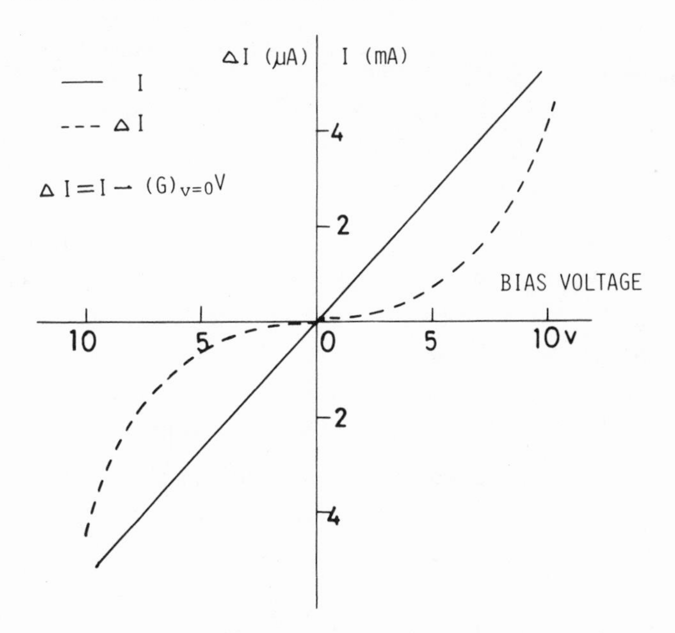

Figure 14. V-I characteristics of film detector deposited on thin glass rod shown in Fig. 12.

7. NOISE PROPERTIES OF DISCONTINUOUS METAL FILM

It is well known that a metal-insulator-metal diode has three kinds of noise, namely, thermal noise, shot noise and 1/f noise. At room temperature, shot noise is negligible compared with thermal noise. A large amount of 1/f noise was observed with the gold film, but 1/f noise was much reduced with the superposed film of platinum-palladium and gold (Okamura et al., 1976; Ijichi et al., 1977).

Noise properties of the superposed film were measured using the detector mount shown in Fig. 13, and the results are shown in Fig.21 . In the low-frequency region, 1/f noise is dominant and depends on the bias voltage as shown in Fig. 22. The experimental results can be summarized in the equation:

$$v_n = 12.8 \times 10^{-7} V_B \sqrt{1/f} \qquad [V/\sqrt{Hz}] \tag{5}$$

where v_n : root-mean-square value of noise voltage at the terminals per 1-Hz bandwidth,

V_B : bias voltage (V),

f : observed frequency (Hz).

In the high-frequency region, more than approximately several hundred kiloherz, the thermal noise will dominate and the results are given by

$$V_n = \sqrt{4kTR} \qquad [V/\sqrt{Hz}], \tag{6}$$

which results in 1.27×10^{-8} $V/\sqrt{Hz}$ for R of 10 kΩ.

Therefore, the noise equivalent power (NEP) of the detector at low-frequency is estimated to be:

$$[NEP]_{LF} = 6.4 \times 10^{-5}\sqrt{1/f} \qquad [W/\sqrt{Hz}], \tag{7}$$

supposing the responsivity of $2 \times 10^{-2} V_B$ V/W. At high frequency, thermal noise dominates and NEP reduces to

$$[NEP]_{HF} = 6.3 \times 10^{-7}/V_B \qquad [W/\sqrt{Hz}]. \tag{8}$$

These results are still too bad, but could be much reduced in the near future by improving the impedance matching.

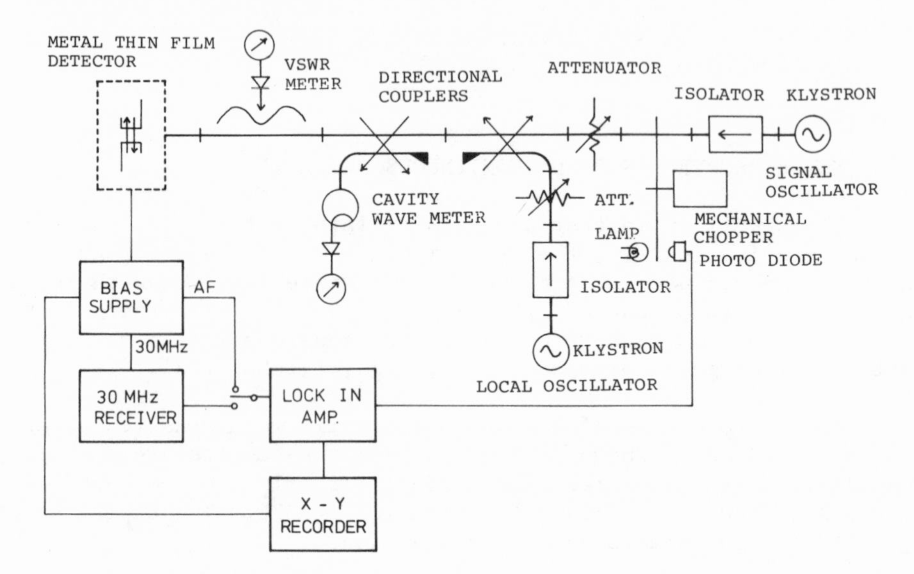

Figure 15. Schematic diagram of measuring circuit.

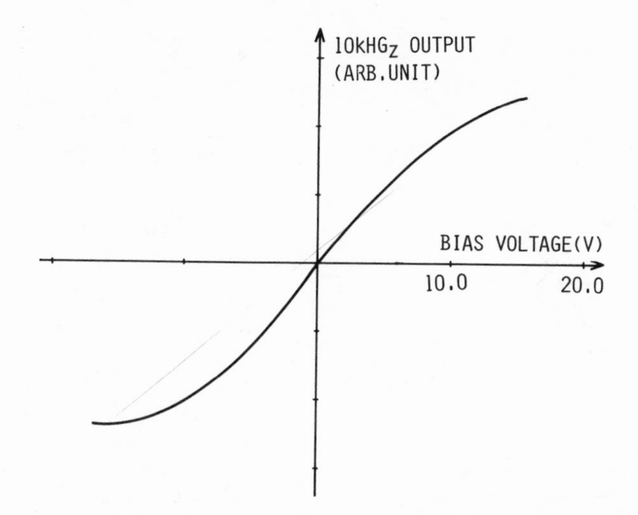

Figure 16. Detection property of 35 GHz millimeter wave by superposed thin film.

As shown in the next section, voltage responsivity of the film has been improved by reducing the dimensions of the film detector, and the optimum responsivity was about 2 V/W at the bias voltage of 5 V. The noise characteristics of this detector were not measured, but if the same noise characteristics were assumed, NEP would be improved as .

$$[NEP]_{LF} = 3.2\times10^{-6}\ \sqrt{1/f} \qquad [W/\sqrt{Hz}] \tag{9}$$

and

$$[NEP]_{HF} = 3.2\times10^{-8}/V_B \qquad [W/\sqrt{Hz}] \tag{10}$$

8. SIZE EFFECT ON DETECTION PROPERTIES

It was found that when the bias voltage is applied to the discountinuous metal film, current does not flow uniformly, but

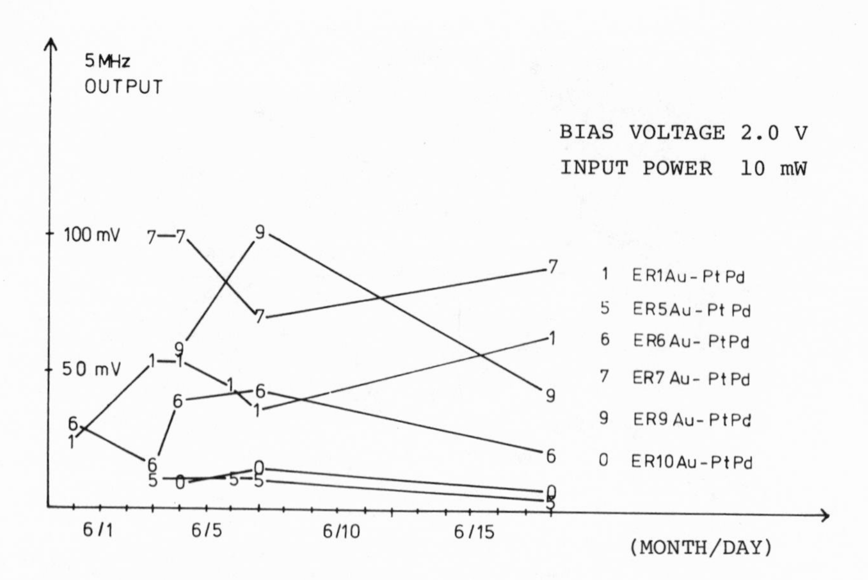

Figure 17. Sensitivity change of superposed thin film detector.

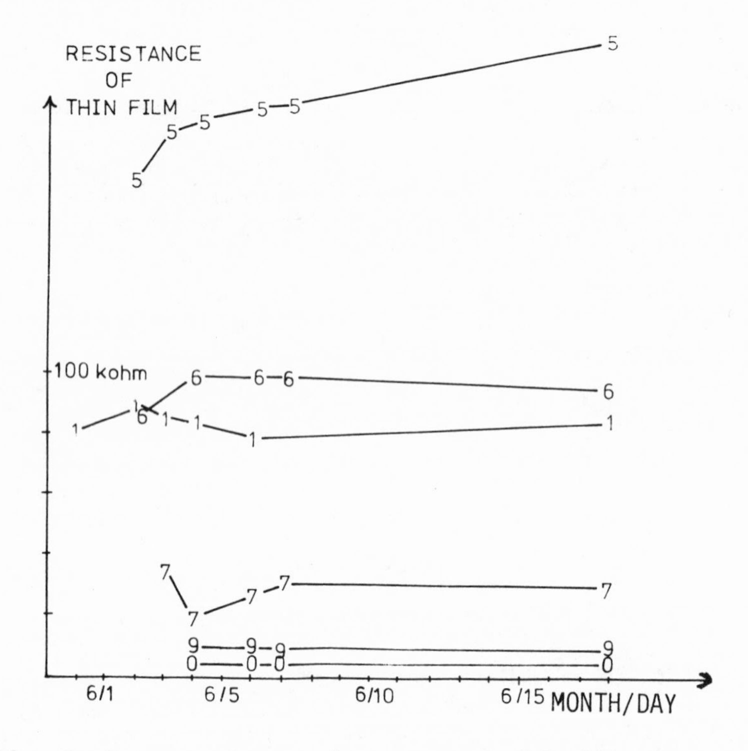

Figure 18. Resistance change of superposed thin film detector.

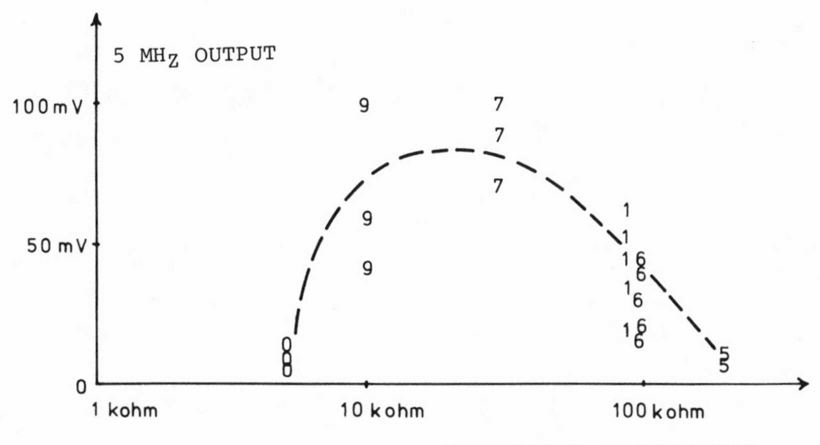

Figure 19. Sensitivity vs. resistance of superposed thin film detector.

flows with a dendriform path. Therefore only the small part of the film along the current path is effective for the detector purpose. Moreover, a slight variation of the evaporated metal of the thin film caused a large amount of resistivity change along the surface (Okamura et al., 1979; 1980). For film detectors of wide area, in many cases, the evaporation is not so uniform and, therefore, the current does not flow in a well-distributed manner. In the worst case, current paths exist just at the device edge, which sometimes locates outside the waveguide opening in the mount shown in Fig.13. This requires thin film detectors of small size.

In addition to the uniformity, small detectors also have the advantage that millimeter wave or optical wave power can be focused on the small area by an appropriate impedance-matching and give it a large rf amplitude, which results in high efficiency of detection.

Suppose a thin film of the length L and the width W. When the voltage and the current at the terminal of the device are V and I, the voltage across a unit length is V/L, and the current for a unit width is I/W. If the film of unit area has a conductance G and its nonlinearity G', namely

$$(I/W) = G(V/L)+G'(V/L)^2, \tag{11}$$

the increment of current, I_d, induced by rf voltage, V_{rf}, would be given by

$$I_d = WG' (V_{rf}/L)^2. \tag{12}$$

The rf power, P_{rf}, is equal to $(GW/1)V_{rf}^2$ where GW/L is the device impedance, then

$$I_d/P_{rf}=G'/(GL). \tag{13}$$

A similar relation for the increment of voltage, V_d, is also given by

$$V_d/P_{rf}=G'/(G^2W), \tag{14}$$

where V_d is assumed to be $I_dL/(GW)$, supposing small nonlinearity.

These equations show that the current sensitivity is inversely proportional to the length, and the voltage sensitivity is inversely proportional to the width, if the film characteristics are fixed.

Figure 23 shows an example of the film detector made by vacuum deposition through a metal mask (Okamura et al., 1981a), and is tested using the detector mount shown in Fig. 24 (Okamura et al., 1981b). The minimum size of the film is approximately 100 μm. The detection experiments were carried out, and the results are summarized in Fig. 25, where the responsivity of the detectors are plotted against the surface resistivity. Results clearly show that the responsivity of the detector becomes better when the size of the film becomes smaller,as shown in Eq. (14).

Film detectors of very small size of a few micrometers were also fabricated (Tamura et al.). The fabrication process is given in Fig. 26. The results will appear in the near future.

9. SUPERCONDUCTING GRANULAR THIN FILMS

Superconducting granular thin films are also considered to be an array of Josephson junctions which can be used as a sensitive detector of microwave and optical wave radiation. As thin film detectors in room temperature, series arrays of Josephson junctions are expected to have higher responsivity and ease in impedance matching. Bertin and Rose reported that the obtained responsivity of tin granular thin films is 104 V/W at X band (Bertin et al.,1971). Hansma and Kirtley have made tin and lead granular thin films and obtained the responsivity of 70 V/W at 36 GHz and 2 V/W for broadband far infrared radiation (Hansma et al., 1974). Ayer and Rose reported the correlated operations of tin granular thin films irradiated by millimeter wave radiation (Ayer et al., 1975).

Granular thin films of tin were also fabricated by the

authors' group (Fujimaki et al., 1981). Tin was evaporated by a tungsten heater onto a glass or fused quartz substrate whose temperature is controlled by the Peltier refrigerator beneath it. The evaporation was stopped when the resistance decreases to a certain value.

Under the lower background pressure from 10^{-5} to 10^{-6} Pa, a sudden decrease of the film resistance to a few hundred ohms happens after few minutes of evaporation. Observation by an electron microscope indicated that tin granules of the size of about a micrometer were formed and are connected to one another by narrow bridges of tin.

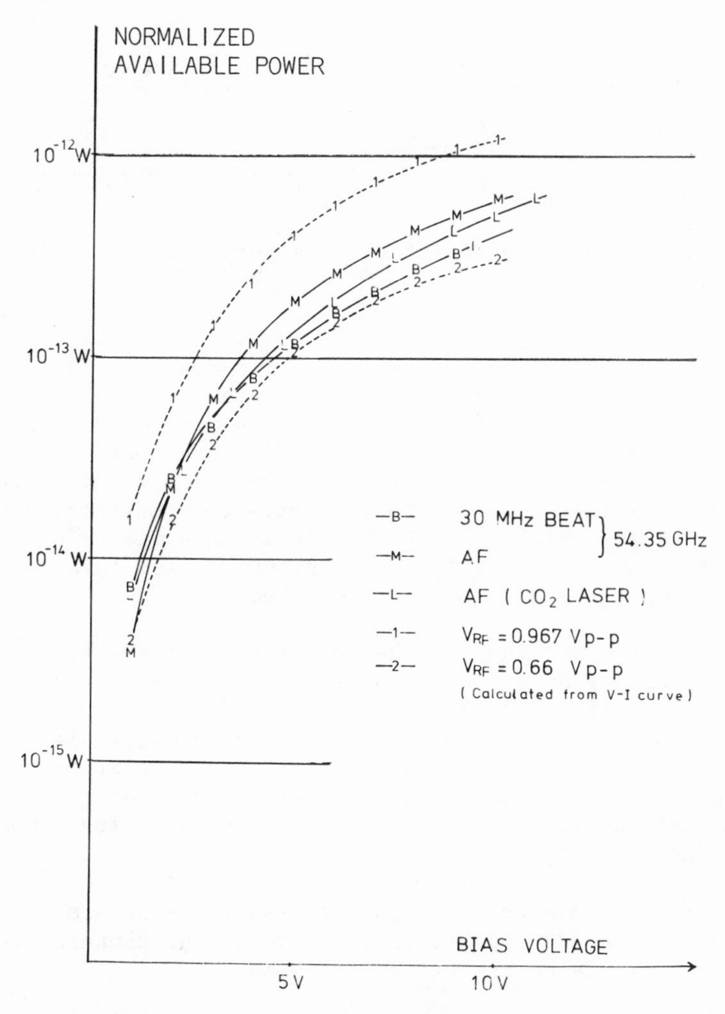

Figure 20. Comparison of various detection and mixing properties of superposed thin film detector.

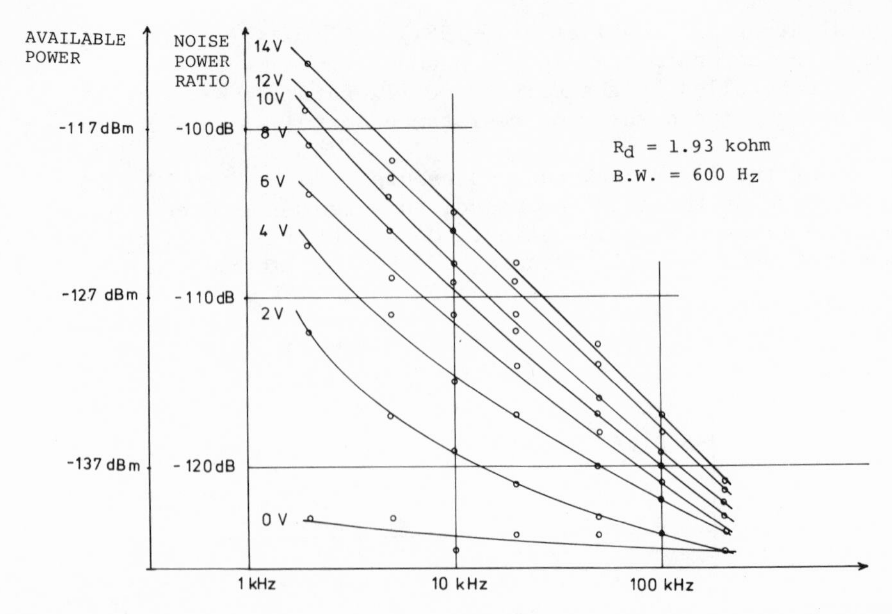

Figure 21. Noise characteristics of superposed thin film detector.

Under the higher background pressure from 10^{-2} to 10^{-3} Pa, the film resistance decreases slowly within about several tens of seconds after a few minutes. A film resistance is more than a few kilo-ohms. The electron microscope image in Fig. 27 shows that tin granules are closely connected to each other. The typical size of these granules is the same as that of the granules deposited under the lower background pressure. Higher electrical resistance with a closely connected structure suggests that each granule is so oxidized that tunneling junctions are formed among them.

Figure 28 shows the V-I characteristics of the films deposited at lower pressure at 3.6 K which is slightly lower than the critical temperature of tin (3.7 K). Many hysteresis and jumps appear, and the measurement of detection voltage is quite difficult. They are, moreover, very sensitive to the temperature fluctuation. Films on glass substrates fluctuate more than films on fused quartz. It is presumed that the glass has lower thermal conductivity compared with that of fused quartz.

For films deposited at higher pressure, V-I characteristics are shown in Fig. 29 and are insensitive to the temperature fluctuation. There is no hysteresis on them.

The granular thin film is mounted transversely in an X band

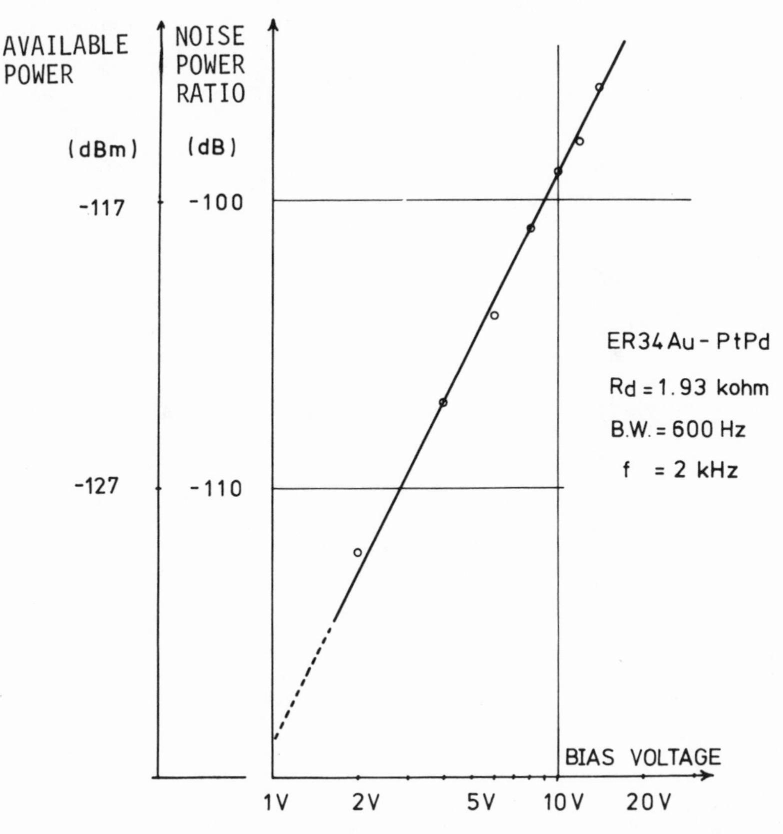

Figure 22. Noise vs. bias voltage of superposed thin film detector.

waveguide which is terminated by a quarter wavelength short. The mount is dipped into liquid helium, and no effort has been made for impedance matching. Therefore, responsitvity defined here is the ratio of the detection voltage to the incident power. Figure 30 shows the characteristics of the bias current versus the detection voltage of 10 GHz, which is measured by lock-in-amplifier. The incident power is 1.7×10^{-12} W, the maximum responsivity is 9×10^{6} V/W and NEP is 3×10^{-13} W/$\sqrt{Hz}$ for the film evaporated at the higher background pressure.

The responsivity values obtained for some films are shown in Fig.31. The detection mechanism is not supposed to be a bolometric one, because the maximum responsivity is obtained for the film of residual resistance ratio, R(4.2 K)/R(RT), of nearly unity; that is, the resistance of the film does not change much with the temperature change. However, the bolometric mechanism is still possible

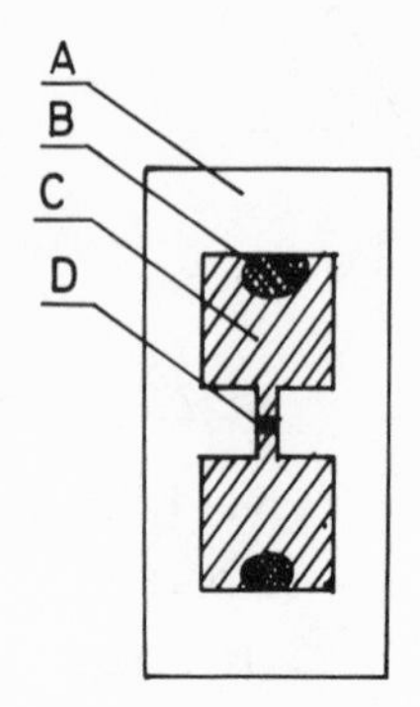

Figure 23. Small film detector fabricated by vacuum deposition through metal mask.

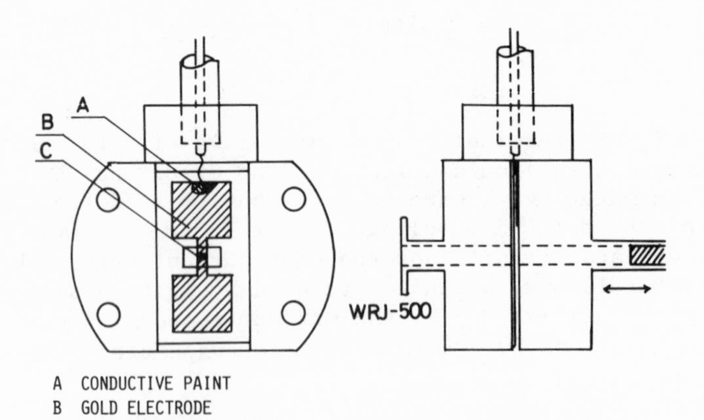

Figure 24. Millimeter wave mount of small film detector.

because the polarity of the detection voltage and the feature of square-law response are common to the Josephson mechanism and the bolometric one. The films with a residual resistance ratio more than unity showed no detection voltage.

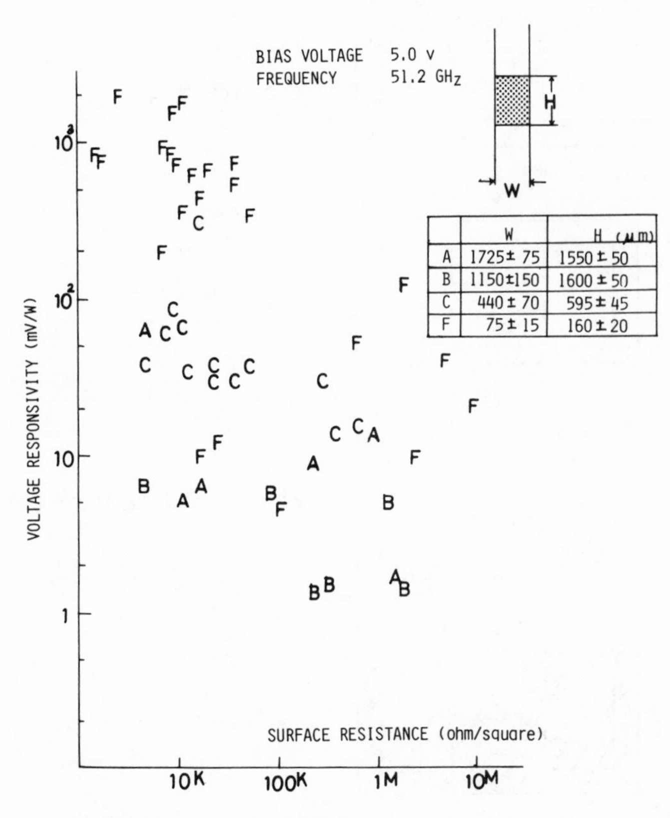

Figure 25. Responsivity vs. surface resistivity for small thin film detectors of various size.

Far infrared detection was carried out at 891 GHz of HCN laser using the film deposited at the higher pressure of 8×10^{-3} Pa with

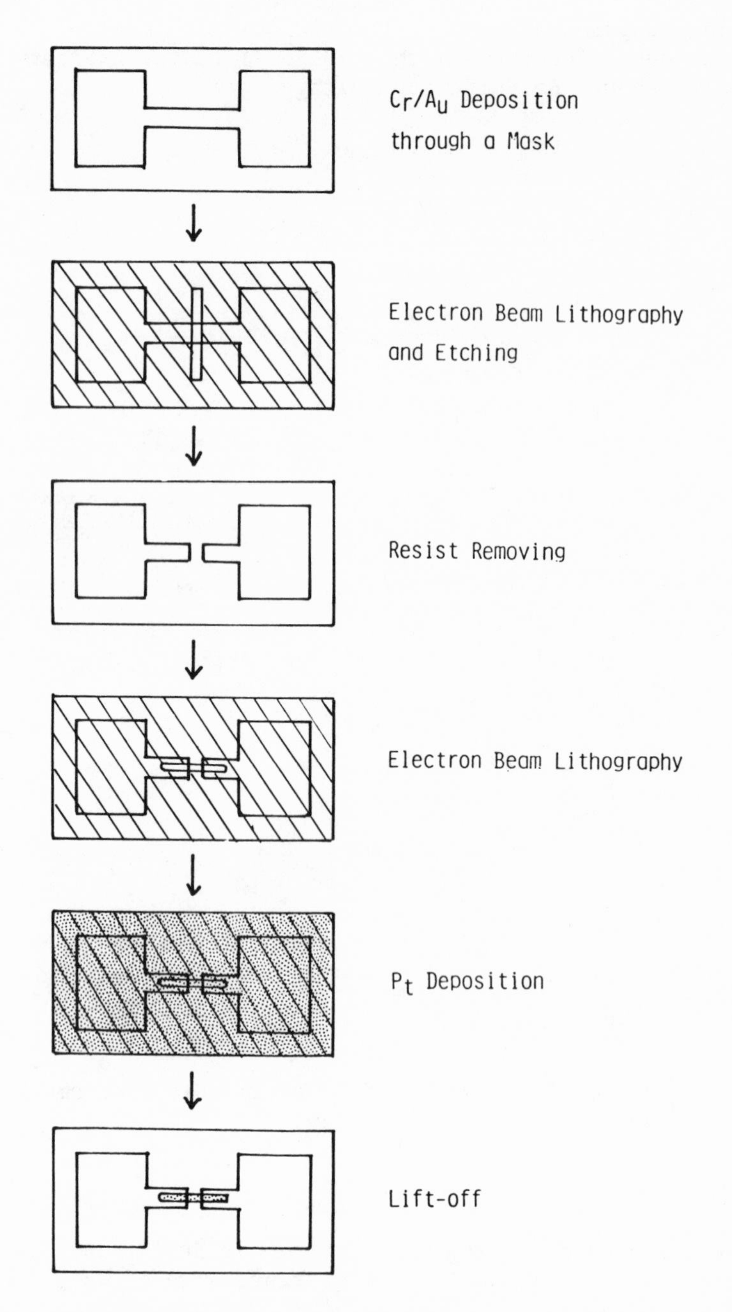

Figure 26. Fabrication process of very small thin film detector of a few micrometers.

Figure 27. Scanning electron microscope image of tin granular thin film evaporated at higher background pressure of $7x10^{-3}$ Pa. The substrate temperature is lowered about 10 K below room temperature.

a surface resistance of 1.3 kΩ. The laser ray is guided into the cryostat by a light pipe and the film is placed in front of the light pipe . Peak detection voltage is 85 μV for an incident power of $3x10^{-7}$ W, and the responsivity is 270 V/W.

The results above show that films evaporated at higher background pressure are supposed to be series and parallel connections of Josephson tunnel junctions and realize high responsivity for higher frequency regions.

10. CONCLUSION

Discontinuous metal film detectors are still under development, but show a bright future for high responsivity detectors in microwave through optical frequency region. Series and parallel

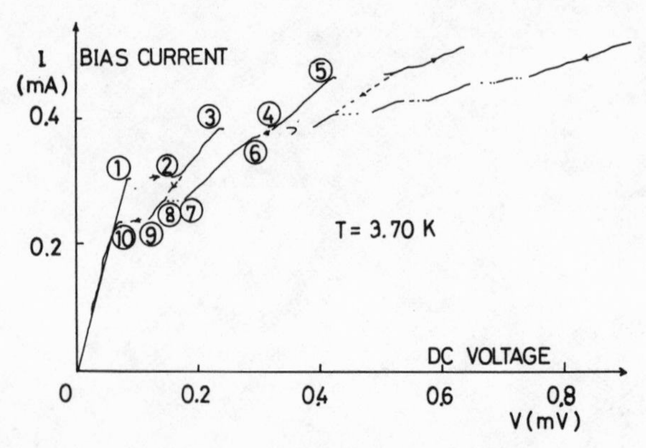

Figure 28. V-I characteristics of tin granular thin film deposited at the background pressure of $1x10^{-4}$ Pa.

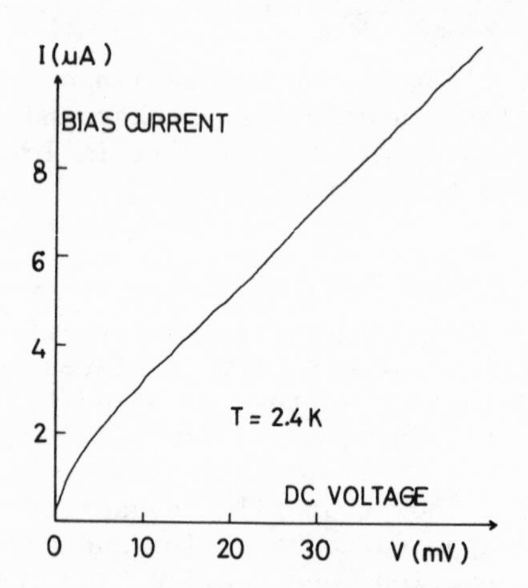

Figure 29. V-I characteristics of tin granular thin film deposited at higher background pressure of $1x10^{-2}$ Pa.

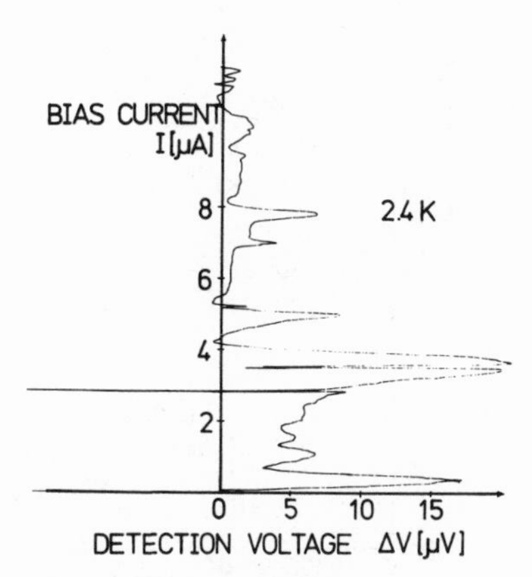

Figure 30. Increment detection voltage vs. bias current. Incident power of 10 GHz is 1.7×10^{-12} W.

connections of tunnel junctions are automatically realized by a simple evaporation technique. The junctions in nature have very small areas, which give a short RC constant, and can respond the higher frequencies.

Especially films of reduced dimensions with metal of high melting point, such as platinum or platinum-palladium alloy, would show good responsivity-more than several volts per watt-and good NEP-less than several 10^{-9} W/$\sqrt{\text{Hz}}$.

The above mentioned characteristics were also confirmed by superconducting granular films, and it was shown that tin granular films deposited in the higher background pressure formed many series and parallel connections of Josephson tunnel junctions, and gave high responsivity, too.

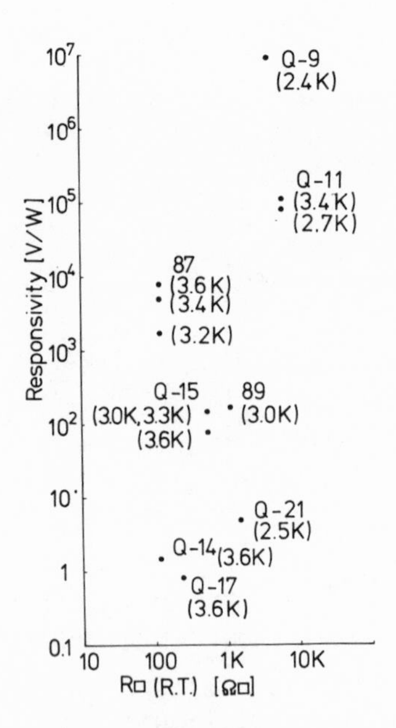

Figure 31. Measured responsivity at 10 GHz vs. film resistance at room temperature, R(RT). Films No. 87 and 89 are deposited onto glass substrate and films from No. Q-9 to Q-21 are on fused quartz substrates.

Finally, the authors want to thank Dr. K. Ijichi, Dr.N.Fujimaki and Dr. H. Tamura for their assistances. They performed experimental and theoretical works in the author's laboratory as graduate students. Dr. Ijichi did the first work on the discontinuous metal films, Dr. Fujimaki did works on the superconducting granular films, and Dr. Tamura fabricated the very small size thin films.

REFERENCES

Abram, R. L. and Gandrud, W.B., 1970, "Heterodyne detection of 10.6 μm radiation by metal-to-metal point contact diode", Appl. Phys. Lett., 17:150.

Ayer, W.J. and Rose, K., 1975, IEEE Trans.,Mag-11: 678.

Bertin, C.L. and Rose, K., 1971, J. of Appl. Phys., 42: 631

Evenson, K.M., 1974, "Frequency measurement in the optical region and the speed of light", Conf. on Precision Meas. on Elec.Mag.

Evenson, K.M., Pay, G.W., and Wells, J.S., 1972, "Extension of absolute frequency measurement to the cw He-Ne laser at 88 THz", Appl. Phys. Lett., 20: 133.

Evenson, K.M., Wells, J.S., Matarrese, L.M. and Elwell, L.B., 1970a, "Absolute frequency measurement of the 28 and 78 μm CW water vapor laser lines", Appl. Phys. Lett., 16: 160.

Evenson, K.M., Wells, J.S. and Matarrese, L.M., 1970b, "Absolute frequency measurements of the CO_2 cw laser at 28 THz (10.6 μm)", Appl. Phys. Lett., 16: 251.

Farris, S.M., Gustafson, T.K. and Weisner, J.C., 1973, "Detection of optical and infrared radiation with dc-biased electron-tunneling metal-barrier-metal diodes", IEEE Trans., QE-9: 737.

Fujimaki, N., Okabe, Y., and Okamura, S., 1981, "Radiation detection with superconducting granular thin films", J.Appl. Phys., 52: 912.

Hansma, P.K., and Kirtley, J.R., 1974, J.Appl. Phys., 45: 4016.

Hartman, T.E., 1962, "Tunneling of wave packet", J. Appl. Phys., 33: 3427.

Hill, R.M., 1969, "Electrical conduction in ultra thin metal films, I. Theoretical, II. Experimental", Proc. Roy. Soc. A., A-309: 377.

Ijichi, K. and Okamura S., 1973, "Characteristics of metal to metal point contact diode in infrared region (10.6 μm)," Paper of Tech. Group in Microwave, IECE Japan, MN73-27

Ijichi, K. and Okamura ., 1977, "A fast detector by discontinuous metal films for the millimeter wave through infrared range". Trans. IECE Japan, 60-C: 170.

Jenning, D.A., Peterson, F.R., and Evenson, K.M., 1975, "Extension of absolute frequency measurement to 148 THz; Frequencies of the 2.0-3.0 μm Xe laser", Appl. Phys. Lett., 26: 510.

Matarrese, L.M., and Evenson, K.M., 1970, "Improved coupling to infrared whisker diode by use of antenna theory", Appl. Phys. Lett., 17: 8.

Nahman, N.S., (NBS, Boulder Lab.), private comm.

Neugebauer, C.A., 1967, "Measurement technique for thin films", 191, Electrochem. Soc.

Nishiura, S, and Uozumi, K., 1975, "Electric resistance of discontinuous metal films", Appl. Phys. Japan, 44: 71.

Okamura, S., Shindo, Y., Nakazato, H., Okabe, Y., and Tamiya, S., 1981a, "Fabrication of discontinuous metal films", National Convention of IECE Japan, No.805.

Okamura, S., Shindo, Y., Nakazato, H., Okabe, Y., and Tamiya, S., 1981b, "Characteristics of fast detectors by discontinuous metal films, II", National Convention of IECE Japan, No.806.

Okamura, S., and Ijichi, K., 1974, "Characteristics of metal to metal point contact diode in infrared region", Ann. Rep. of Engineering Res. Inst., Fac. of Engineering, Univ. of Tokyo, 33: 121.

Okamura, S. and Ijichi, K., 1976, "A fast detector by discontinuous metal film for the millimeter through optical frequency range", IEEE Trans., IM-25: 437.

Okamura, S., Hino, T., Okabe, Y. and Tamiya, S., 1979, "Fast detector by discontinuous metal films", Paper of Tech. Group of Microwave, IECE Japan, MW79-62.

Okamura, S., Shindo, Y., Okabe, Y., Tamiya, S. and Hino, T., 1980, "Characteristics of fast detectors by discontinuous metal films, II", National Convention of IECE Japan, No.701.

Sakuma, E. and Evenson, K.M., 1974, "Characteristics of tungsten-nickel point contact diode used as laser harmonic-generator mixers", IEEE J. Quant. Elect. QE-10, 559.

Simmons, J.G., 1963a, "Generalized formula for the electrical tunneling effect between similar electrodes separated by a thin insulating film", J.Appl. Phys., 34: 1793.

Simmons, J.G., "Electric tunnel effect between dissimilar electrodes separated by a thin insulating films", J.Appl. Phys., 34:2581.

Tamura, H., Tamiya, S., Okabe, Y., and Okamura, S., not published.

Twu, B., and Schwarz, S.E., 1975, "Properties of infrared cat-whisker antennas near 10.6 μm", Appl. Phys.Lett., 26: 672.

Wang, S.Y., Heiblum, M., Gustafson, K., and Whinnery, J. R., 1977, "An integrated metal-oxide-metal device: A new integrated device with the widest band known", Laser Engineering and Applications

MICROWAVE RADIOMETRY FOR MEASUREMENT OF WATER VAPOR

D.C. Hogg, F.O. Guiraud, J.B. Snider,
M.T. Decker, and E.R. Westwater

NOAA/ERL/Wave Propagation Laboratory
Boulder, Colorado 80303

The amount of tropospheric water vapor is highly variable in both time and space; therefore the effect of vapor on certain observing systems must be reckoned with in real time. Examples may be encountered in determining heat balance in climatology, in cloud seeding experiments, in mesoscale meteorology, and in atmospheric absorption at millimeter wavelengths for radio astronomy and space communications. At microwave wavelengths, the vapor has strong refractive (Bean and Dutton, 1966) as well as absorptive properties. Therefore radio waves traversing the troposphere undergo a varying delay which is evidenced as a changing phase shift. Since varying delay is location dependent, paths through the atmosphere that are widely separated, as in Very Long Baseline Interferometry (VLBI), introduce uncorrelated phase shifts. In VLBI, this effect limits the quality of the interference fringes and therefore the resolution of the radio interferometer (Hinder and Ryle, 1971). However, if an independent measurement of integrated water vapor can be obtained on the paths, real time correction for these phase shifts can be made.

Application of dual-channel microwave radiometry (Guiraud et al., 1979) to measurement of water vapor integrated along an earth-space path shows that the resulting phase shifts can be determined (Hogg et al., 1981 a) to about 10° in the microwave range. The amount of water vapor is found to vary with periodicities from months down to minutes (Hogg et al., 1981 b). The design, implementation, and products of a radiometric instrument for this purpose are the subjects of this chapter. Although emphasis is given to measurement of integrated vapor (for purposes of correcting for excess radio path length), the simultaneous measurement of liquid in clouds is also discussed.

1. FACTORS INFLUENCING DESIGN OF THE INSTRUMENT

Six design considerations that result in high stability, sensitivity, and reliability in the instrument are discussed in the following paragraphs; they are:

1. Two frequencies, 20.6 and 31.6 GHz, are used; this ensures that the integrated vapor can be measured on occasions when liquid-bearing clouds are on the path (see Sections 2 and 6).

2. 20.6 GHz, which primarily senses the vapor, is removed from peak of the absorption line (22.2 GHz); these results in a measurement of integrated vapor that is relatively independent of pressure, (Westwater, 1978), i.e., independent of the particular distribution of water vapor with height (see Sections 2 and 6).

3. The electronics of the radiometer, and the antenna and feed, are all housed in a benign enviroment (see Section 3).

4. The antenna, an offset paraboloid with a hybrid-mode feed, results in high-quality radiation patterns that minimize the effect of extraneous sources of noise; the antenna aperture is devoid of blockage and the beam fully steerable(see Section 3).

5. The antenna, operates with the same beamwidth at the two frequencies; this allows for proper operation when liquid-bearing clouds are present, and leads to improved accuracy at low elevation angles.

6. The radiometer is triple-switched, two reference loads being provided; these results in continuous unattended calibration and high stability. The AGC function is accomplished in a computer (see Section 4).

2. FREQUENCY CONSIDERATIONS

It is common practice to use a frequency at the peak of an absorption line to measure the amount or density of a substance. In particular, when radiometric techniques are used, the "signal-to-noise" ratio is largest at that frequency; provided the pressure and temperature are constant, using this frequency can result in a good measure of the amount.

But when the pressure and temperature change, as occurs with height for vapor in the troposphere, operation at the line center is not optimal. The physical reasoning behind this argument is indicated schematically in Fig. 1 by the pressure-broadening of the rotational line at 22.2 GHz. As shown in Fig. 2, both at 20.6 and at 24.4 GHz the change in absorption caused by pressure broadening is minimized, but at the expense of dropping to about three-quarters

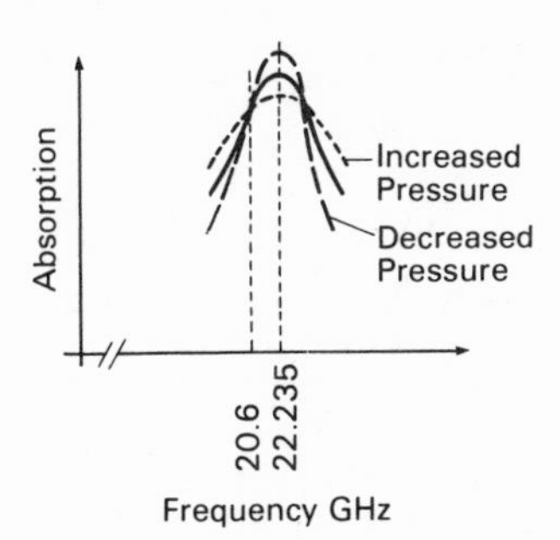

Figure 1. Absorption by water vapor at different pressures showing variation on the peak of the line and invariance at 20.6 GHz.

of the peak absorption. Calculations on the improvement in accuracy of determining atmospheric vapor using 20.6 GHz(rather than 22.2 GHz), calculated from typical profiles measured by radiosondes, are given by Westwater (1978).

The other frequency in the dual-channel system, 31.65 GHz, is situated in a transmission window of the troposphere. At this frequency, the absorption coefficient of water vapor, under standard conditions, is lower than the value at 20.6 GHz by a factor of about 1.8. Conversely, absorption by liquid water at 31.65 GHz is about 2.2 times greater than that at 20.6 GHz. From an operational point of view, 31.65 GHz is desirable because it is centered in a band allocated to passive space research.

Although we have chosen 20.6 GHz as our operating frequency, to circumvent possible interference from future satellite-communication bands, a relocation of the center of the passive band to 24.4 GHz might be required. To maintain a comparable degree of separation of liquid and vapor effects, (so the instrument would still measure vapor accurately in the presence of liquid-bearing

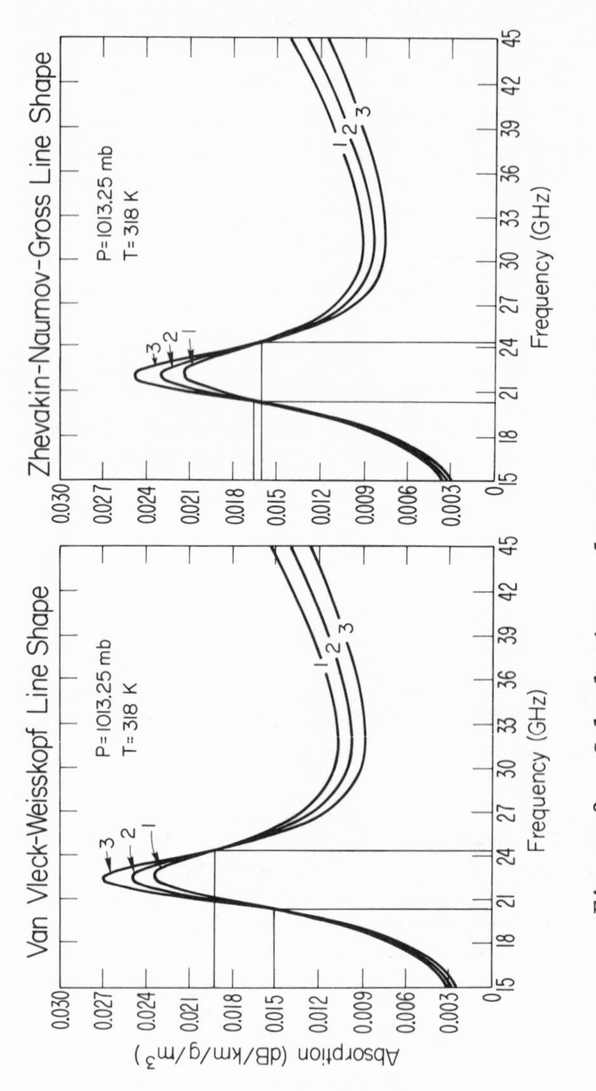

Figure 2. Calculations of water-vapor mass absorption coefficient for ±10% change on linewidth constant $\Delta\nu_o$;
1: linewidth increased by 10%,
2: original linewidth,
3: linewidth decreased by 10%.

clouds), the higher frequency channel would then also have to be relocated; a frequency of 35 GHz may be reasonable for this liquid-cloud-sensing channel.

The two brightness temperatures are incorporated in a pair of simultaneous equations to retrieve the amounts of vapor and liquid. In section 6, the method of retrieval and the coefficients in these equations are discussed in more detail.

3. PHYSICAL CONFIGURATION

A cross section of the physical layout of the system is shown in Fig. 3. The electronics for both the 20.6 and 31.6 GHz radiometers are located in a single package, and the feed horn for the offset paraboloid is mounted in a wall of the box containing the electronics. This arrangement provides a benign environment, in particular essentially constant temperature, for both the electronics and the antenna. The mini-computer, an LSI 11/02, is also located inside the trailer. A heat pump (not shown) is used to stabilize the temperature of the trailer housing.

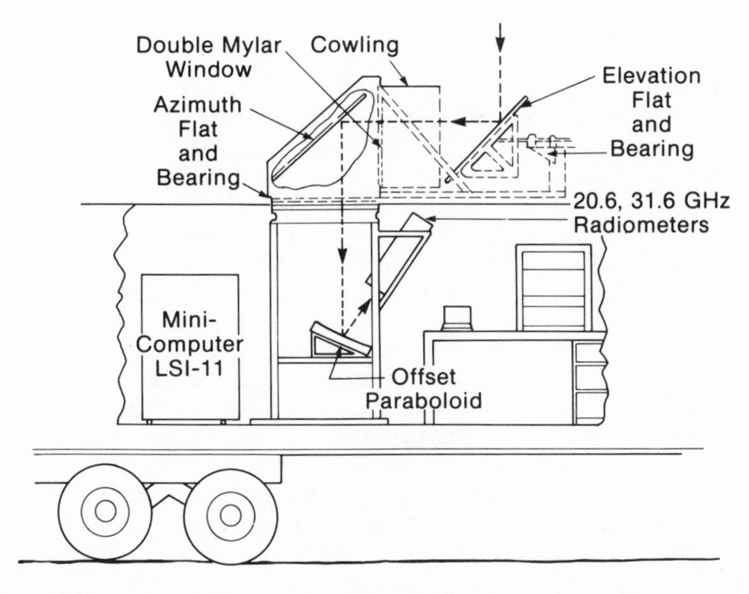

Figure 3. Side view of a steerable dual-channel radiometer showing the mechanical layout; a central ray from the zenith is shown as dashed lines with arrows.

The offset paraboloid and feed horn are of the same design as discussed by Hogg et al.(1979a); the configuration is such that the aperture is unblocked, and the corrugated (multi-mode) feed such that equal beamwidths are produced at 20.6 and 31.6 GHz. The performance of this feed-reflector combination is discussed in detail in the above reference. It remains to provide full steerability for the beam without introducing blockage to the radiation.

The roof of the trailer, although structurally strong, is further braced by supports surrounding the offset paraboloid as indicated in Fig. 3. Above these supports is incorporated a bearing which is 75 cm in diameter. On this bearing, in turn, is mounted a structure (rotatable with the bearing) that serves as a support for two flat reflectors. The flat immediately above the bearing is fixed on the rotatable structure; it is referred to as the azimuth flat in Fig. 3. The flat on the outrigger of the structure is mounted on an independent bearing that rotates in a plane orthogonal to the azimuth bearing. When this flat is rotated, the antenna beam scans in elevation, thus this reflector is referred to as the elevation reflector in Fig. 3. Full-sky coverage by the antenna beam can therefore be achieved by suitable rotation of the two bearings. (Modes of scan are discussed further in later sections). A photograph of the trailer and external elements of the antenna system is given in Fig. 4.

The azimuth flat and bearing are protected with a weather-tight cover fitted with a multilayer window (see Fig.3). The window is formed of two mylar sheets each 0.05 mm thick; this window is, in turn, shielded from the weather by a cowling whose primary function is to prevent rain from wetting the window. A double window is used to help prevent condensation on the outer surface during humid conditions. An external heater on the floor of the cowling is added if condensation becomes objectionable.

In this configuration, the microwave window is protected from wetting by rain, and the elevation flat is the only component left subject to the elements. The physical rationale for this design is that in the case of the reflector, radiation reflects (in part) from a water layer, whereas with a wetted window, radiation must pass through. To demonstrate the magnitude of the effects, Blevis(1965) calculated absorption, at normal incidence, for water layers of given thicknesses of both a reflector and a window; Fig. 5 shows the results for a frequency of 24 GHz. For a layer thickness of 0.1 mm, for example, the wet window absorbs about 50 times more than the wet reflector, and the brightness temperature is about 30 times higher. These ratios do not necessarily apply if the water forms rivulets or beads on either surface. As far as the reflector is concerned, a wetting surface (such as flat paint) is useful in preventing beading and formation of rivulets.

Figure 4. Side view of a steerable dual-channel radiometer at a weather service forecast office.

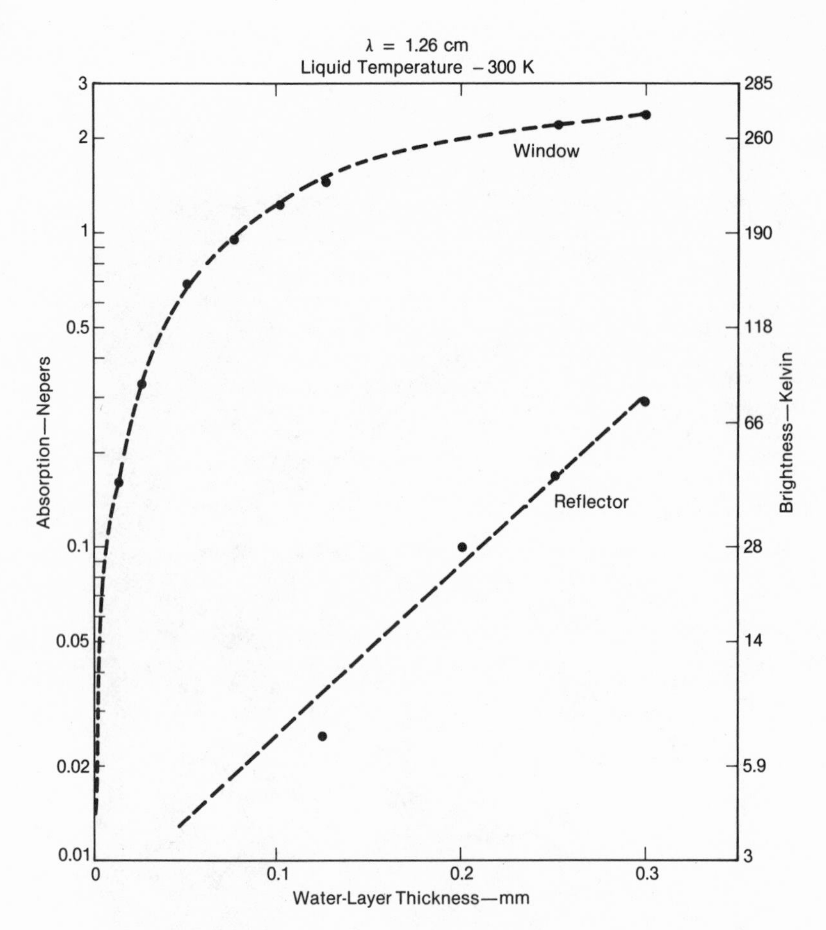

Figure 5. Plots showing computer brightness temperatures generated by a water layer on a transmission window, and on a flat reflector.

The performance of the overall antenna system was checked in a rather elementary way by measuring the far-field patterns of the main beam at 20.6 GHz. The azimuth and elevation half-power beamwidths are both 2.3°. Since previous tests (Hogg et al., 1979a) on this type of antenna at both frequencies were satisfactory, and since both of the flat reflectors were constructed and measured to be of high precision, in-depth measurements of antenna patterns were deemed unnecessary.

Mobility of the instrument is achieved by using a trailer as both a small laboratory and an antenna platform. However, because of clearance restrictions on many highways, the flat-reflector assembly is removed from the azimuth bearing and stored in the trailer for long-distance transport. Trips from Colorado to California, and to Virginia for the Army Research Office, which partially supported the construction, have been satisfactory in every respect.

4. ELECTRONICS PACKAGE AND COMPUTER

4.1. Electronics Package

The radiometers are of the Dicke switching type and are operated in an off-balance mode. Both radiometers, including all associated electronics, are mounted in the same 46 x 18 x 61 cm enclosure. The power supplies and controls are excluded from the package for convenience, and because of temperature considerations. Table 1 lists the characteristics of the radiometers. The degradation in sensitivity with gain control applied is real; this is a direct result of a scaling factor having been applied. The improvement we obtain in absolute accuracy overshadows this degradation. We also apply signal integration external to the radiometers, which improves the sensitivity further.

TABLE 1
RADIOMETER CHARACTERISTICS

Center Frequency	20.6 GHz	31.6 GHz
Bandwidth	>1.5 GHz	>1.5 GHz
Receiver Noise Temperature	680 K	725 K
Sensitivity w/o AGC	0.05 K	0.06 K
Sensitivity w AGC	0.26 K	0.26 K

The radiometer design can best be understood by following the block diagram, Fig. 6. Emission signals are focused into a wideband hybrid-mode feed by an offset 50 cm x 50 cm paraboloidal surface (not shown in the block diagram). By selecting a proper combination of focal length and horn feed design, different-sized areas of the surface are seen by the horn, a larger area for the lower frequency and a smaller one for the higher frequency. The resulting

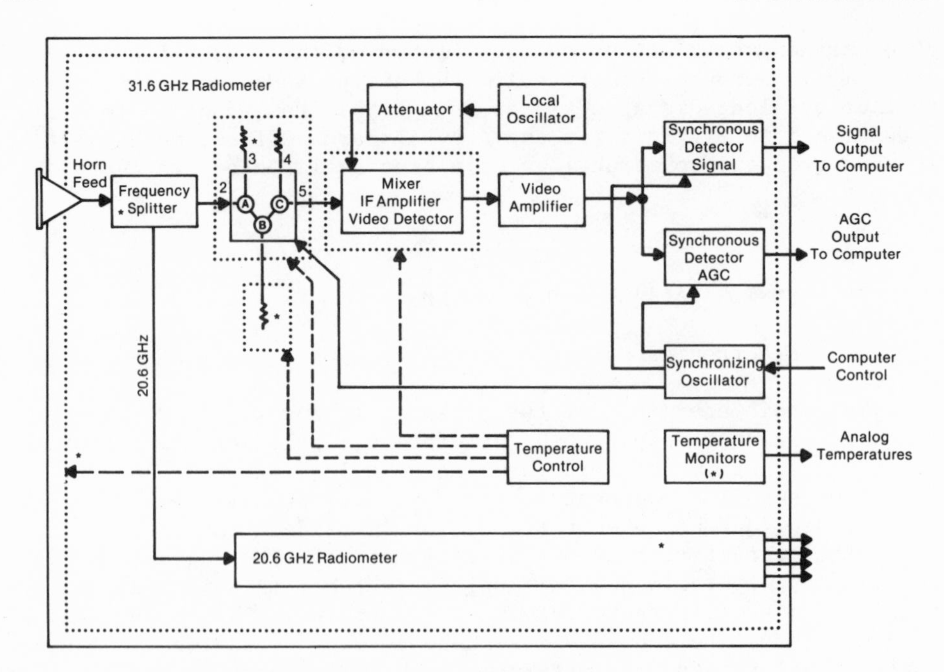

Figure 6. Electrical layout of a radiometer; temperature is monitored at points marked by asterisk.

combination produces patterns from the parabolic surface that have near-equal beamwidths as discussed in Section 3. Following the antenna feed, the two frequencies are split and passed to the two radiometers. The frequency splitter is a modified orthogonal polarization coupler. As a polarization coupler, one polarization is taken from a sidewall port, and the other from the through port. For our application, the through port has been stepped down in size to the higher frequency waveguide. The 20.6 GHz signal is taken from the sidewall port and the 31.6 GHz signal from the through port. The 20.6 GHz and 31.6 GHz signals therefore have orthogonal polarizations; this is not a major problem for our application. The Dicke switch is a three junction, five-port, ferrite device; it requires slightly over one microsecond to settle after being switched and can be toggled at a 2 kHz rate. The design of the radiometers extends the normal Dicke design by adding a second source to the switching sequence; this second source provides a different noise-power level than the first. The source we refer to as the reference load is a temperature controlled, low VSWR waveguide termination; the temperature is controlled at 45°C, a level higher than the expected ambient will reach. The second source, the hot load, is also a temperature controlled waveguide termination, at 145°C. Junctions A and B are used to connect in sequence the antenna, the hot load and the reference loads to the mixer through junction C. Junction C is not switched and functions as an isolator.

The mixer is a balanced design using selected low noise Shottky-barrier diodes. Incorporated in the mixer package is a series of thin-film wideband amplifiers with a video detector at the output. A 50 MHz high-pass filter is added in the IF section to reduce mixer shot noise; to maximize sensitivity no other filtering is added. The high side of the band pass of the amplifiers exceeds 1.5 GHz. Sufficient gain is provided by an external video amplifier to produce a 2 volt peak-to-peak video signal at the input to the synchronous detectors are required to make use of the triple switching. One is gated to detect the power difference between the antenna and reference load; the other is used to detect the difference between the reference and the hot load. The voltage produced by the second synchronous detector is in effect a monitor of any changes in gain and is used as an automatic gain control. This output is referred to as the agc channel. The output from the first synchronous detector will be referred to as the signal channel.

The synchronizing oscillator controls the switching sequence of the ferrite switch and the gating of the synchronous detectors. The sequence of the switch is never changed. The gating of the detectors is periodically changed to establish reference levels and to determine channel sensitivities. The programs for these modes are stored in a PROM memory chip in the synchronizing circuitry. The programs are selected by the computer and once these values are established they are stored and used by the computer.

The dotted lines surrounding areas in the block diagram,(Fig.6), are areas under temperature control; thermistors are used as the sensing elements. A thermistor is connected in a bridge network forming an input to an integrated-circuit voltage comparator. The output of the comparator controls a solid state relay, which in turn applies an AC voltage to resistive elements in the area requiring temperature control. We have already mentioned that the hot load and reference load are temperature controlled. We also include the ferrite switch block with the reference load; this ensures that the associated waveguide losses of the switch do not modify our reference level. We are able to maintain a 0.1°C temperature control for the reference load and hot load for weeks on end. A mounting plate supporting the mixer-amplifier unit is also temperature controlled near 42°C. This control is less effective since no insulation is used, but it greatly reduces the burden placed on the agc. Finally the whole electronic package is temperature stabilized near 37°C.

Linearized thermistors are used to monitor the temperatures of the hot load, reference load, and antenna waveguide. These measurements are all critical to accurate reduction of the brightness measurements. Other temperatures such as the interior of the package, room temperature, and several others, are also monitored as an indication of the general health of the radiometers, but are not used in the reduction of the measurements.

4.2 Computer System

Data from the dual-channel radiometer are processed in real-time in an LSI-11/02 minicomputer. In addition, the computer system controls many radiometer functions including data sampling rates, averaging intervals, output data rates and initiation of periodic radiometer self-calibrations. In the present model, antenna position is not controlled by the computer; however, this capability can be readily added if desired.

The LSI-11/02 minicomputer operates with 32,000 16-bit words of active semiconductor memory; inactive system sofware and radiometer data are stored on magnetic disks. The magnetic disk drive unit contains both a fixed and a removable disk. Each disk has two recording surfaces with a storage capacity of 4,762 512-word blocks. System software is stored on the fixed disk and radiometric data are archived on the removable disk.

Communication with the computer is through a teletype terminal. Output data are also printed out periodically on the teletype terminal.

Computer software supports three radiometer modes of operation:

1. Fixed azimuth and elevation angles,
2. Variable azimuth or elevation scans with the remaining axis fixed
3. Continuous 360 degree azimuth scans at a fixed elevation angle.

The system is capable or unattended, continuous operation in modes 1 and 3. Mode 2, which is employed to measure the spatial distribution of vapor and liquid in clouds, requires an operator to be present. Operating mode is changed either by activating the appropiate computer program or, for mode 2, by modifying the input parameters to a single program. In all modes, the start of operation is selfprompting, i.e.,the operator enters input parameters into the teletype terminal in response to questions from the computer. Thereafter, program execution is continuous until stopped by the operator.

Processed data outputs, path-integrated water vapor and path-integrated liquid, are continuously displayed on a two-channel strip-chart recorder. The recorder display is scaled to indicate directly total water vapor in cm and total liquid in mm. In addition, the output data are printed on the computer teletype at certain intervals. The print-out also includes a simple plot of water vapor and total liquid versus time for ready identification of trends. Unprocessed radiometric data are written on a removable magnetic disk for later processing if desired. Table 2 summarizes the sample rates, input data averaging times, and output data update intervals for the three operating modes. Output data are averaged over the internvals shown.

A 1.0-s radiometer time constant is used in all modes of operation. In order to ensure an adequate number of independent samples rate is limited to 0.5 deg/s. At the maximum scan rate, five 1-s samples are obtained in 2.5 degrees of antenna rotation.

TABLE 2

SUMMARY OF DATA-PROCESSING CHARACTERISTICS FOR STEERABLE DUAL-CHANNEL RADIOMETER

Mode of Operation	Sample Rate (Hz)	Averaging Time(s)	Output Interval Strip-Chart	Printer*	Disk
1	10	10	10 s	10 min	2 min
2	10	1	1 s	5 min	5 s
3	10	1	1 s	2.5 deg	2.5 deg

* A print-on-demand feature offers printouts at 2 min, 5 s and 15 s intervals, for modes 1, 2, and 3 respectively.

5. CALIBRATION

We use the two temperature controlled loads to establish the relationship between the voltage output from the radiometers and input brightness. The outputs are linear with brightness. Junction B in Fig. 6 is the first common point for all the signals. All signals must be transferred to or from this junction for our determination. Since the waveguide leading to the reference load is at the same temperature as the reference load, the emission generated by the guide exactly compensates for the quantity attenuated; this is not the case for the signals coming from the antenna and hot load. For the transmission path to the antenna we estimate the attenuation because it is extremely difficult to make an adequate attenuation measurement. By monitoring the physical temperature of the antenna line, we account for both the added emission and the attenuation effects on the signal. The waveguide from the hot load is temperature controlled either in the hot load or in the ferrite switch; the emission is therefore unchanging. We thus determine a factor with which we can multiply the indicated hot-load temperature to determine its effective temperature at junction B.

A value for this factor is found from a series of "tipping-curve" measurement made on cloudless days when the emission from the atmosphere is unchanging. An initial calibration is made by first placing a high quality absorber in front of the antenna; the output voltage from the radiometers and the physical temperature of the absorber are noted. The antenna is then directed toward the zenith, and the output voltage is again noted. The zenith brightness will range in value between 10 K and 40 K depending on the amount of vapor present; initially a value of 20 K is assumed. Using this assumption and the temperature of the absorber as the two points to define a temporary calibration, a series of measurements are made at various elevation angles. These emission measurements are related to absorption (τ) by the expression

$$\tau = \ell n(T_m - 2.9)/(T_m - T_b)) \tag{1}$$

where the absorption is in nepers, T_m is the mean radiating temperature for the atmosphere, T_b is the brightness at the antenna input, and 2.9 is a constant representing the cosmic background, all in kelvins.

For calibration purposes, a relative value of absorption is all that is needed. T_m is estimated at 270 K. We now plot the absorption measurements against the number of atmospheres included in each measurement (secant of the zenith angle). This new curve is also linear and must pass through the origin, i.e., no atmosphere--no absorption. We slide the plot up or down, forcing it through zero, if a poor assumption has been made in the initial two-point calibration. The value of the adjusted curve at one air mass is the true relative zenith absorption. Typical tipping curves showing measurements of absorption at both frequencies versus number of atmospheres are given in Fig. 7. The true brightness is found by solving Eq.(1) for T_b. In turn, the hot load multiplying factor is found,thus establishing an absolute calibration for the radiometers.

6. RETRIEVAL METHODS

6.1 Statistical Inversion

In deriving total precipitable water vapor and integrated cloud liquid from radiometric observations, statistical retrieval methods are commonly employed (Guiraud et al., 1979; Westwater and Guiraud, 1980; Grody et al., 1980; Staelin et al., 1976). We will outline this method below, first in a form that is appropriate in the case

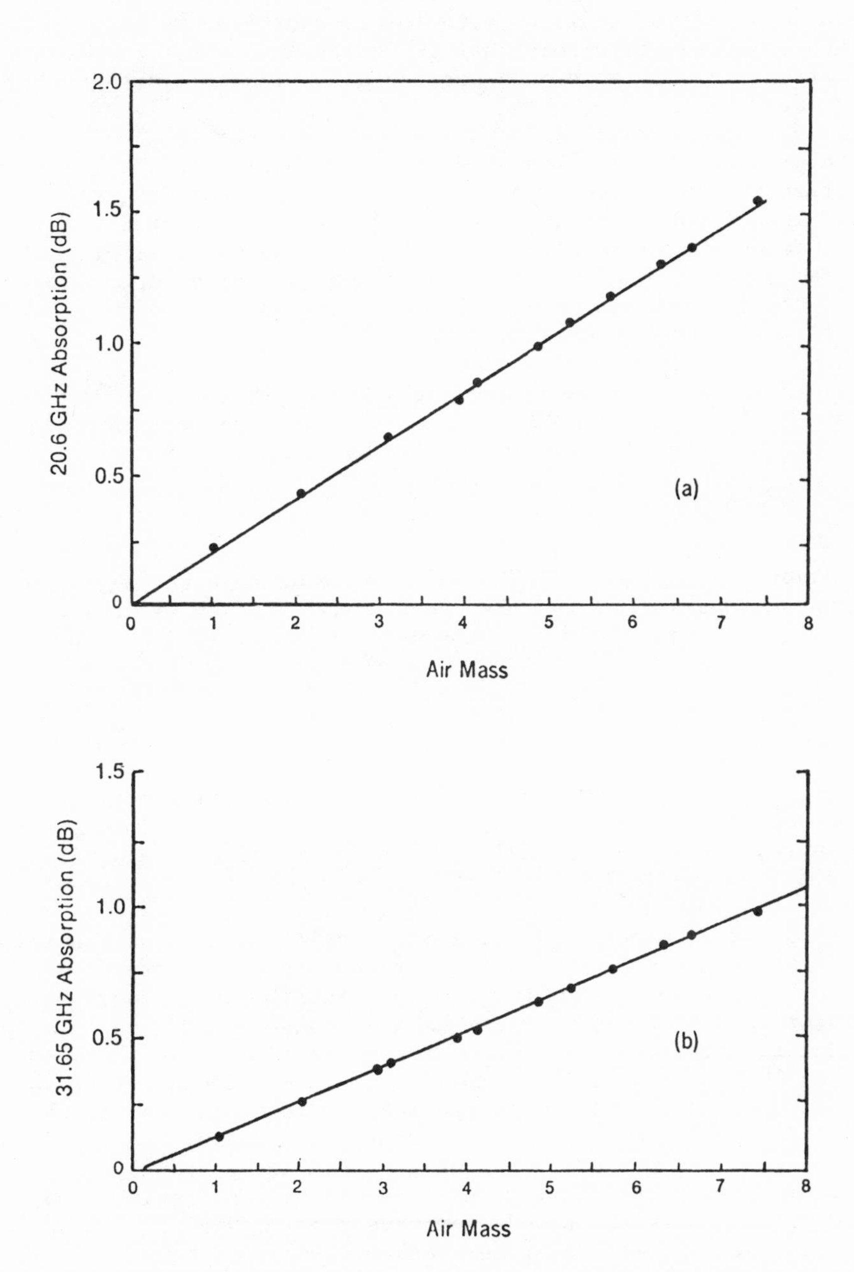

Figure 7. Measured absorption at (a) 20.6 GHz and (b) 31.65 GHz versus elevation on a clear day at Denver, Colorado; these tipping curves are used in calibration of the instrument.

where nonprecipitating clouds with low attenuation are involved, and then in an adaptive form that is designed to treat relatively high attenuation due to clouds bearing a considerable amount of liquid water.

Under conditions of low attenuation (Optical depth $\lesssim 1$), total absorption can be derived from atmospheric emission (Hogg and Chu, 1975). This absorption can, in turn, be directly related to corresponding amounts of integrated water vapor V and cloud liquid L. As discussed by Staelin (1966) measurements of low attenuation at a vapor-sensitive frequency and a liquid-sensitive frequency allow separation of the two water phases.

If the total atmospheric transmission at frequency ν is written as $\exp(-\tau_\nu)$, then measurements of microwave brightness temperature $T_{b\nu}$ can be converted to absorption τ_ν by

$$\tau_\nu = -\ell n[(T_{mr} - T_{b\nu})/(T_{mr} - T_{bb})], \tag{2}$$

as in equation (1), where T_{bb} is the cosmic background "big-bang" brightness temperature equal to 2.9 K, and T_{mr} is an estimated "mean radiating temperature". For nonprecipitating clouds, we can write

$$\tau_\nu = - \kappa_{V\nu} V + \kappa_{L\nu} + \tau_{d\nu} \tag{3}$$

where $\kappa_{V\nu}$ and $\kappa_{L\nu}$ are path-averaged mass absorption coefficients of vapor and liquid, and $\tau_{d\nu}$ is the dry absorption. Since microwaves interact only weakly with ice clouds (Staelin et al., 1975), their effect is neglected here.

In the usual experimental situation neither the mass absorption coefficients $\kappa_{V\nu}$ and $\kappa_{L\nu}$ nor the dry absorption τ_ν are known since they depend on the unkown spatial distributions of pressure, temperature and humidity . As an example Table 3 shows calculations of the climatological variation of these quantities over a two-month period at Oklahoma City, Oklahoma. Note, in particular, that the liquid attenuation coefficients are subject to variations of the order of 25-30%. Because of this variation, deriving V and L from dual-frequency measurements by simple matrix inversion is not optimal, and we use instead statistical inversion. The application of this technique to ground-based microwave remote sensing is thoroughly discussed by Westwater and Decker (1977) and Decker et al. (1978). For completeness we briefly outline the method below.

TABLE 3

CLIMATOLOGICAL VARIATION OF PARAMETERS USED IN DERIVING VAPOR AND LIQUID FROM DUAL-CHANNEL RADIOMETRIC OBSERVATIONS. BASED ON OKLAHOMA CITY RADIOSONDE OBSERVATIONS DURING APRIL-MAY 1975-1977. ν_ℓ = 20.6 GHz, ν_u = 31.65 GHz

	$\tau_{d\ell}$ nepers	τ_{du} nepers	$\kappa_{V\ell}$ nepers per cm H_2O	κ_{Vu} nepers per cm H_2O	$\kappa_{L\ell}$ nepers per cm H_2O	κ_{Lu} nepers per cm H_2O	$\frac{\kappa_{V\ell}}{\kappa_{Vu}}$	$\frac{\kappa_{Lu}}{\kappa_{L\ell}}$	$T_{mr,\ell}$ K	$T_{mr,u}$ K
Average	0.01320	0.02336	0.04257	0.02417	0.8289	1.8567	1.7761	2.2538	277.8	275.4
Std.Dev.	0.00033	0.00059	0.00034	0.00127	0.2423	0.4978	0.0926	0.0492	5.1	5.7
$\frac{\text{Std.Dev.}}{\text{Average}}$ x100	2.51	2.51	.79%	5.26%	29.23%	26.81%	5.25%	2.18%	1.8%	2.1%

The linear statistical inversion algorithm may be written (Deutsch, 1965) as

$$\hat{\underset{\sim}{p}} = \langle \underset{\sim}{p} \rangle + \langle \underset{\sim}{p}' \underset{\sim}{d}'^T \rangle \langle \underset{\sim}{d}' \underset{\sim}{d}'^T \rangle^{-1} \underset{\sim}{d}' \tag{4}$$

where $\hat{\underset{\sim}{p}}$ is the estimate of the m-dimensional parameter vector $\underset{\sim}{p}$, $\underset{\sim}{d}$ is the $\tilde{n}$-dimensional data vector, $\langle \cdot \rangle$ refers to ensemble average, $\underset{\sim}{d}^T$ is the transpose of the vector d, and the primed quantities refer to departures from the average, i.e., $\underset{\sim}{d}' = \tilde{d} - \langle \underset{\sim}{d} \rangle$. The matrix inverse $\langle \underset{\sim}{d}' \underset{\sim}{d}'^T \rangle^{-1}$ is assumed to exist.

In our applications of (4) to determine V and L from measurements of brightness temperature $T_{b,\nu}$, the two-component data vector d is chosen to be the total absorption τ_ν, ν = 20.6 GHz, 31.65 GHz. In using (2), the quantity τ_ν is derived from $T_{b,\nu}$ and a climatological average mean radiating temperature. The ensemble over which the averages and covariances in (4) are calculated is based on a history of radiosonde profiles; clouds are inserted into this ensemble following the modeling scheme described by Decker et al. (1978). Brightness temperatures are calculated from the radiosondes using a standard water-vapor absorption model (Waters, 1976) which accounts for the empirical excess attenuation related to the continuum absorption (Deepak, 1980; Gaut and Reifenstein, 1971; Hogg, 1959); instrument noise is simulated by assuming Gaussian noise. Numerical values of the coefficients in (4) for selected climatologies are given in Table 4(a) for vapor and in Table 4(b) for liquid. For these tables, the optical depth τ is in nepers, and an instrumental noise level of 1.0 K is assumed.

The predicted rms accuracies in the radiometric measurement of vapor and liquid for the conditions of Oklahoma City, OK are shown in Fig. 8(a). Note, that for cloud liquid much in excess of 3 mm, retrieval accuracies of vapor are marginal. As shown below, adaptive retrieval methods can increase the range of cloud liquid over which accurate retrievals can be obtained. However, because of the low frequency of occurrence of nonprecipitating clouds whose liquid in the zenith direction exceeds 3 mm, ordinary statistical retrieval methods are adequate most of the time.

6.2 Adaptive Statistical Inversion

Linear statistical inversion techniques, described in Section 6.1, have led to accurate measurements of water vapor during clear conditions or during conditions of light clouds (Guiraud et al., 1979). However, retrievals during periods in which clouds with high liquid content are present, are not always satisfactory (Westwater and Guiraud, 1980). To extend the range of cloud liquid under which the radiometer can operate, we developed an adaptive method of retrieving both vapor and liquid. The method is adaptive in the

TABLE 4(a)

	$\hat{V} = a_o + a_{20.6} \cdot \tau_{20.6} + a_{31.65} \cdot \tau_{31.65}$ (cm)				
	a_o	$a_{20.6}$	$a_{31.65}$	T_{mr}(K) 20.6 GHz	T_{mr}(K) 31.65 GHz
Sterling VA Jul - Nov	-.05662	30.429	-12.754	281.04	279.90
Sheridan CA Nov. - Apr	-.03980	30.408	-13.363	272.99	271.13
Oklahoma City OK Apr - May	-.02067	29.623	-12.593	277.8	275.4
Denver CO All Year	-.00111	26.966	-11.772	268.49	265.47

TABLE 4(b)

	$\hat{L} = b_o + b_{20.6} \cdot \tau_{20.6} + b_{31.65} \cdot \tau_{31.64}$ (cm)		
	b_o	$b_{20.6}$	$b_{31.65}$
Sterling VA Jul - Nov	-.01285	-.51291	.85769
Sheridan CA Nov - Apr	-.01716	-.33190	.67706
Oklahoma City OK Arp - May	-.01034	-.44446	.75298
Denver CO All Year	-.00950	-.22866	.56300

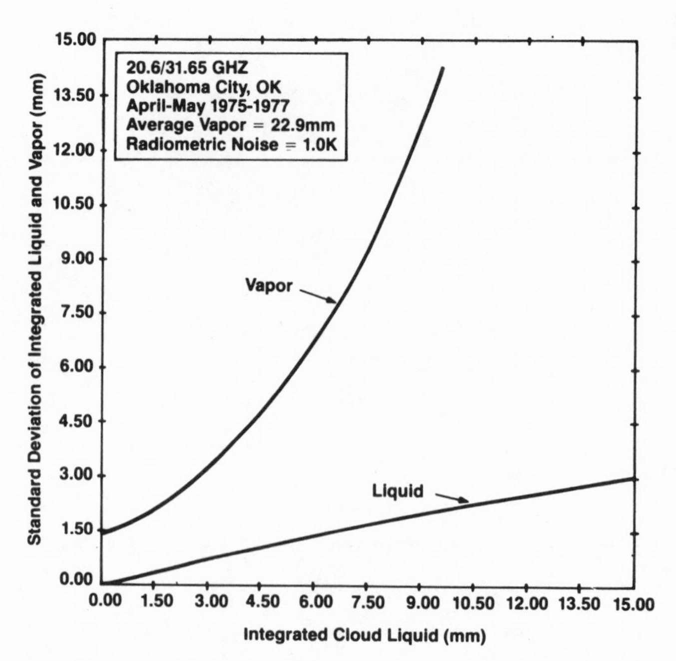

Figure 8(a). Predicted rms error in vapor and liquid retrieval from a dual-frequency radiometer, as a function of cloud liquid.

sense that the retrieval coefficients are themselves functions of the amounts of (inferred) liquid and vapor that are present. This method is described below, following Westwater and Guiraud (1980).

For a linear random process with a joint Gaussian probability distribution, linear statistical inversion (or estimation) yields a minimum expected mean square error (Swerling, 1966). Hence in the linear Gaussian case, (3) is the optimal of all estimators, linear or not. However, calculated frequency distributions of cloud liquid content L depart markedly from a Gaussian character (Westwater et al., 1976). For a large percentage of the time, thin clouds with low values of cloud-liquid density occur; much more infrequently, thicker and more dense clouds are present. In addition, at the higher cloud liquids, rain is much more likely to occur. However, we concern ourselves with nonprecipitating clouds. Nevertheless, the marked departures from Gaussian statistics suggest that retrievals which use (3) can be improved.

As discussed above (see also Table 3), the liquid attenuation coefficient κ_L is subject to a rather large climatological variation of the order of 30%. As a consequence of this variation, (3) contains multiplicative errors and hence is nonlinear. If we express κ_L as an average $<\kappa_L>$ plus a departure from the average κ'_L, then (3) may be written

$$\tau_\nu = \kappa_{\nu V} \cdot V + \kappa_{\nu L} \cdot L + \tau_{d\nu} \approx <\kappa_{\nu V}>V + <\kappa_{\nu L}>L + \kappa'_{\nu L} L + <\tau_{d\nu}> \quad (5)$$

where we have neglected the small variation in $\kappa_{\nu V}$ and $\tau_{d\nu}$. We recall that if

$$\underset{\sim}{d}' = A \underset{\sim}{p}' + \underset{\sim}{\varepsilon} \quad (6)$$

where A is a known n x m matrix and ε is a random zero-mean measurement error, then one way to derive (4) is to minimize the quadratic form

$$Q(\underset{\sim}{p}') = (\underset{\sim}{d}' - A \underset{\sim}{p}')^T S_\varepsilon^{-1}(\underset{\sim}{d}' - A \underset{\sim}{p}') + \underset{\sim}{p}'^T S_p^{-1} \underset{\sim}{p}' \quad (7)$$

with respect to $\underset{\sim}{p}'$ (Strand and Westwater, 1968). In (6), n is the number of measurements, and m is the dimension p'. In (7), S_ε and S_p are the covariance matrices of the vectors $\underset{\sim}{\varepsilon}$ and $\underset{\sim}{p}$. For the moment, consider (5) for a fixed value of L; then the equation can be approximately written in the form of (6):

$$\underset{\sim}{d} = \underset{\sim}{\tau} - <\underset{\sim}{\tau}_d> + \underset{\sim}{\varepsilon} = <\underset{\sim}{\kappa}_V> \; V + <\underset{\sim}{\kappa}_L> \; L + (\underset{\sim}{\kappa}'_L L + \underset{\sim}{\varepsilon}) \quad (8)$$

where $<\underset{\sim}{\kappa}> = (\kappa_{\nu 1}, \kappa_{\nu u})^T$, $<\underset{\sim}{\tau}_d> = (\tau_{d1}, \tau_{du})^T$, and the term $(\underset{\sim}{\kappa}'_L L + \underset{\sim}{\varepsilon})$ may be regarded as the effective noise in $\underset{\sim}{d}$. For the case in which the measurement error ε and $\underset{\sim}{\kappa}'$ are uncorrelated, the effective noise covariance matrix S_ε^{eff} is

$$S_\varepsilon^{eff} = L^2 <\kappa'_L \kappa'^T_L> + S_\varepsilon \, , \quad (9)$$

and thus the effective noise in determining V is greater than the experimental noise by a term that depends on the product of L^2 and the covariance matrix describing climatological fluctuations in $\underset{\sim}{\kappa}_L$. Consequently, retrieval coefficients deriving V from measurements of $\underset{\sim}{\tau}$ should depend on the value of L. Our method of inserting this dependence of the retrieval coefficients is described below.

We first constrain every profile in the ensemble to contain the same amount of liquid L. Within cloud layers we choose some realistic distribution of cloud liquid density $\rho_L(h)$; our distributions are chosen according to the method given by Decker et al. (1978). For an arbitrary distribution of $\rho_L(h)$ we normalize the liquid absorption coefficient $\alpha_L(h)$ to be consistent with L:

$$\alpha_{\nu L}(h) = \frac{\kappa_{\nu L}(h)\ \rho_L(h)}{\int_0^\infty \rho_L(h)\ dh} L . \tag{10}$$

In this way, every profile in the ensemble has its original spatial distribution of temperature and gaseous absorption, and a distribution of liquid absorption that is consistent with the temperature profile, but is now constrained to the desired value of L. Over this constrained ensemble we calculate brightness temperatures, add to them random noise, then apply statistical inversion (see Section) 6.1) to derive retrieval coefficients for this specific amount of L. By incrementing L over a range of values we derive a set of adaptive coefficients to estimate V:

$$\hat{V} = a_0(L) + a_1(L)\ \tau_{20.6} + a_2(L)\ \tau_{31.65} .$$

Adaptive coefficients for the climatology of Oklahoma City are shown in Figs. 8(b) and 8(d). The behavior of the coefficients (a_0, a_1, a_2) as a function of L is qualitatively explained as follows. For L = 0 we are solving two equations for one unknown; consequently, we derive a least squares solution with positive a_1 and a_2. For intermediate values of L say $0.02 \leq L \leq 0.5$ cm, we are solving two equations for two unknowns, and the resulting coefficients are close to those derived by statistical inversion (see Table 3). Finally, for $L \gtrsim 0.5$ cm, the effective error resulting from large liquid drives the coefficients a_1 and a_2 to zero, and the solution approaches the climatological mean. The predicted rms retrieval error for vapor as a function of liquid content is shown in Figure 8(c). This figure, compared with figure 8(a), indicates the theoretical reduction in error achieved by the adaptive coefficient method.

In a similar manner, we can derive coefficients to infer L from τ if V is fixed. However, since the climatic variation of the vapor mass absorption coefficients κ_V is much less than that of liquid, the sensitivity to V is also much less for this case; this is seen in Figure 8(d), where the liquid retrieval coefficients are shown as a function of V.

To use the above two sets of coefficients to retrieve V and L, and iterative procedure can be used. We first estimate $L^{(0)}$ using nonadaptive statistical inversion. Then we use adaptive coefficients appropriate to this $L^{(0)}$ to infer $V^{(0)}$. Next the coefficients corresponding to $V^{(0)}$ are used to infer $L^{(1)}$, and the iteration is continued until $V^{(n+1)} - V^{(n)}$ and $L^{(n+1)} - L^{(n)}$ are

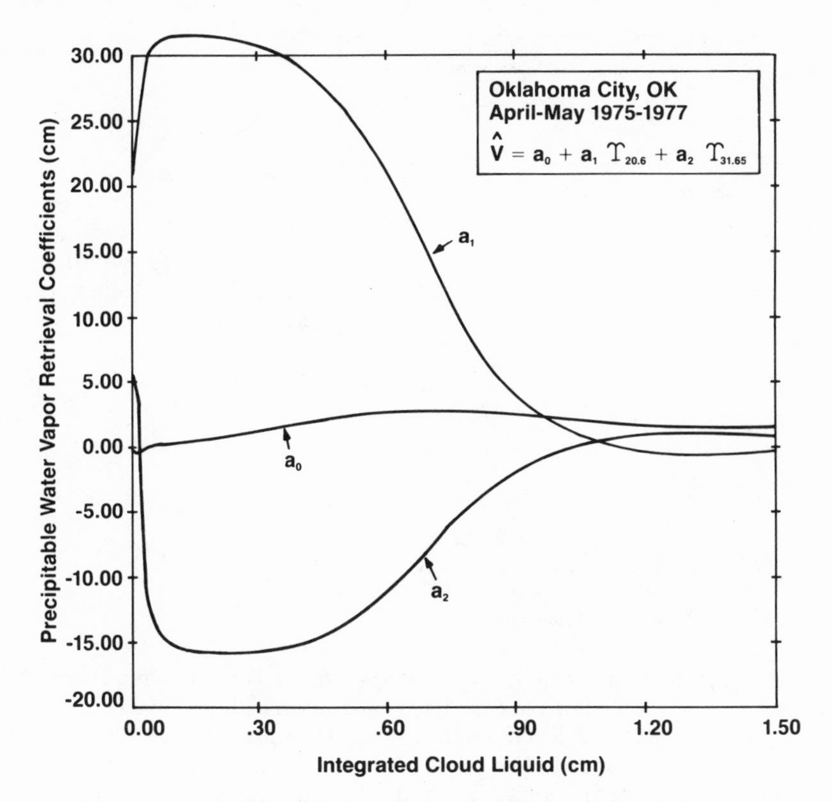

Figure 8(b). Adaptive water vapor retrieval coefficients, as a function of cloud liquid.

sufficiently small. Experience on data has indicated that one of two iterations are sufficient for this method.

In section 7, examples of retrievals are given using the methods outlined above.

7. MEASUREMENTS OF WATER VAPOR

The integrated amount of water vapor in the zenith direction is called the total precipitable water vapor (PWV). This amount is used routinely in weather forecasting, being especially relevant to prediction of precipitation. Vapor is measured conventionally by a hygristor in a radiosonde package that is borne aloft by a balloon. These soundings, made twice daily at stations of the National Weather

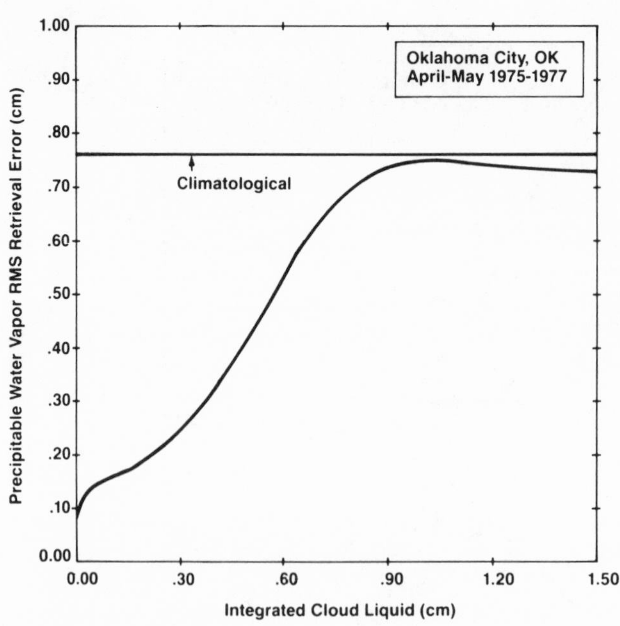

Figure 8(c). Theoretical rms error in vapor retrieval using the adaptive method.

Service (NWS), are integrated to obtain the PWV; it takes something more than an hour to complete a radiosonde profile to altitudes above the tropopause. Although the precision of radiosondes in measurement of PWV is not especially good, it is an accepted standard; operationally the error would exceed the inherent precision. However, radiosondes are the only available instruments with which long-term comparisons of PWV can be made.

7.1 Long-Term Measurements

Comparisons of PWV measured by a dual-channel radiometer with amounts obtained by operational radiosondes have been made over a 6-month period at the NWS facility, Stapleton Airport, Denver, CO (Guiraud et al., 1979).

A typical analog plot of total precipitable water vapor above Denver, CO, taken over a 2-week period, is shown in Fig. 9. The solid record is radiometric measurement, and the triangles are total precipitable water vapor from integrated radiosonde data obtained at 12-h intervals. The agreement is quite good. Of the differences that do exist on individual reading, we do not know how much is due to the radiometric system and how much is error stemming from the radiosonde measurement. It is clear, however, that during certain 12-h intervals (e.g., August 8, P.M. to August 9, A.M.), a considerable amount of vapor passes overhead unobserved by the sondes.

Further verification of performance is obtained by plotting precipitable water vapor amounts measured by the radiometer and the

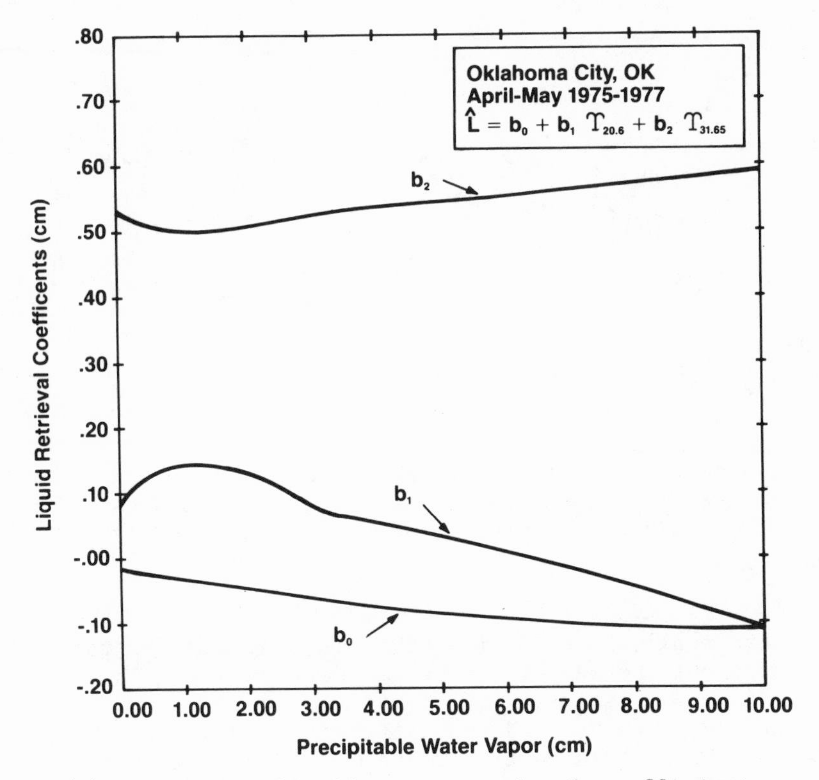

Figure 8(d). Adaptive liquid water retrieval coefficients, as a function of the amount of water vapor.

radiosonde directly against one another, as in Fig. 10. In this 6-month sample, the rms difference between the two is 1.7 mm, some part of this being attributable to each measuring device. Although assignment of an error in operational-radiosonde measurements is difficult, estimates by experienced operators and researchers, for typical midrange humidity and temperature, place $\overline{(V_S - V_A)^2}$ at $(1.5\ \text{mm})^2$, where V_S is the sonde-measured precipitable water vapor and V_A is the actual value. Thus, assuming no correlation between radiosonde and radiometric errors, one obtains

$$\overline{(V_R - V_A)^2} = 1.7^2 - 1.5^2 = 0.65$$

or 0.8 mm for the rms error of the radiometric measurement. However,

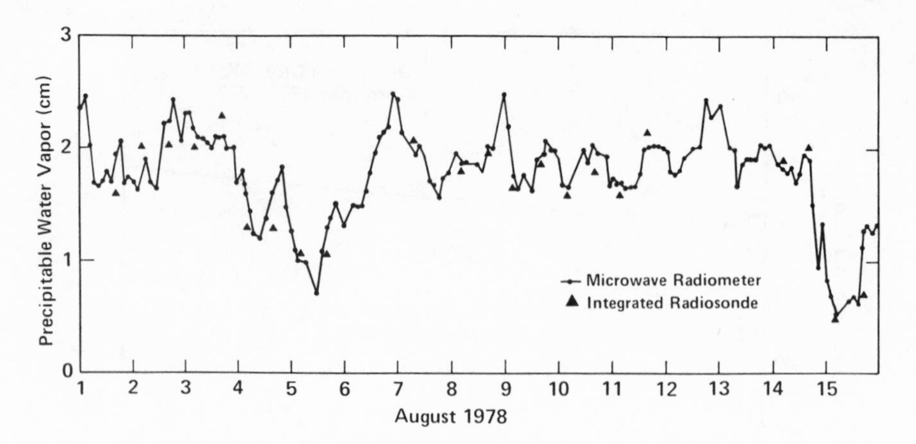

Figure 9. Time-series of precipitable water vapor measured by dual-channel radiometry compared with amounts obtained by radiosonde; Denver, Colorado, August 1978.

all that can really be said with confidence is that the radiometric values are about the same as or better than those of the radiosonde.

It should be emphasized that all data for the 6-month period, regardless of weather conditions, are plotted in Fig.10 for those times when total precipitable water vapor from the radiosonde was available. This remark points to the fact that many cases involving clouds are included.

7.2 Short-Term Measurements

First evidence of the overall stability of a dual-channel system was obtained in December 1978, as shown in Fig. 11. The atmosphere became very dry, the precipitable water vapor decreasing from 0.5 to about 0.2 cm during the early morning. Then for an interval of about two hours, 8 to 10 A.M., the atmosphere was remarkably stable. The data over this interval, which had been recorded in the form of 2-minute averages, was analyzed; a root-mean-square deviation of 0.007 cm was obtained. As discussed in Section 8, this corresponds to a change in radio path length of less than 0.05 cm. With this order of sensitivity, it was realized that the instrument was capable of providing useful corrections for phase shifts due to the troposphere that affect VLBI and other systems used in tectonic metrology.

It is also possible to determine the magnitude of the fluctuations and their associated periods by way of spectral analysis (Hogg et al., 1981 b). The amount of water vapor varies greatly with time.

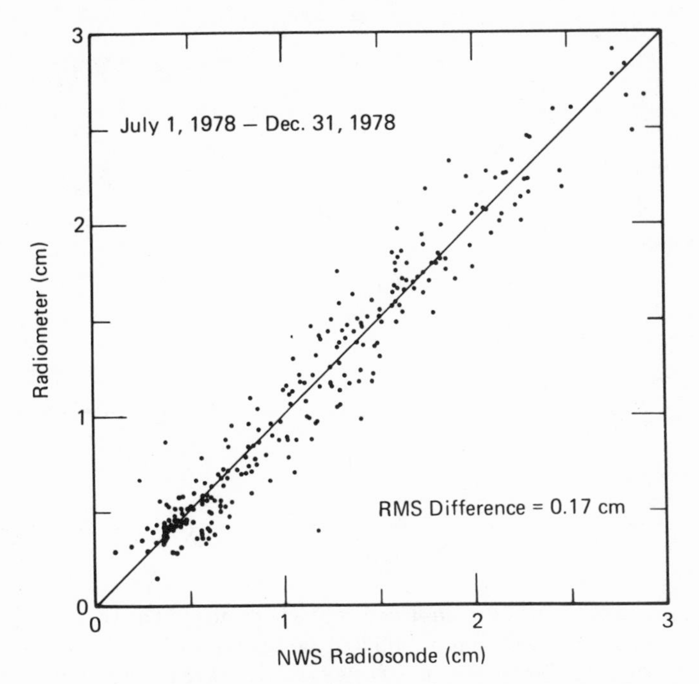

Figure 10. Plot showing the differences between precipitable water vapor measured by dual-channel radiometer and radiosonde during a 6-month period at Denver, Colorado.

For synoptic weather forecasting, one obtains this quantity by integrating in situ measurements from balloon-borne radiosondes conventionally launched every 12 hours ; but the periodicities of the variation span a wide scale. For example, seasonal and yearly changes are significant, especially in temperate zones. Likewise, large air masses on the synoptic scale produce variations with periods of the order of a week. On the other hand, weather fronts give rise to variations on the order of hours. In addition, there are cloud-sized patches of vapor that produce periods of minutes, as well as very-small-scale patches, caused by turbulence, that move with the wind and can give rise to periods on the order of seconds.

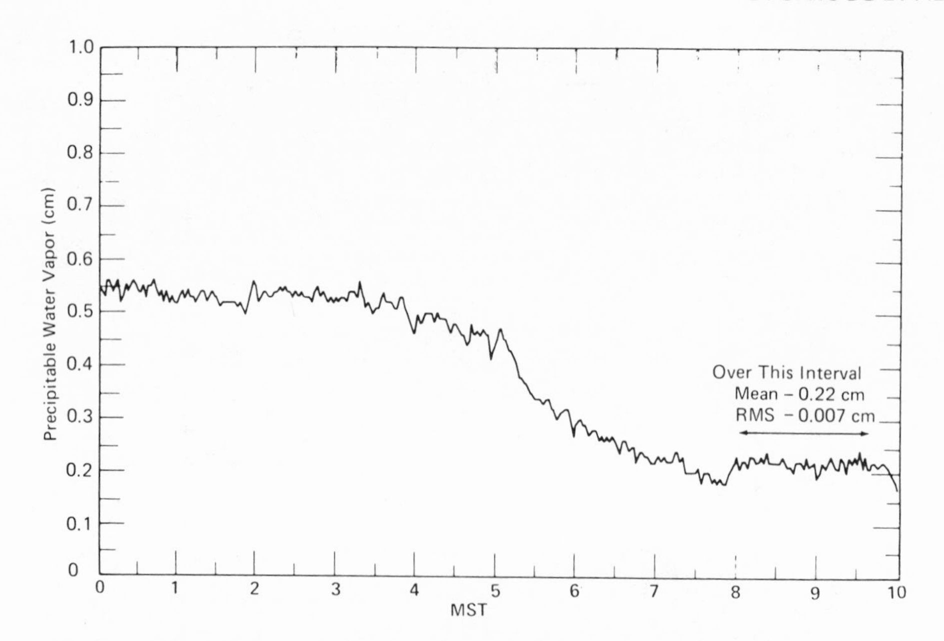

Figure 11. Radiometrically measured total-precipitable water vapor at Denver, Colorado, WSFO, December 13, 1978; the measurements are 2-minute averages on a clear, very dry day.

Figure 12 is a time series that shows how the PWV varies during time intervals of less than a week. For example, on day 273, the PWV decreased to half-value from midnight to midday, and then regained half of that decrease by evening. Periods of much less than an hour are observed on several occasions; for example, the fine-grained low-amplitude fluctuations near noon of day 269. The time constant of the instrument is 2 minutes for these data. An analysis of very-short-term fluctuations is given by Orhaug (1965).

Examples of spectra are shown in Fig. 13; curve a corresponds to the time series of Fig. 12. The abscissa is in frequency units (F) of cycles per day; therefore, $F = 10^2$, for example, applies to a period of about 15 minutes. The length of the individual data samples was 34 hours, and five of these were averaged to produce the spectrum of curve a. The weather was clear with broken clouds; the average PWV for the period was about 1.2 cm, and the average wind was about 8 m/s. We detrended the time-series data by taking first differences before removing the mean. The spectrum, computed by Fourier transform, is adjusted by taking into account the effect of differencing the time series. Differences between the spectra of the detrended data and data with only the mean removed are observed only at the lowest frequencies for nonstationary samples.

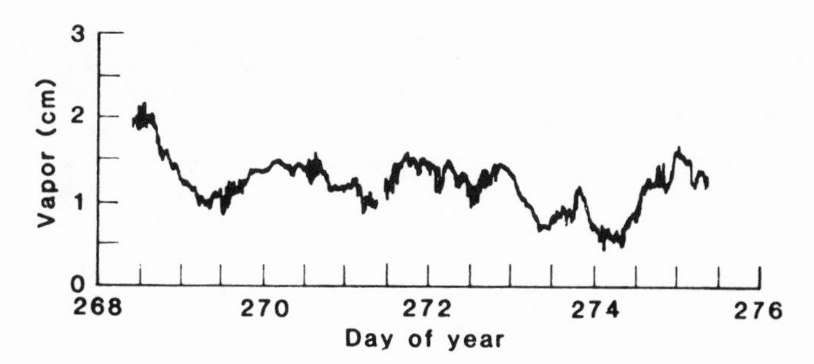

Figure 12. Microwave radiometric measurement of total PWV for a 7-day period in 1978 at the Weather Service Forecasting Office, Denver, Colorado.

Another spectrum, taken under much drier conditions, is curve b in Fig. 13; three 34-hour data samples were averaged. Visual comparison indicates that this power spectrum is lower than curve a by almost a factor of 10.

A matter of considerable interest in the theory of atmospheric structure is the slope of the power spectrum. Although the shape is found to change considerably with weather conditions, one can fit the slopes of spectra such as those in Fig. 13 with relationships of the form $S(F) = KF^{-x}$. However, it is clear from Fig. 13 that the curves can be more satisfactorily represented by two straight lines rather than by one. Such being the case, the behavior of the PWV spectrum appears to change at about $F = 10$, corresponding to a period of about 2.5 hours.

A concern in power spectrum measurements is the signal-to-noise ratio over the frequency range of interest. Low frequency noise in a microwave radiometric instrument is generated by effects such as changes in the temperatures of components, gains of amplifiers, and lack of stability in power supplies. To determine the noise performance of the dual-channel system, the microwave antenna (which normally is permanently attached to generate a beam in the zenith) is disconnected and replaced by an enclosed microwave termination of known temperature. Under these conditions, the radiometer is operated over a long period of time, and data are recorded in the usual way. The spectrum of the background noise of the instrument is then obtained by the same technique, and length of interval, as that used in generating the PWV spectra. The result (Fig. 14) shows that the most objectionable noise components occur at the lower frequencies, near two to three cycles per day. A comparison of curve a of Fig. 13, the "moist" condition, with the noise spectrum of Fig. 14 shows

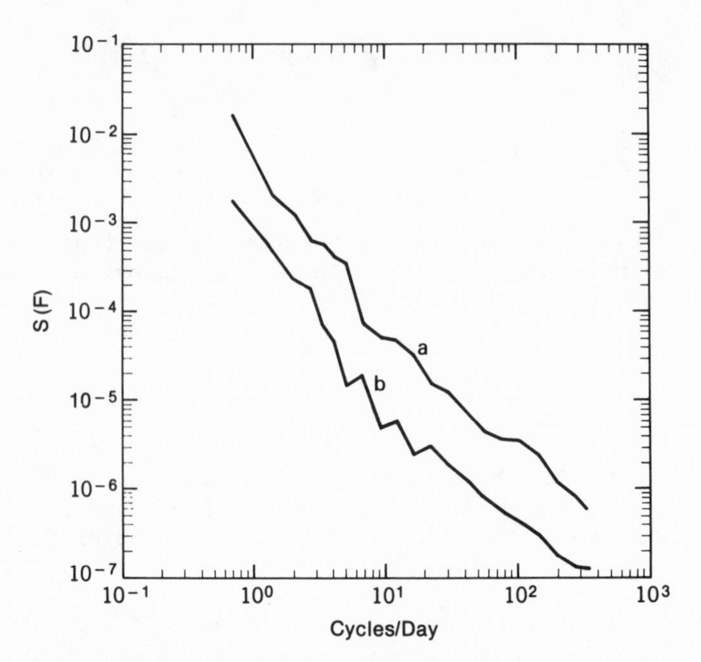

Figure 13. Curve a is an example of the spectrum of PWV measured at the Weather Service Forecasting Office, Denver, CO, over five 34-hour periods; the average PWV for the period is 1.2 cm. Curve b is the corresponding spectrum for a drier condition over three 34-hour periods; the average PWV is 0.35 cm. The units of the ordinate are square centimeters per cycle per day.

that the data spectrum exceeds the noise spectrum by two orders of magnitude over most of the frequency range. In curve b, the "dry" condition, the signal-to-noise ratio is less than dB, especially at the high frequencies.

7.3 Other Comparative Measurements

Another way of establishing the credibility of an instrument of given design is to construct two of them and compare the measured data from the two instruments in the same environment. Such a

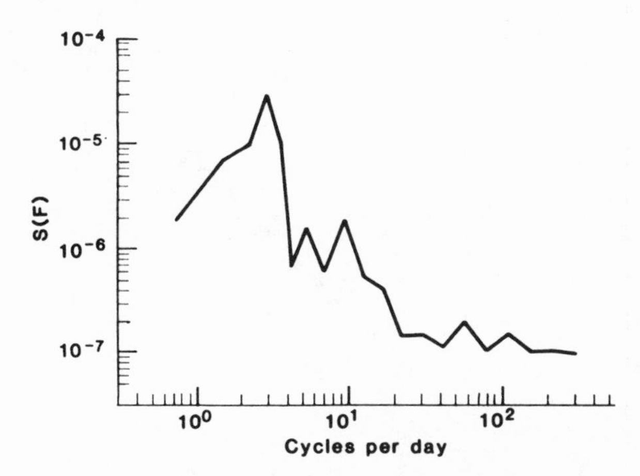

Figure 14. A 34-hour measurement of the spectrum of the background noise of a dual-channel microwave radiometric system; the ordinate is lower by a factor of 10^3 than in Fig.13 The units of the ordinate are square centimeters per cycle per day.

comparison has been made during part of the summer of 1981 between two dual-channel radiometers, with vertically-oriented beams, situated side-by-side at Denver, CO. Figure 15 shows a plot of precipitable water vapor measured by the two instruments, one as ordinate and the other as abscissa. These sets of points happen to be taken at the times of radiosonde launches; the individual points are 2-minute averages. The root-mean-square difference between the two sets is about 0.8 mm, the same number as obtained by the estimate discussed in Section 7.1 in connection with the data of Fig. 10. It is clear from visual inspection of Fig. 15 that the agreement is quite good; therefore, the two instuments measure the same amount of water vapor.

The steerable-beam capability permits real-time comparison with other instruments measuring water vapor, such as those on satellites. An example of a comparison with measurements by the infrared (IR) VAS system on the synchronous GOES-E satellite is shown in Fig. 16. Both instruments are oriented to observe the same path, in this case from the orbital location to Denver. During the relatively clear morning, three measurements were obtained by the I.R. instrument,

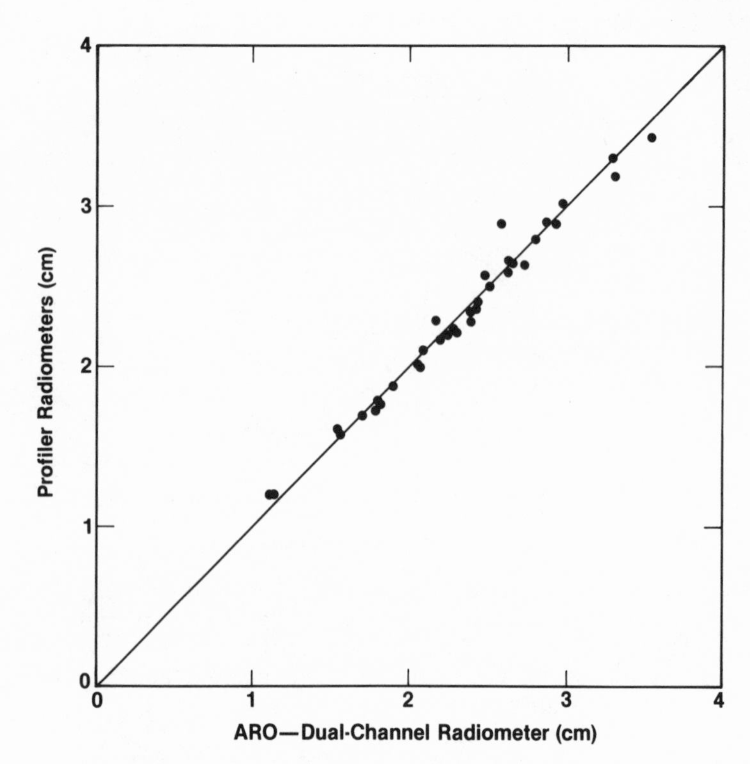

Figure 15. Side-byside measurements of precipitable water vapor by two dual-channel radiometers; Stapleton Airport, Denver, CO, June 15 to July 16, 1981.

as shown by the solid dots in the figure. The amounts of path-averaged water vapor calculated by multiplying the PWV obtained from the Denver radiosondes by the secant of the zenith angle are shown by the two X's near 0500 and 1700; the recording of the microwave radiometer is the full line. The agreement between the I.R. data and the microwave measurement is surprisingly good. However, during the afternoon, cloudy conditions set in, including occurrence of rain as shown by the path liquid measured by the microwave instrument, and further I.R. data could not be retrieved. As a matter of interest, the liquid occurring at 1330 was a heavy discrete cell some ten miles from the radiometer site whereas the showers at 2000 and 2200 occurred at the site.

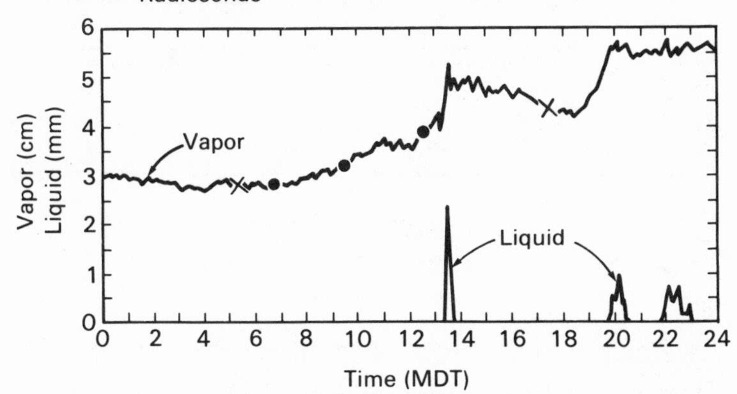

Figure 16. Integrated water vapor measured by dual-channel radiometry on a slant path to the GOES-E satellite compared with VAS infrared data initially reported from the satellite (solid dots) and with the Denver radiosonde measurements (X's).

If the atmosphere happens to be horizontally homogeneous, and if the radiometric instrument is properly levelled, the radiometer, with the antenna beam at a fixed elevation angle, should produce a constant output over a full rotation in azimuth. Under the above assumptions, the constancy of the output is some measure of the quality of the antenna radiation pattern, especially when the elevation is low. Figure 17 shows three full azimuth scans, with elevation angle 20°,taken in California on March 10,1981; each rotation is completed in about 20 minutes. The integrated vapor is about 3.5 cm at all azimuths, a fairly dry day, with corresponding PWV of 1.2 cm. Although the 3.5 cm average appears to be well preserved over the one-hour period of measurement, certain repeating characteristics show up in specific azimuth sectors, e.g., the fluctuations over the 120-140° sector in Fig. 17; the origin of these is not known.

7.4 Comparison of Retrievals Using Statistical and Adaptive-Statistical Methods

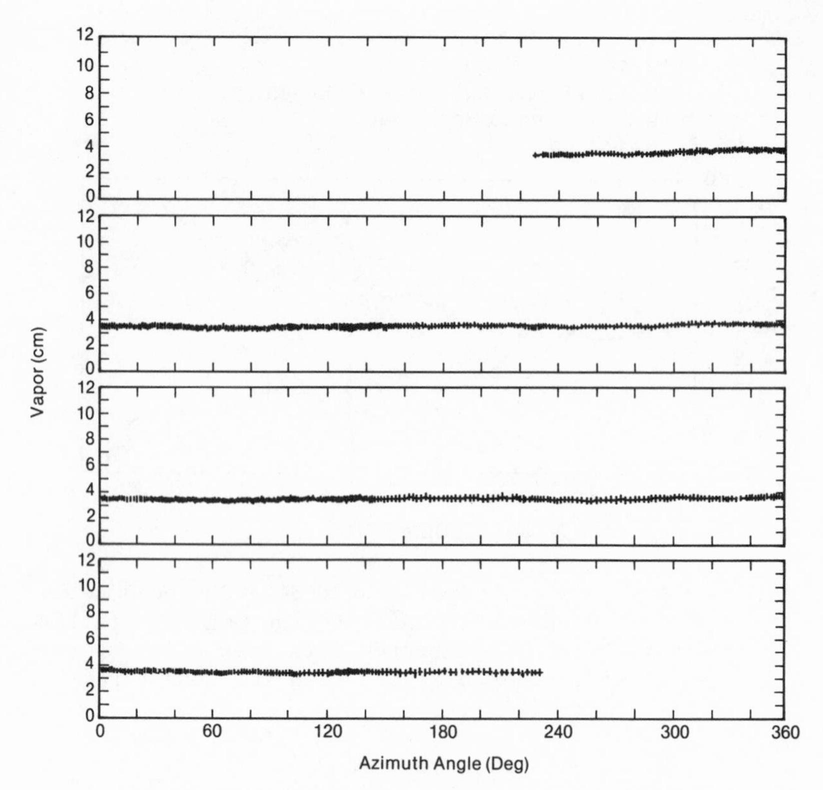

Figure 17. Azimuthal scans showing integrated water vapor on slant paths at an elevation of 20° during a day when the atmosphere was quite homogeneous; Auburn, CA, March 10, 1981.

A dual-channel radiometer was operated at Lawton, OK, from March 29,1979 to June 8, 1979 as a part of a Severe Storms and Mesoscale Experiment (SESAME). The radiometer was intended to provide a continuous monitor of total precipitable vapor as it flowed from the Gulf of Mexico into the SESAME monitoring area. In contrast to the range of atmospheric conditions encountered during the previous operation of this radiometer at Denver, CO (Guiraud et al., 1979), there were many periods during which clouds with high liquid content were present; frequently these conditions developed into severe storms and rain. Retrievals of vapor during these high-liquid conditions were performed, using both statistical (see Section 6.1) and adaptive statistical methods (see Section 6.2.). In all, a total of 15,579 dual-channel brightness temperatures were available. Because of the high likelihood of rain, we did not attempt to retrieve vapor when the 30-GHz brightness temperature exceeded 250 K. A total of 38 data pairs exceeded the above stated thresholds, leaving 15,541 pairs to be analyzed by our retrieval techniques.

We present the data in two ways, the first is a case study of a 7-day period for which (a) there were periods of high liquid content and (b) the Fort Sill and the Oklahoma City soundings were in reasonable agreement. The case selected is shown in Figs. 18 and 19, in which nonadaptive statistical and adaptive retrievals are shown. During periods of heavy liquid the advaptive retrieval yields vapor estimates substantially lower than those of the non-adaptive method,but during periods of clear to low liquid content, the retrievals are practically identical. In all likelihood, the vapor estimates during times of high liquid, such as the spikes of liquid greater than ∿1 cm, exceed the true value. We obtained some evidence for this by integrating the Oklahoma City (OKC), May 3, 0600 CST sounding, assuming saturation conditions. This gave an upper bound of vapor content of 4.0 cm. During the 18-hour period of high liquid variations on day 123, the value of 4.0 was exceeded four times.

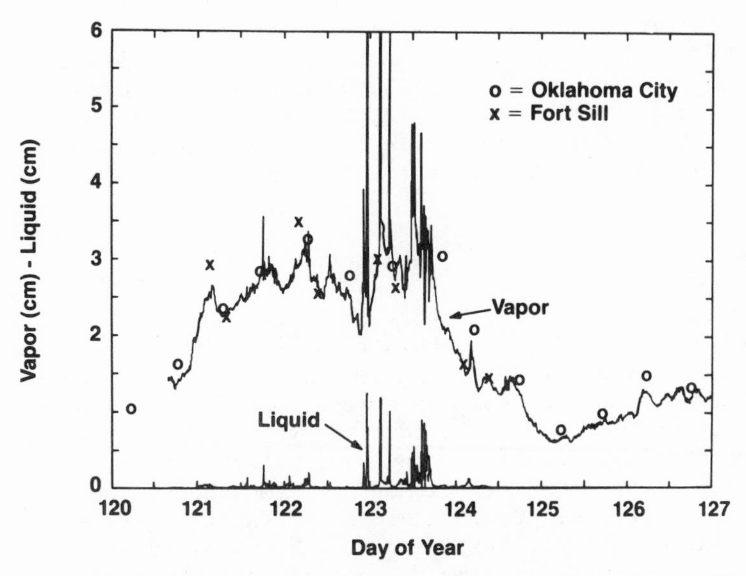

Figure 18. Nonadaptive statistical retrieval of vapor and liquid. Data were taken on April 30 - May 7,1979, during SESAME experiment.

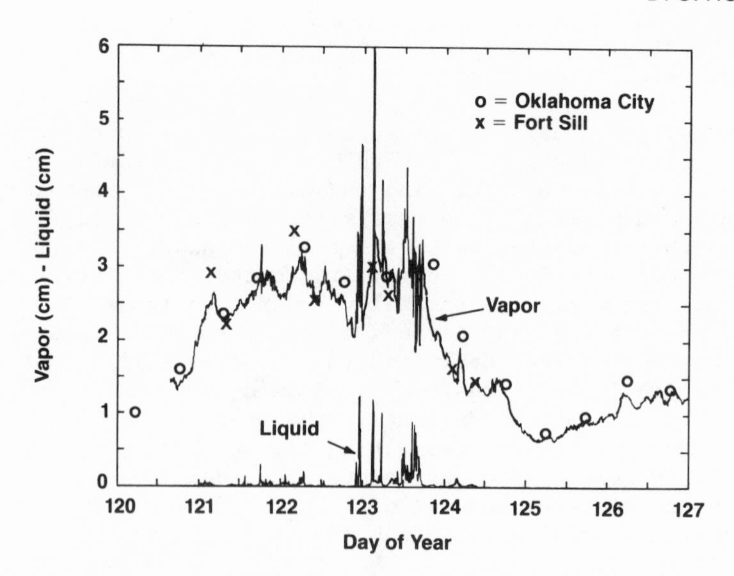

Figure 19. Adaptive retrieval of vapor and liquid. Conditions are as in Fig. 18.

For the complete set of 15,491 observed data we compared PWV obtained by statistical and adaptive statistical methods for values of inferred liquid ranging from 0.0 to 1.5 cm. The rms differences are shown in Fig. 20. For liquid less than about 0.2 cm, there is practically no difference between the methods. For liquids greater than 0.2 cm, the rms differences between retrievals dramatically increase, and there is a large bias between the adaptive method and the other method. We believe this is because the adaptive method decreases the tendency to overestimate vapor during high-liquid situations.

7.5 Long-Term Measurements of Absorption

In section 5, a tipping curve, obtained by scanning the antenna beam in elevation angle, was shown to be useful for determining total absorption in the zenith, for a given amount of precipitable water vapor measured simultaneously by a radiosonde. This type of measure ment is made under cloudless conditions to ensure that liquid does ńot contaminate the results in any way. If the measurement is carried out many times over a matter of months, atmospheres are encountered in which the amount of precipitable water vapor covers a considerable range of values. In temperate zones, the amount in high during a warm and humid summer, and low on cool and dry winter days.

Measurements taken in this way, at Denver, CO, are shown in

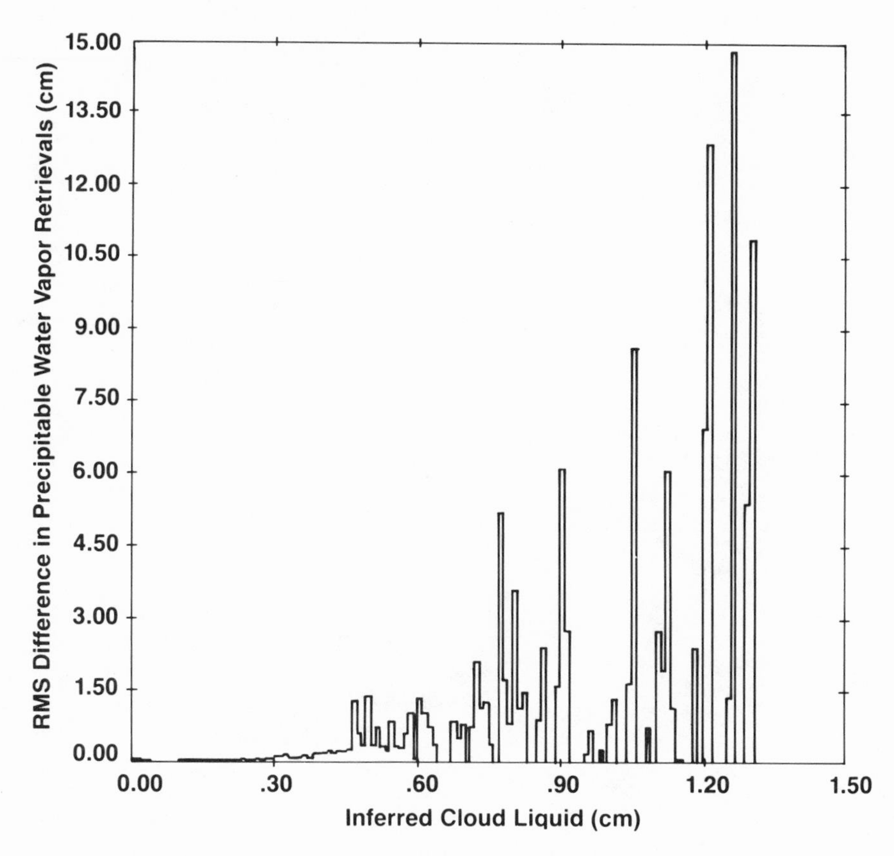

Figure 20. Rms difference between adaptive and physical retrievals of vapor. Data from SESAME experiment, March 29 to June 8, 1979; 15.541 data points.

Fig. 21 (Hogg and Guiraud, 1979) for the frequencies 20.6 and 31.6 GHz. Although Colorado has a relatively dry climate, precipitable vapor up to about 2.5 cm is encountered. Fits to a straight line, and to curves of higher power, have been made to the data. For linear fit, the standard error of measurement is 0.031 dB at 20.6 GHz and 0.019 dB and 31.65 GHz. However, fitting with a polynomial results in standard errors of 0.028 and 0.015 at 20.6 and 31.65 GHz respectively. The polynomial fit therefore show some improvement over the linear fit, especially at 31.65 GHz in the transmission 'window' where the 'continuum' absorption is most influential. The scatter in the data is caused in part by lack of horizontal homogeneity in the atmosphere, and in part by non-coincidence in radiosonde flight time and absorption measurement; the latter contributes about 1 mm rms error on the abscissa in Fig. 21. The error introduced by

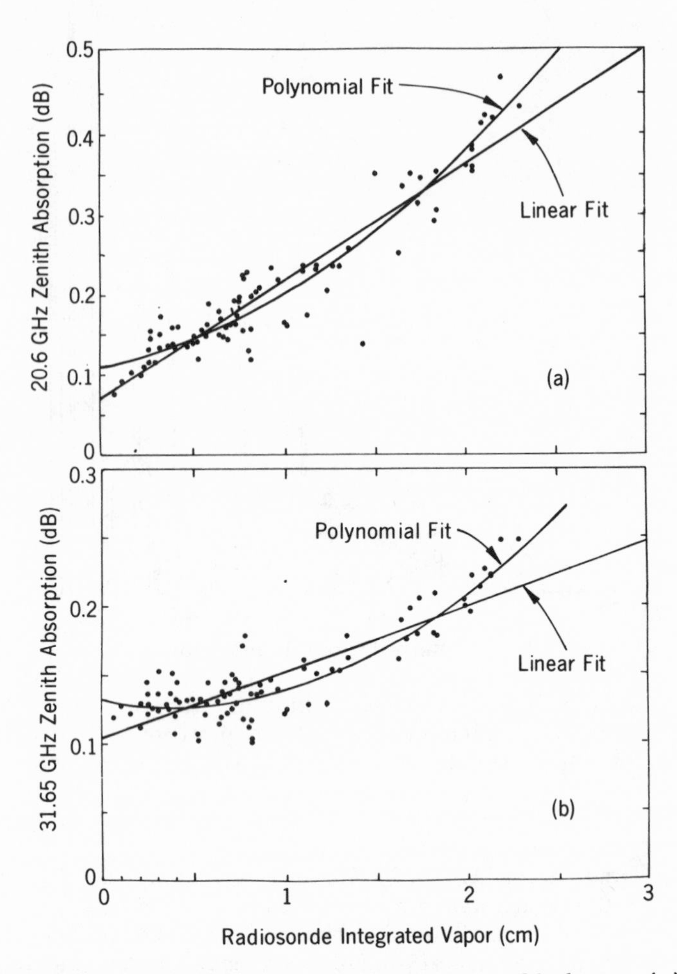

Figure 21. Total absorption in the zenith at 20.6 GHz (a) and 31.65 GHz (b) plotted against precipitable water vapor measured by radiosondes; standard errors at 20.6 GHz and 31.65 GHz are 0.031 and 0.019, 0.028 and 0.015 (dB) for the linear and polynomial fits respectively. The slight downward trend in (b), from 0 to 0.5 cm, is caused by an inverse temperature effect in absorption by oxygen.

the radiometric instrument is <0.01 dB which is about one-tenth of the maximum peak-to-peak deviations. The accuracy of humidity measurement by radiosonde is known to be questionable; the hygristors used are especially erratic at relative humidities below 20%. Estimates place the accuracy of measurement of precipitable vapor at about 1.5 mm which does not materially alter the conclusion regarding the nonlinear fit in Fig. 21.

The behavior of the 31.65 GHz data of Fig. 21(b) at low values of vapor is curious; the total absorption decreases slightly with increasing vapor. This effect is believed caused by the inverse dependence of oxygen absorption on temperature, that is, as low precipitable vapor normally occurs during cool temperatures, the absorption by oxygen is higher then than during high-humidity conditions. The equation describing the nonlinear fit in Fig. 21 (b) is

$$A = 0.13 - 0.026\ V + 0.032\ V^2.$$

The microwave absorption in this transmission window of the atmosphere appears to have a quadratic dependence on the water-vapor content. This matter is of importance in remotely sensing water vapor in the atmosphere, because if the absorption process is not fully understood, accurate values of vapor are not retrieved from radiometric observations.

8. VARIATIONS IN RADIO PATH LENGTH

It is changes in phase at given wavelengths, generated along paths through the atmosphere, that are troublesome in interferometers and related systems. For measurements of water vapor, such as those discussed above, to be useful in correcting for these effects, the integrated vapor, V, must be converted to phase at the operating wavelength of the interferometer. This procedure has recently been discussed (Hogg et al., 1981a) in some detail.

By utilizing temperature, pressure, and water vapor, the refractive index of the air (Thayer, 1974), and therefore the total excess path length, L, can be determined. The contribution of water vapor to this length, L_v, can be dealt with separately and is given by

$$L_v = 6.5\ V \tag{11}$$

to good accuracy; L_v and V are in the same units, say centimeters. When the measured accuracies for V determined by the measurements discussed in Section 7, are applied to eq. (11), one obtains

$\Delta L_v \leq 0.52$ cm (long-term)

$\Delta L_v = 0.046$ cm (short-term)

If these values are, in turn, applied to wavelengths such as those of the very large array of NRAO at Soccoro, NM, one obtains the following table for a lower bound on the measurable phase shift $\Delta\phi = \frac{2\pi\Delta L_v}{\lambda}$, in degrees:

Wavelength (cm)	$\Delta\phi$(long-term)	$\Delta\phi$(short- term)
21	9	1
6	31	3
2	94	8
1.2	156	14

These results apply to a near-zenith beam-pointing configuration.

Since the density of dry air contributes some fifty times more than water vapor to excess path length (but is much less variable in time and space), it is of interest to develop a simple approximate relationship for overall excess path length, L_t. By forsaking the conventional procedure of dissociating the so-called wet and dry terms of the refractive index, one obtains a result that depends only on the precipitable water vapor and the total surface pressure; for a zenith-pointing situation:

$$L_t(\text{cm}) = 0.2276\ P_s + 6.277\ V \tag{12}$$

where P_s is the total surface pressure in millibars and V the precipitable water vapor in centimeters. By comparing lengths computed from this equation with those obtained by numerical integration of radiosonde data, agreement to a root-mean-square difference of 0.18 cm has been found (Hogg et al., 1981a). Evidently, good measurements of overall phase shift are therefore obtainable by the combination of an instrument that measures vapor accurately, along with a good barometer that measures to an accuracy of order 0.1 mbar. By continuance of the argument that the temperature and pressure fields in the atmosphere are often well stratified, one is tempted to speculate that application of eq. (12) to the geometry of a given slant path would be successful. In such a case, the vapor V would be measured by a steerable dual-channel radiometer with beam oriented along the path under consideration.

REFERENCES

Bean, B.R., and Dutton, E.J. 1966, Radio Meteorology, NBS Monograph No.92, U.S. Government Printing Office, 393

Blevis, B.C. 1965, IEEE Trans. Antennas and Propagation, AP-13:175-176.

Decker, M.T., Westwater, E.R., and Guiraud, F.O., 1978, J.Appl. Meteorol., 17: 1788-1795.

Deepak, A., 1980, Atmospheric Water Vapor, edited by A. Deepak, T.D. Wilkerson, and L.H. Ruhnke, Academic Press, New York.

Deutsch, R., 1965, Estimation Theory, Prentice Hall, Englewood Cliffs, N.J., 54-71

Guat, N.E., and Reifenstein, E.C. III 1971, Environmental Res. and Tech. Rep., 13.

Grody, N.C., Gruber, A., and Shen, W.C., 1980, J. Appl. Meteorol., 19:986-996.

Guiraud, F.O., Howard, J., and Hogg, D.C., 1979, IEEE Transactions on Geoscience Electronics, GE-17, 4:129-136.

Hinder, R., and Ryle, M., 1971, Mon. Not. R. Astr. Soc., 154:229-253

Hogg, D.C., 1959, J. Appl. Phys., 30:1417-1419.

Hogg, D.C., and Chu, Ta-Shing, 1975, Proceedings of the IEEE, 63, 9:1308-1331

Hogg, D.C., Guiraud, F.O., Howard, J., Newell, A.C., Kremer, D.P., and Repjar, A.G., 1979a, IEEE Trans. on Antennas and Prop.,AP-27, 764-771.

Hogg, D.C., and Guiraud, F.O., 1979b, Nature, 279: 408-409.

Hogg, D.C., Guiraud, F.O., and Decker, M.T., 1981a, Astron. and Astrophys., 95: 304-307.

Hogg, D.C., Guiraud, F.O., and Sweezy, W.B., 1981b, Science, 213 : 1112-1113.

Orhaug, T.A., 1965, Transaction 300, Chalmers University of Technology, Sweden.

Staelin, D.H., 1966, J. Geophys. Res., 71: 2875-2881.

Staelin, D.H., Cassel, A.L., Kunzi, K.F., Pettyjohn, R.L., Poon, R. K.L., and Rosenkranz, P.W., 1975, J. Atmos. Sci., 32:1970-1976.

Staelin, D.H., Kunzi, K.F., Pettyjohn, R.L., Poon, R.K.L., Wilcox, R.W., and Waters, J.W., 1976, J. Appl. Meteorol., 15: 1204-1214.

Strand, O.N., and Westwater, E.R., 1968, J. Assoc. Comput. Mach., 15: 100-114.

Swerling, P., 1966, J. Soc. Ind. Appl. Math., 14: 998-1031.

Thayer, G.D., 1974, Radio Sci., 9,10: 803.

Waters, J.R., 1976, Absorption and emission by atmospheric gases, in Methods of Experimental Physics, 12B, edited by M.L. Meeks, chap. 2.3., Academic, New York.

Westwater, E.R., 1978, Radio Sci., 13, 4:677-685.

Westwater, E.R., and Decker, M.T., 1977, Inversion Methods in Atmospheric Remote Sounding, edited by A. Deepak, Academic Press, New York, 395-427.

Westwater, E.R., and Guiraud, F.O., 1980, Radio Sci., 15: 947-957.

Westwater, E.R., Decker, M.T., and Guiraud, F.O., 1976, NOAA Tech. Rep., ERL 375-WPL 48, NTIS 262-421, Nat. Tech. Infor. Serv. Springfield VA.

FAR INFRARED METAL MESH FILTERS
AND FABRY-PEROT INTERFEROMETRY

Kiyomi Sakai

Osaka University, Faculty of Engineering
Department of Applied Physics
Osaka 565, Japan

and

Ludwig Genzel

Max-Planck Institute fur Festkorperforschung
Stuttgart, Federal Republic of Germany

ABSTRACT

The use of metal meshes is becoming increasingly important for applications in the far infrared. This paper reviews the important aspects of metal meshes, metal mesh filters, Fabry-Perot interferometers and their applications. The article includes the following: an introductory description and historical review of far-infrared metal meshes, the optical properties of metal meshes, beam splitters and filters based on inherent optical properties of meshes, Fabry-Perot interferometers, multimesh filters, applications of metal mesh filters and Fabry-Perot interferometers for far-infrared lasers, for plasma diagnostics and for astronomy.

1. INTRODUCTION

The far infrared (FIR) applications of metal mesh filters and Fabry-Perot interferometers (FPI's) are becoming increasingly important in recent years. The metal mesh (electroformed type) was originally introduced into the FIR region, in 1962, by Genzel and co-workers (Renk et al., 1962) as nearly lossless reflectors for the FPI.

In the ultraviolet, visible and near infrared up to 20 μm wavelength, thin metal coatings or multiple dielectric layers are used as low-loss reflectors for interference filters and FPI's. On the other hand, in the microwave region, wire gratings had already been used for reflectors of a FPI by Lewis and Casey (Lewis et al., 1951, 1952; Casey et al., 1952) early in the 1950's. In the FIR region, however, desirable reflectors were not available until 1962. Before Lewis and Casey realized the microwave FPI, the electromagnetic properties of the wire grating had been studied by many investigators since the day of Herz and Dubois. (Larsen,1962) Through 1959 into 1960, Culshaw (Culshaw, 1959,1960) used regularly perforated metal plates as reflectors for a 6 mm FPI. The scaling of this idea for the operation in FIR indicated the applicability of metal meshes.

After the invention of the metal mesh, many studies have been performed on the optical properties of mesh itself, the metal mesh FPI, the multimesh interference filters and their applications. The FIR properties of metal meshes have been investigated experimentally first by Renk and Genzel (1962) followed by Ulrich et al. (1963), Vogel and Genzel (1964), Sakai et al. (1969), and many others (see, for instance, Sakai et al.,1978; Ressler et al., 1967; Romero et al., 1973; Lecullier et al., 1976; Belland et al., 1980; Lamarre et al., 1981). In theoretical work, much attention has long been devoted to the one-dimensional wire grating and very little on the two-dimensional structure. A historical review and bibliography of early investigations can be found in Larsen. An extensive calculation has been presented by Saksena et al.(1969) Apart from their analysis based on the induced current and reradiation,the equivalent circuit model by Markovitz (1951) has been used quite often in the diffraction free region, because of the relative simplicity of the approach. More recently, based on a least-square method (Davies, 1973) rigorous calculations involving the diffraction region have been presented by Beunen et al. (1981). As to the two-dimensional structure, the approach of Chen (1973) and Lee (1971) or McPhedran and Maystre (1977) was found to account successfully for the experimental data in spite of its complexity. Brief descriptions on the features of respective theoretical approaches and the comparison of experimental and theoretical values have been made recently (1981). In this paper,however, we use principally the equivalent circuit and transmission line model.

Ulrich (1967a) introduced the "capacitive mesh" having structure and optical properties complementary to the normal "inductive mesh". He constructed multimesh interference filters (1967b,1968) using the capacitive mesh, inductive mesh or meshes having other structures. These studies have been followed by other investigators (See, Holah et al., 1972, 1974, 1980a, 1980b; Varma et al., 1969; Grenier et al.,1973; Whitcomb et al., 1980; Baldecchi et al., 1974; Daris,(1980).

As early as in 1962, E.E.Bell (1966) introduced the inductive mesh as a beam splitter in a Michelson type Fourier transform spectrometer (FTS). Mitsuishi et al. (1963) showed that wirecloth metal meshes are excellent lowpass filters. Sakai et al. (1969, 1978) showed that the inductive meshes are useful as wide or narrow band-pass filters. Ulrich et al. (1963) and Sakai et al. (1969) measured FIR spectra with a tunable FPI in a very simple optical arrangement. Gornik et al. (1976) operated a tunable FPI in liquid helium for the detection of radiation from InSb Landau states. After the invention of optically pumped FIR lasers by Chang and Bridges (1970) in 1970, numerous papers reported the use of the metal mesh FPI for selecting and measuring the various laser wavelenghts. In addition to the laser work, the FPI, multimesh lowpass filters, and wide and narrow bandpass filters play important roles in the field of plasma diagnostics and also for the observations of astronomical objects.

Brief or relatively detailed descriptions of this subject are found in the review articles by Genzel and Sakai (1977),and Holah and Smith (1977) or in a book by Moller and Rothschild (1971).

2. OPTICAL PROPERTIES OF METAL MESHES IN THE FAR INFRARED REGION

2.1. Electromagnetic Properties of the Wire Grating

Before going into the FIR properties of metal meshes, we describe briefly the electromagnetic properties of the wire grating. In the FIR region, it is useful only as a polarizer and not important for practical FPI or filters. However, a discussion of its properties will be very helpful for an understanding of the properties of metal meshes.

2.1.1.Wire Gratings in Free Space (Moller and Rothschild, 1971)

The grating wire can have either a circular, elliptical, or rectangular cross section. The grating is essentially transparent for electromagnetic waves with an electric vector perpendicular to the wires. The parallel polarized wave is partly reflected, partly transmitted and partly absorbed. The important situation is this parallel polarized case. Casey and Lewis (1952) give a formula for the absorptance and for the reflectance of a wire grating under the restriction that $\lambda>2g$ and $a<<\lambda$, where λ is the wavelength of electromagnetic wave in free space, g the grating constant and a the wire radius assumed large compared to the skin depth. The absorptance A is represented by

$$A = \frac{g}{\pi a} \left(\frac{c}{\sigma\lambda}\right)^{\frac{1}{2}} R, \qquad (1)$$

Where R is the power reflectance, σ the conductivity of the wire, and c the velocity of light. With the conditions $\lambda>2g$ and $a<<\lambda$, the equation of reflectance is still complicated. In the limiting case $\lambda \gg g$, it is expressed in a simpler form. The reflectance R and the transmittance T are given by (Renk et al., 1962; Casey et al., 1952)

$$R = 1 - \left(\frac{2g}{\lambda} \ln \frac{g}{2\pi a}\right)^2 = 1 - \tan^2 \phi$$

$$T \simeq 1 - R = \left(\frac{2g}{\lambda} \ln \frac{g}{2\pi a}\right)^2 = \tan^2 \phi \tag{2}$$

where ϕ is the phase with respect to the reflected wave.

Here we describe briefly the transmission line theory in order to use it effectively for the calculation of multimesh filters. The simplest equivalent circuit is shown as Fig. 1 (a) for $\lambda \gg g$ and $g \gg a$. The amplitude reflection coefficient Γ of this transmission line is given by

$$\Gamma = \frac{Z - Z_o}{Z + Z_o} \tag{3}$$

where Z_o means the impedance of free space, JX_o ($j = \sqrt{-1}$) the impedance of the wire grating and Z the total impedance given by

$$Z = JX_o \ // \ Z_o = \frac{jX_o Z_o}{Z_o + jX_o} \tag{4}$$

Substituting Eq. (4) into Eq. (3), one obtains

$$\Gamma = - \frac{Z_o}{Z_o + j2X_o} = \frac{Z_o}{\sqrt{Z_o^2 + 4X_o^2}} \exp j \left[\pi - \arctan(2X_o/Z_o)\right] \tag{5}$$

From this expression, writing $\Gamma = |\Gamma| e^{i\phi}$ one can obtain the amplitude reflectance $|\Gamma|$ and the phase jump ϕ on reflection as

$$\left.\begin{aligned} |\Gamma| &= \frac{Z_o}{\sqrt{Z_o^2 + 4X_o^2}} \\ \phi &= \pi - \arctan \frac{2X_o}{Z_o} \\ \tan\phi &= - \frac{2X_o}{Z_o} \end{aligned}\right\} \tag{6}$$

The power reflectance R is then represented by

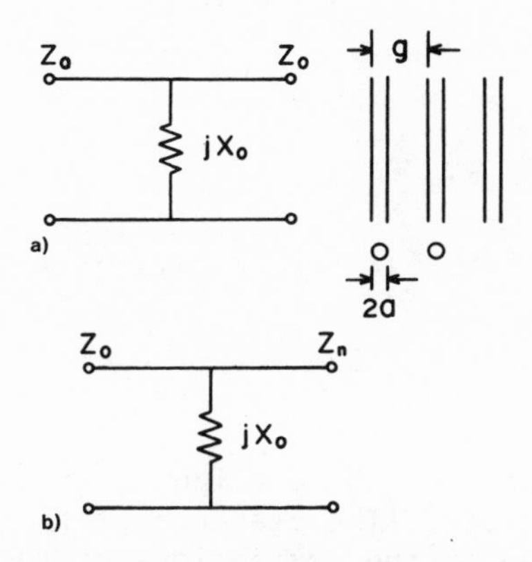

Figure 1. Equivalent circuit representation for a wire grating having grating constant g and wire radius a : (a) grating in empty space, (b) grating on the surface of a dielectric film. (After Ulrich et al., 1963).

$$R = \Gamma\Gamma^* = |\Gamma|^2 = \frac{Z_o^2}{Z_o^2 + 4X_o^2} \tag{7}$$

where R is the power reflectance

When the shunt impedance in the equivalent circuit is purely imaginary, the energy dissipation does not occur, so that one obtains the power transmittance as

$$T = 1 - R = \frac{4X_o^2}{Z_o^2 + 4X_o^2} \tag{8}$$

The quantity X_o/Z_o is, in general, a function of the wavelength and the grating parameters. For a grating with wire radius a, under the conditions that $\lambda >> g$ and $a << g$, it is represented as (Marcuvitz, 1951)

$$\frac{X_o}{Z_o} = \frac{g}{\lambda} \ln \frac{g}{2\pi a} \tag{9}$$

Substituting Eq. (9) into Eq. (7) or Eq. (8) yields again Eq. (2).

2.1.2. Wire Gratings on the Surface of a Dielectric Medium (Moller and Rothschild, 1971)

This case occurs when the grating or the mesh is glued or deposited on a substrate. The presence of the substrate requires a slight modification of the expressions of the transmittance and reflectance. In considering this problem, we neglect the absorption of the substrate. We consider here also the case $\lambda >> g$. Denoting the refractive index of a dielectric medium with n, the wavelength in the medium is $\lambda_n = \lambda/n$ and the impedance of the medium is $Z_n = Z_o/n$. When grating is imbedded in a dielectric medium, Eq. (9) can be used in the form

$$\frac{X_n}{Z_n} = \frac{g}{\lambda_n} \ln \frac{g}{2\pi a} \tag{10}$$

for a wire grating with wire radius a. From Eq. (10) it follows that

$$X_n = Z_o \frac{g}{\lambda} \ln \frac{g}{2\pi a} = X_o \tag{11}$$

This implies that the absolute impedance of the grating is independent of n. It can be assumed that this holds approximately also for the case in which the grating is on the surface of a dielectric medium. In fact, this problem was trated theoretically by Balakhanov (1966) and he found that the assumption is valid for $\lambda >> g$. In conclusion, the absolute impedance of the grating is not affected by the presence of a dielectric medium and the shunt impedance of the grating sees a transmission line with characteristic impedance Z_o on the free side and sees a line of impedance Z_n on the side of the dielectric medium (Ulrich et al., 1963). The equivalent circuit for this case is shown in Fig. 1 (b). When the wave propagates from the left to the right, the amplitude reflection coefficient Γ of this transmission line is given by the same equation as Eq. (3), replacing Z with

$$Z = jX_o // Z_n = \frac{jX_o Z_n}{Z_o + jX_o} \tag{12}$$

When the wave propagates from the right to the left, Γ is expressed by

$$\Gamma = \frac{Z - Z_n}{Z + Z_n} \tag{13}$$

with the same Z as Eq. (4). Therefore, the quantities corresponding to those from Eq. (5) to Eq. (8) can be calculated by use of Eqs. (3) and (12) or Eqs. (13) and (4). Hence, the reflectance and transmittance R' and T' for the dielectrically backed grating can be calculated from the equivalent circuit of Fig. 1(b) and they are expressed as

$$\frac{T'}{T} = \frac{1 - R'}{1 - R} = \frac{n}{R + (n+1)^2\, T/4} \tag{14}$$

for the waves propagating to the right side or to the left side.

2.2. FAR INFRARED PROPERTIES OF METAL MESHES

2.2.1. Introductory Description of Inductive and Capacitive Meshes

We first describe the construction of inductive and capacitive meshes briefly. The structures of the inductive mesh and the capacitive mesh are shown in Fig.2 (a) and (b). The parameters g, a and t indicate the mesh periodicity, half of the strip breadth and the strip thickness, respectively. The inductive meshes are made of cooper, nickel, siler or gold. The inductive mesh is generally free-standing and it is formed on a cross-ruled grating of glass by an electroforming process. Various kinds of inductive meshes are commercially available (e.g. Buckbee-Mears Co., St.Paul, Minnesota or Dainippon Screen, Kyoto). The capacitive mesh is

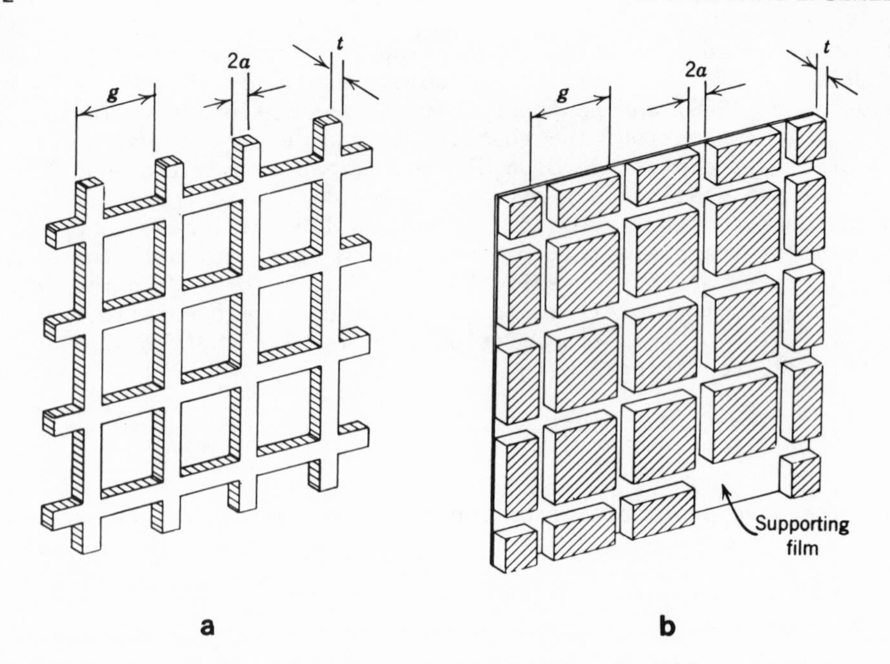

Figure 2. Schematic of normal metal meshes and their geometrical parameters : (a) inductive mesh, (b) capacitive mesh. (After Ulrich 1967a)

composed of a metal such as copper, aluminum, nickel or gold, supported in most cases by a thin transparent dielectric Mylar film of typically 2.5 μm thickness. The capacitive mesh is fabricated by either of the following two methods. The first method (Ressler et al., 1967; Romero et al., 1972) is simple, that is, one obtains the capacitive mesh by the direct vacuum deposition of a metal on a substrate film of Mylar, masking it with an inductive mesh (Fig. 3). The thickness of the metal layer formed by this method, however, does not exceed approximately 0.3 μm. The second method (Ulrich et al., 1969) can prepare a thicker metal layer. It again starts with the vacuum deposition of a thin conducting layer of copper on the Mylar substrate. This primary layer is electroplated (See Fig. 4) up to the required thickness of approximately 1 μm (several skin depths). The metal layer is coated with light sensitive lacquer (Kodak Photo Resist KPR3, thinned 1:1 with Kodak Ortho Resist Thinner) and then a photographic negative of the desired pattern is contact-printed on it (see Fig. 5). Either an inductive mesh

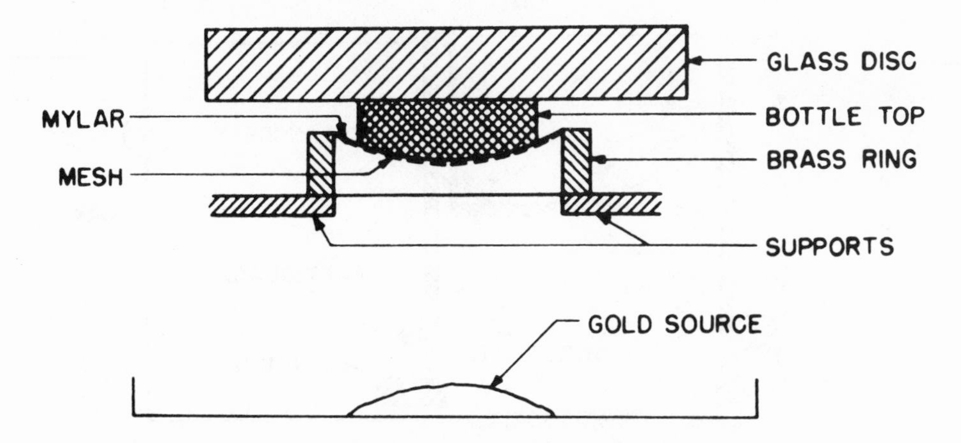

Figure 3. Diagram of vacuum evaporation configuration used in construction of capacitive meshes. (After Romero et al., 1972).

or a photographic plate having a pattern of fretwork is used. After the printing of the desired pattern, via the process of development (in Kodak Ortho Resist Developer KOR) and the etching (using a solution of approximately 50 g $FeCl_3.6H_2O$ per liter of water), one obtains a capacitive mesh. The capacitive mesh can be prepared in the laboratory but it is commercially available now (e.g. Buckbee-Mears).

By use of the photoetching process just mentioned, it is possible to fabricate meshes other than the capacitive one. Up to the present the cross-shaped meshes (Ulrich, 1968; Tomaselli, 1981) with such patterns as in Fig.6(a), (b) have been developed. As in case of ordinary meshes, they are called "inductive cross" and "capacitive cross" mesh, respectively. This new structure gives rise to a new aspect for the bandpass filter and the bandstop filter. The description of these meshes will be given in sections 5. 2. 3 and 5. 2. 4. Except for these sections, this paper deals with the normal inductive and capacitive meshes.

2.2.2. General Properties of Inductive and Capacitive Meshes

The FIR properties of metal meshes are of interest and important for many applications. In this section we shall outline the theoretical considerations of the general properties of the meshes following Ulrich's paper (1967 a). Here the meshes are considered to be very thin and made of perfectly conducting metal, and the influence of the dielectric film supporting the capacitive mesh is neglected. Their optical properties; the power transmittance and reflectance, T and R, the phase shifts upon reflection and transmission, ψ_Γ and ψ_τ, and the loss mechanisms by absorption

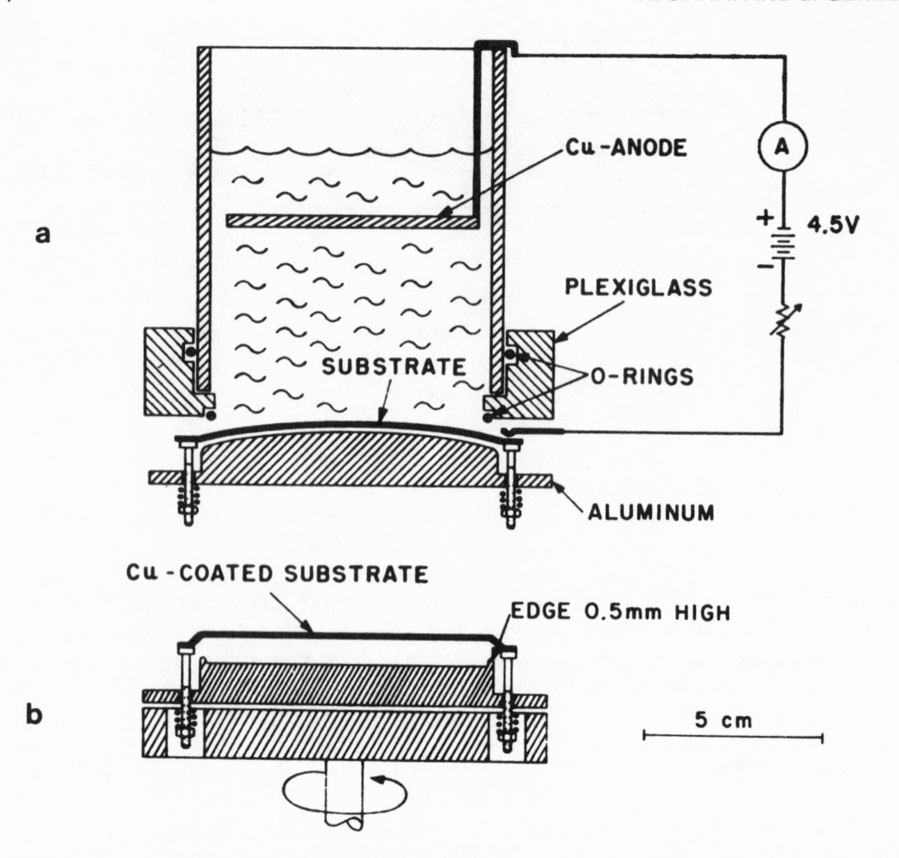

Figure 4. Electroplating process : (a) a copper evapolated Mylar film forms the bottom of the electrolytic cell ; electrolyte (188 g/liter of $CuSO_4$ $5H_2O$ and 74 g/liter of H_2SO_4), (b) Cleaning and KPR (light sensitive lacquer) coating process. (After Ulrich 1969).

and diffraction, A and D, depend upon t, g and a and upon the wavelength λ. The equation of energy balance is expressed as

$$R + T + A + D = 1 \tag{15}$$

Diffraction does not appear in the region $\lambda/g \geq 1 + \sin\theta$, where θ is the incidence angle. Now we will employ the"normalized" frecuency $\omega = g/\lambda$ for convenience, since the relevant expressions involve λ/g rather than λ. For most applications the frequency range $\omega<1$ is used. Therefore, the diffraction is neglected since the condition of normal incidence is in general satisfied. In addition, the absorption is very small and it can be almost neglected. Then only reflection and transmission remain, and the energy balance may be

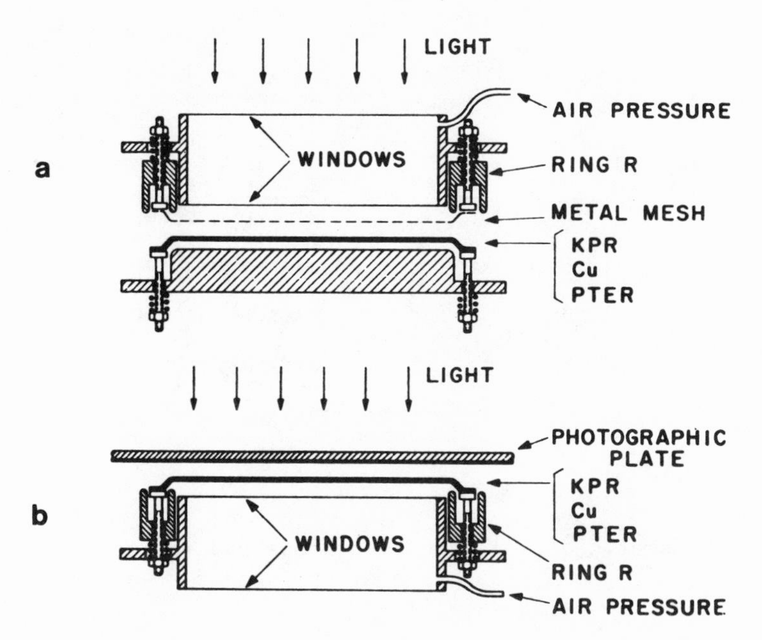

Figure 5. Process for contact printing a negative on the KPR layer: (a) an inductive mesh serves as the negative, (b) a photographic plate is the negative. (After Ulrich 1969).

approximated by

$$|\tau(\omega)|^2 + |\Gamma(\omega)|^2 = 1. \tag{16}$$

where τ and Γ are the amplitude transmittance and reflectance, respectively and where $|\tau\ (\omega)|^2 = T$ and $|\Gamma\ (\omega)|^2 = R$. By using the boundary condition at the position of a mesh, namely that the tangential component of the electric field is continuous on both sides, we obtain the relation

$$\tau(\omega) = 1 + \Gamma(\omega) \tag{17}$$

As a consequence of Eqs. (16) and (17), the phases $\psi_\Gamma(\omega)$ and $\psi_\tau(\omega)$ are related directly to the power transmittance $\tau(\omega)^2$ which can be measured most easily among all optical properties of a mesh,

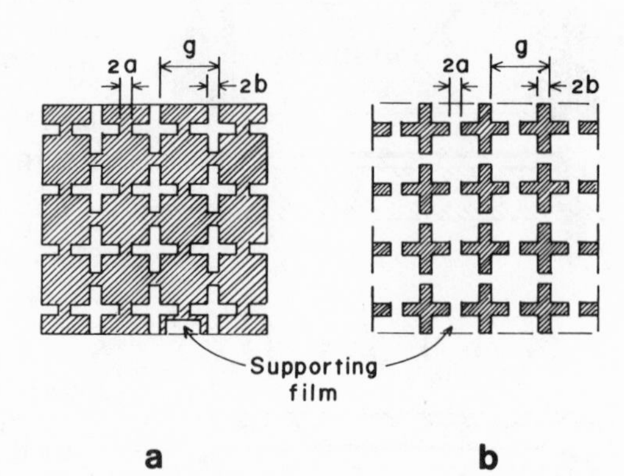

Figure 6. Schematic of (a) inductive cross and (b) capacitive cross meshes and their geometrical parameters, shaded areas are metallic. (After Tomaselli et al., 1981).

$$\sin^2\psi_\Gamma(\omega) = |\tau(\omega)|^2 \quad (18a)$$

$$\sin^2\psi_\tau(\omega) = 1 - |\tau(\omega)|^2 \quad (18b)$$

When the frequency varies, the points of the vectors Γ and τ move around the circles shown in Fig 7. At all frequencies, the reflected and the transmitted wave are seen to be 90° out of phase. Under the above-mentioned idealizations these considerations apply to both types of meshes in the range $\omega<1$.

The comparison of complementary meshes can be made using the electromagnetic equivalent of Babinet's principle. Ulrich has arrived at the following relations between the various quantities

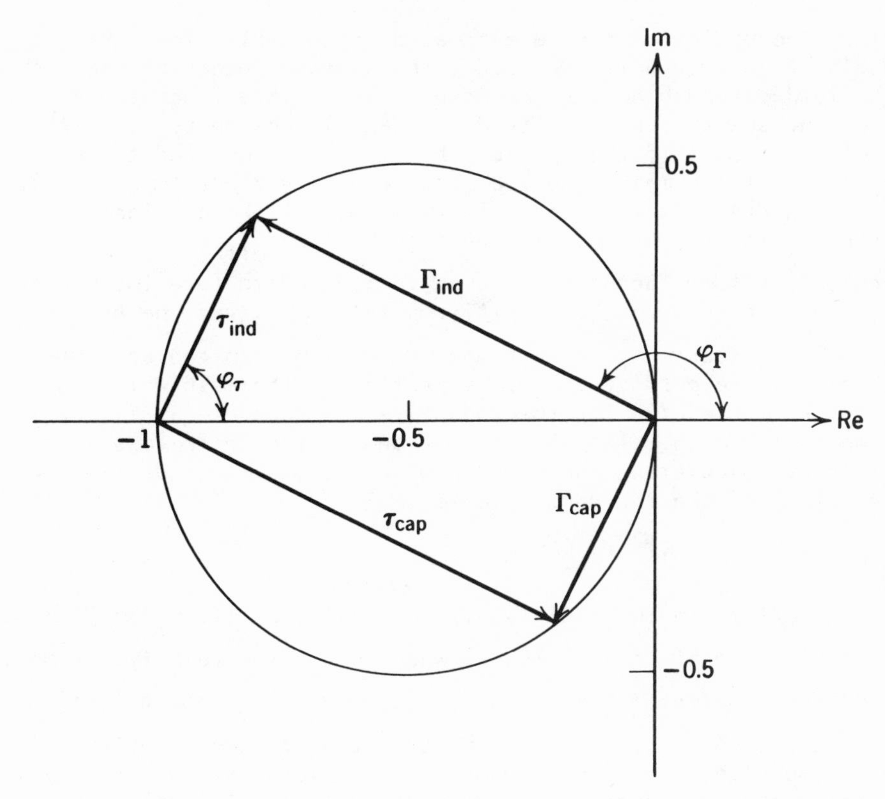

Figure 7. The locus of the amplitude reflection (Γ) and transmission (τ) coefficients of thin, lossless, inductive and capacitive meshes in the complex plane. Valid for $g/\lambda<1$. (After Ulrich 1967a).

of capacitive (c) and inductive (i) mesh;

$$\tau_i(\omega) + \tau_c(\omega) = 1 \tag{19}$$

$$|\tau_i(\omega)|^2 + |\tau_c(\omega)|^2 = 1 \tag{20}$$

$$\tau_i(\omega) = -\Gamma_c(\omega) \text{ and } \tau_c(\omega) = -\Gamma_i(\omega) \tag{21}$$

Up to now the absorption of the meshes has been neglected. Though its value is assumed to be very small in practice, an estimation of the value is important. The absorption may result from ohmic losses arising from surface currents flowing in the metallic parts of the meshes. The ohmic losses are estimated in the following way: for an incident wave of unit amplitude, the average change of the magnetic field-strength across the mesh is 2Γ . This is caused by an average surface current density $J = c\Gamma/4\pi$ (in cgs units) on either side of the mesh. If the skin depth is small compared to the thickness t of the mesh, the real part of the surface impedance is $\rho = 1/\sigma\delta$, where σ is the conductivity of mesh. The dissipated power per unit area of the mesh becomes $P_d = 2\rho J_{eff}^2 = 2\rho\eta J^2$. A dimensionless form-factor η has been introduced to take into account the difference between the spatial rms value J_{eff} and the average value $\bar{J}$ of the current. As a rough estimation, $1/\eta$ can be equated to that relative part of the cross-section of the mesh which may conduct current: $\eta = g/2a$ for the inductive mesh and $\eta = 1/(1-2a/g)$ for capacitive mesh. The absorptance due to the ohmic losses is obtained by calculating the portion of P_d to the power density $P_o(= c/4\pi)$ of the incident wave as

$$A = |\Gamma|^2\eta(c/\lambda\sigma)^{\frac{1}{2}} \qquad (22)$$

The term $(c/\lambda\sigma)^{\frac{1}{2}}$ in Eq. (22) is also expressed as $(10^7\mu_r/c\sigma\lambda)^{\frac{1}{2}}$ in m.k.s. units (Sakai et al., 1969), where μ_r is the relative permeability of the mesh metal. We usually use $\mu_r = 1$ even for a ferromagnetic metal such as nickel assuming that the frequency of the FIR wave is too high for magnetic domains to follow. We further assume that the bulk dc conductivity σ of the metal is still valid in the FIR region. The estimation of the absorptance under the conditions $|\Gamma|^2 = 1$, $\lambda = 100\ \mu m$ and $2a/g = 0.3$ shows that it does not exceed 1%. As the absorptance is in inverse proportion to the square root of the wavelength, it becomes smaller in the longer wavelength region.

2.2.3. Measurements of Inductive and Capacitive Meshes

Information of FIR properties of meshes is quite important not only for many applications but also for theoretical analysis. Out of numerous measurements on inductive meshes (Renk et al., 1962; Ulrich et al., 1963; Mitsuishi et al., 1963; Vogel et al., 1964; Sakai et al., 1969, 1978; Ressler et al., 1967; Romero et al., 1973; Lecullier et al., 1976; Belland et al., 1980; Lamarre et al., 1981) some aspects of power transmittance and reflectance of the meshes are shown in Fig.8 and in Fig.9. Figure 8 (a) and (b) (Sakai et al., 1969, 1978) show how the values of T $(=|\tau(\omega)|^2)$ and $R(=|\Gamma(\omega)|^2)$ change as function of the parameter 2a/g. The measurements were made for unpolarized radiation at normal incidence for T and at 15° incidence for R, both with a convergent beam having an angle of 15°. The dependence of R and T on 2a/g coincides qualitatively with the theoretical prediction by Marcuvits (1951), in the $\lambda > g$ region, for a one=dimensional wire grating parallel to the polarized wave.

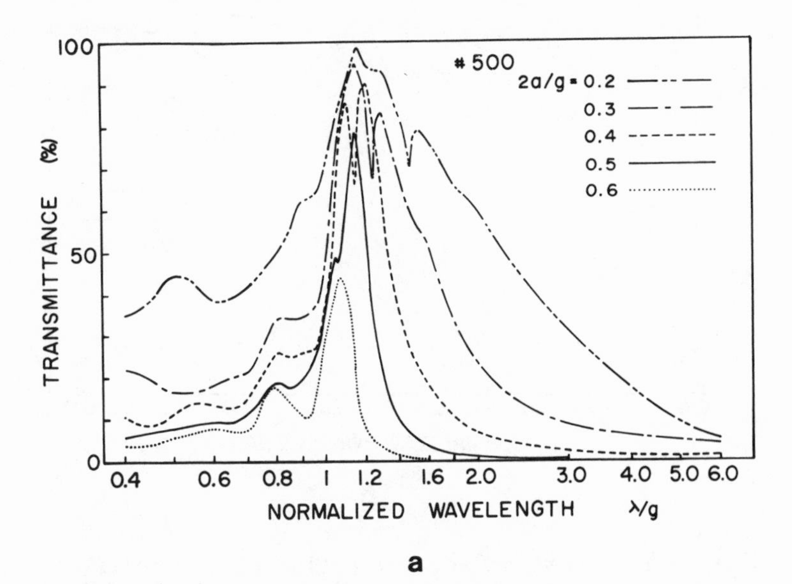

a

Figure 8 (a). FIR transmission and reflection measurements of inductive meshes : transmittance of #500 meshes (500 lines/inch) with 2a/g = 0.2, 0.3. 0.4, 0.5 and 0.6 as a function of /g (After Sakai et al., 1969, 1978.Reflectance for 2a/g = 0.2, 0.5 and 0.6 meshes were newly measured. The results are added and the data of 2a/g=0.25 in the original paper is removed)

Figure 9 (a) and (b) (Vogel et al., 1964) shows the dependence of R and T on the polarization of the incident wave and on the wire direction. All measurements at skew incidence were made at an angle of 45° . Comparing these results, prominent variations are not seen in the $\lambda>g$ region for all arrangements. Here we should mention two noticeable features in Fig. 8 (a). One is the principal transmission maximum at $\lambda\simeq g$, and the other is a sharp dip in transmission. Lamarre et al.(1981) have stated that the origin of the transmission maximum is due to diffraction modes of the first order starting to propagate freely. They have further shown from their measurements with a collimated beam that the wavelength of the transmission maximum λ_o has the relation with the incidence angle θ such as $\lambda_o \simeq g$ $(1 + \sin\theta /2)$, analogous to but slightly different from the expression by Saksena et al. (1969) which is $\lambda_o = g \ (1 + \sin\theta)$.

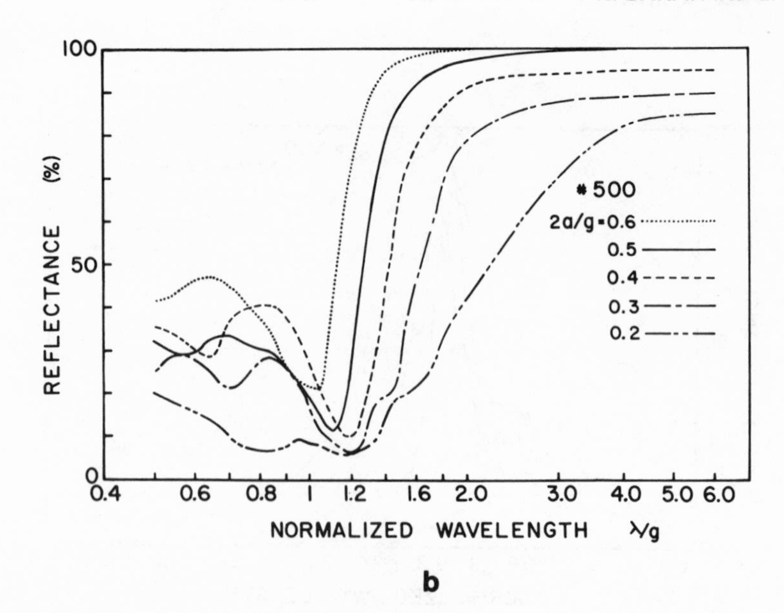

Figure 8(b) FIR transmission and reflection measurements of inductive meshes : reflectance of #500 meshes with 2a/g = 0.2, 0.3, 0.4, 0.5, and 0.6 as a function of λ/g. (N.B. a/g in the original paper corresponds to 2a/g in this review paper) (After Sakai et al., 1969, 1978. Reflectance for 2a/g = 0.2, 0.5, and 0.6 meshes were newly measured. The results are added and the data of 2a/g = 0.25 in the original paper is removed).

McPhedran and Maystre (1977) have pointed out that the cause of sharp dips in attributable to anomalous diffraction effects associated with the passing-off (the transition to evanescent wave) of the propagating spectral orders formed by the mesh, analogous to Wood's anomalies occurring with one-dimensional gratings. Lamarre et al.(1981) also have presented a relation for the wavelength at the dip λ_d,

$$\lambda_d/g = 1.73 - 2.87\ (a/g). \tag{23}$$

Values for 1-R-T, obtained by use of the results of Fig. 8 (a) and (b), is shown in Fig.10. Since the absorptance is negligibly small, the values almost represent the amount of diffraction. Figure 11

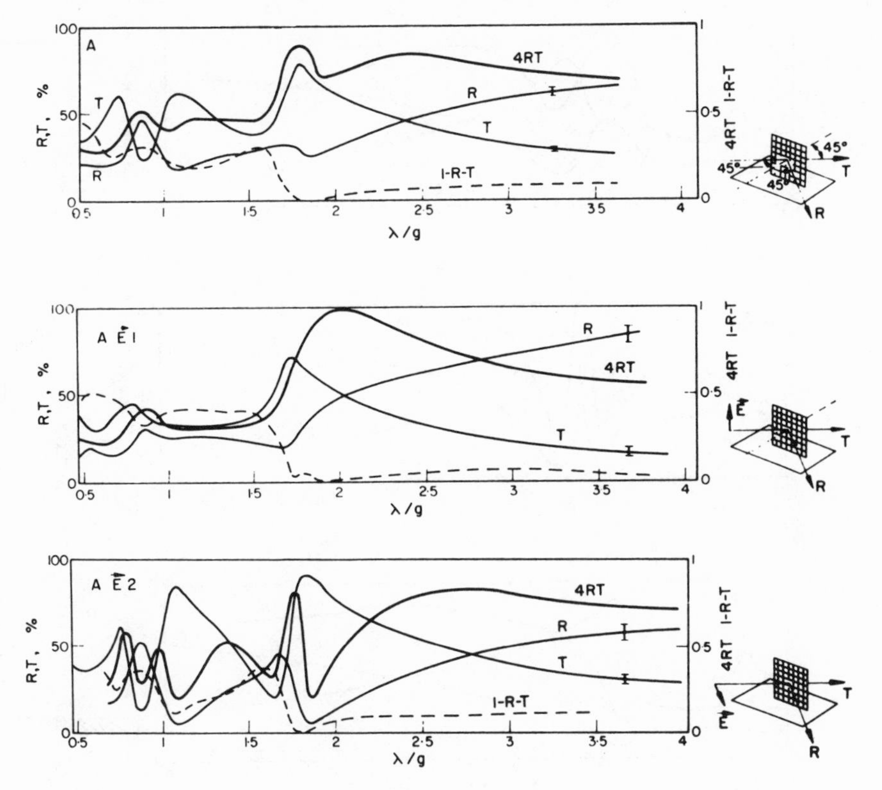

Figure 9 (A)FIR transmittance (T), reflectance (R), 4RT and 1-R-T for polarized or unpolarized beam at 45° incidence. Wires with horizontal and vertical directions; upper, beam not polarized; center, beam polarized with E vector perpendicular to the plane of incidence; lower, beam polarized parallel with the plane of incidence. (After Vogel and Genzel, 1964)

shows the change of phase of reflection obtained using the measured transmittance Fig. 8 (a) and Eq. (18a). The phase change is also obtained directly from the position of the transmission maxima of a FPI of known mesh separation. The general form of the curves in Fig.11 is in good agreement with the experimental results of Lecullier and Chanin (1976).

The optical properties of capacitive meshes have been measured by several investigators (Ulrich, 1967a; Ressler et al., 1967; Romero et al., 1972; Holah et al., 1972). The results of Ulrich (1967a) are shown in Fig.12 together with his results for inductive meshes.

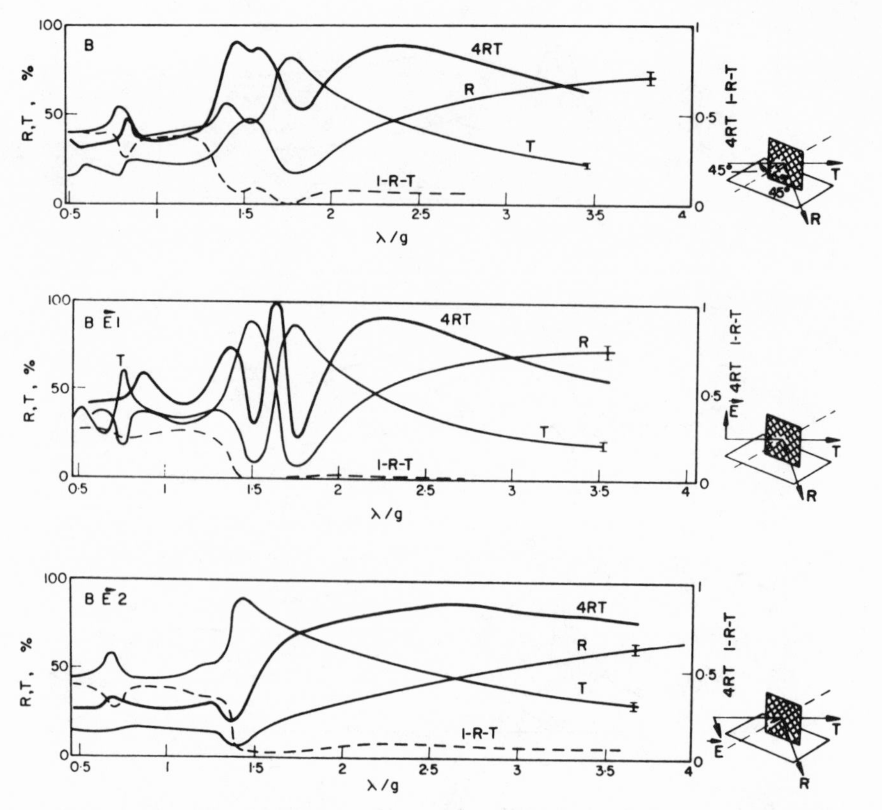

Figure 9(B) FIR transmittance (T), reflectance (R), 4RT and 1-R-T for polarized or unpolarized beam at 45° incidence. Wires inclined about 45° ; upper, beam not polarized ; center, beam polarized perpendicular to the plane of incidence ; lower, beam polarized parallel with the plane of incidence. (After Vogel and Genzel,1964).

The parameters of these meshes are listed in Table I. It is clear from the figure that the power transmittance of capacitive meshes is complementary to that of inductive meshes as Eq. (20) indicates. No anomalous feature can be seen in it. A Fresnel representation of amplitude reflectance for both types of practical meshes are represented in Fig.13.

2.2.4. Equivalent Circuit Representation of Metal Meshes

An equivalent circuit representation of the optical properties of meshes offers the advantage of allowing the use of the well-developed transmission line theory for the calculation of filters

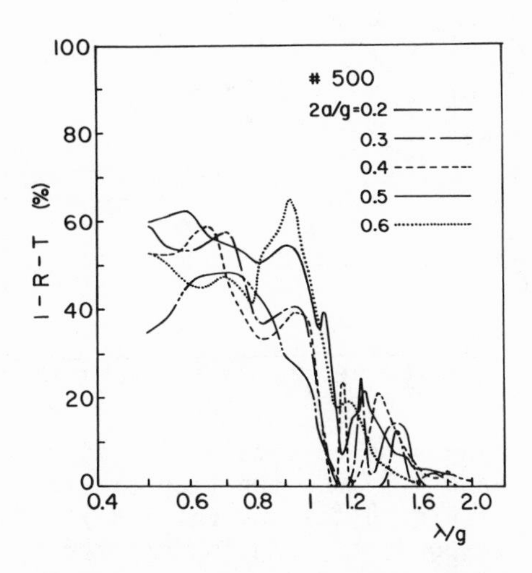

Figure 10. 1-R-T for #500 meshes with 2a/g = 0.2, 0.3, 0.4, 0.5 and 0.6 as a function of λ/g in the range $0.5 \leq \lambda/g \leq 2$. R and T in Fig.8 were used for calculations. (N.B. a/g, in the original paper corresponds to 2a/g in this review paper) (After Sakai et al.1969).

composed of two or more meshes stacked coherently. The general equivalent circuit for a thin mesh can be expressed by a four-terminal transmission line shunted with a lumped admittance Y (ω) which represents the mesh (see Fig.14). The characteristic admittance of the transmission line is $Y_o = 1/Z_o$. The representation of a mesh in free space by a four-terminal network is possible only in the region ω<1. The transmission line theory shows that the shunt Y (ω) gives rise to a voltage reflection coefficient (equal to

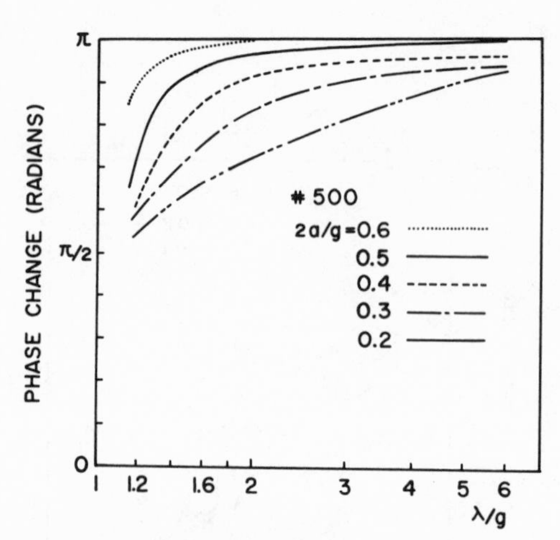

Figure 11. Phase changes on reflection for #500 meshes with 2a/g = 0.2, 0.3, 0.4, 0.5 and 0.6 as a function of λ/g. Calculation was made on the basis of Fig. 8 (a) and Eq. (18a). (N.B. a/g in the original paper corresponds to 2a/g in this review paper) (After Sakai et al., 1969).

the amplitude reflectance)

$$\Gamma(\omega) = \frac{Y(\omega) / 2Y_o}{1 + Y(\omega) / 2Y_o} \tag{24}$$

and a voltage transmission coefficient (equal to the amplitude transmittance)

$$\tau(\omega) = \frac{1}{1 + Y(\omega) / 2Y_o} \tag{25}$$

These expressions are consistent with Eq. 17 and, therefore, the

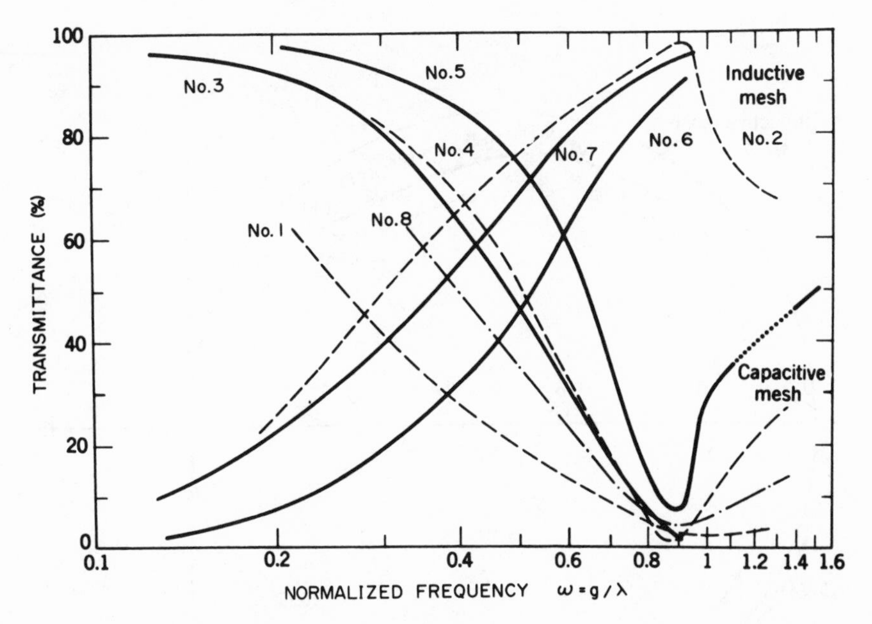

Figure 12. Transmittance of capacitive and inductive meshes. The parameters of the meshes are given in Table I. (After Ulrich 1967a).

value of $Y(\omega)$ is uniquely determined if $\Gamma(\omega)$ or $\tau(\omega)$ is known. Thus it is possible to represent the optical properties $\Gamma(\omega)$ and $\tau(\omega)$ of any thin mesh by the equivalent circuit Fig.14 in the region $\omega<1$. Generally the accuracy of the representation depends on the number of components used in the circuit. Furthermore we must take care that $Y(\omega)$ does not have a negative real part. For a lossless mesh, its admittance becomes a pure susceptance.

$$\frac{Y}{Y_o} = \frac{jB(\omega)}{Y_o} \tag{26}$$

with real $B(\omega)$. The quantity $B(\omega)/Y_o$ is determined empirically from the measurements of $\tau(\omega)$ or $\Gamma(\omega)$. For the capacitive susceptances $B>0$, τ_c and Γ_c are located in the lower half of the circle in the Fresnel representation. For the inductive susceptances $B<0$, τ_i and Γ_i are found in the upper half. The Eq. (21), connecting the optical properties of two complementary meshes, can be

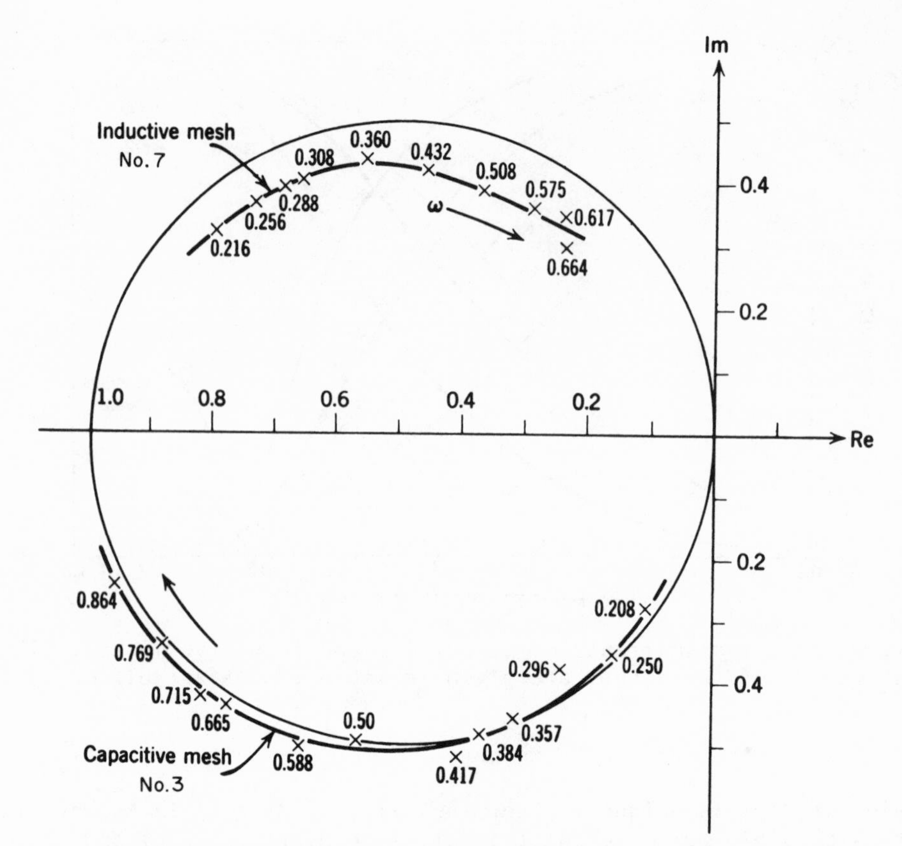

Figure 13. Measured amplitude reflection coefficients of the capacitive mesh No.13 and the inductive mesh No. 7. The numbers on the curves are the normalized frequency $\omega = g/\lambda$. (After Ulrich 1967a).

rewritten in the equivalent circuit representation as

$$\frac{B_c(\omega)}{Y_\circ} \quad \frac{B_i(\omega)}{Y_\circ} = -4 \tag{27}$$

where $B_c(\omega)$ is the susceptance of the capacitive mesh and $B_i(\omega)$ that of the inductive mesh. All information about $|\tau(\omega)|^2$, $|\Gamma(\omega)|^2$, $\psi_\tau(\omega)$ or $\psi_\Gamma(\omega)$ is involved in $B(\omega)/Y_\circ$. Inversely, $B(\omega)/Y_\circ$ can be determined from each of these functions.

Table I. PARAMETERS OF THE MEASURED CAPACITIVE AND INDUCTIVE MESHES

Mesh			Dimension, in μm				Equivalent circuit parameters*		
No.	Type	Material	g	a	t	a/g	2Z	ω_o	$2\bar{R}/Z_o$
1	Cap	Cu on Mylar	368	17.7	5	0.051	0.274	1.0	≤0.02
2	Ind	Ni+	368	17.0	14	0.046	3.00	1†	0.02§
3	Cap	Cu on Mylar	250	40.5	5	0.161	0.70	0.96	≤0.02
4	Cap	Cu on Mylar	473	72.5	5	0.153	0.73	0.96	0.001§
5	Cap	Cu on Mylar	342	68.5	6	0.200	1.19	0.945	0.002§
6	Ind	Cu+	216	28.8	7	0.133	1.33	1†	0.004§
7	Ind	Ni+	216	13.5	12	0.0625	2.42	1†	≤0.04
8	Cap	Cu on Mylar	368	35	7	0.095			

* Refer to Table II-B. They were determined by adaptation of the calculated value of $|\tau|^2$ to the measurements.
\+ Product of Buckbee Mears Co.
† Chosen arbitrarily.
§ Ohmic losses only, calculated from Eq. (31) for $\lambda = 0.5$ mm and $\sigma = 0.25\ \sigma_{bulk}$

(Originally cited from Ulrich (1967a). Above table was cited from Moller and Rothschild (1971).

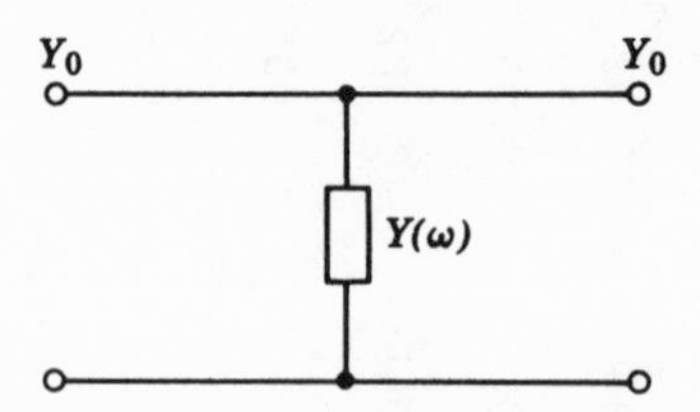

Figure 14. General equivalent circuit representation of a thin mesh in free space by a transmission line shunted with a lumped admittance $Y(\omega)$. (After Moller and Rothshield, 1971).

A first approximation of the synthesis of networks is to use a single element, i.e., a constant capacitance C for the capacitive mesh and a constant inductance L for the inductive mesh. The optical properties resulting from these simplest circuits are given in Table II-A. The numerical values of C and L can be obtained by fitting the calculated $|\tau(\omega)|^2$ to the measured power transmittance. In Fig. 15 the calculated $|\tau(\omega)|^2$ (dashed curve, $C/Y_o = 2.85$) is compared with the experimental values (dots). From this comparison, it is clear that the single element approximation is valid only in the very low frequency region ($\omega<<1$) and it fails near $\omega \sim 1$. The advanced circuit representation is to employ two elements. The results based on the values $C/Y_o = 2.85$ and $L/Z_o = 0.318$ are represented in the form of solid line in Fig.15. We see that this two-element approximation fits closely to the measured data over nearly the whole region of $\omega<1$. From the preceeding general considerations it is sure that the other optical properties such as $|\Gamma(\omega)|^2$, ψ_τ or ψ_Γ are represented correctly by the same equivalent circuit. Ulrich (1967a) has furthermore attempted to represent the inductive and capacitive meshes by resonant circuits given in Table II-B, adding a resistance R in order to involve losses. In this case the normalized admittance Y/Y_o and other optical properties including the absorptance A are expressed in terms of a normalized resonant frequency ω_o, a generalized frequency $\Omega = \omega/\omega_o - \omega_o/\omega$ and a normalized impedance $\hat{\hat{Z}} = \omega_o L/Z_o = Y_o/\omega_o C$. The fitting of the equivalent circuit to the transmission curves in Fig. 12 has produced the values of the parameters as in shown in the last three columns in Table I. The calculated curves

TABLE II-A

SINGLE-ELEMENT EQUIVALENT CIRCUITS FOR THIN CAPACITIVE AND INDUCTIVE MESHES, AND THEIR OPTICAL PROPERTIES

Type of mesh	Capacitive	Inductive
Equivalent circuit	$Y_\circ$ C $Y_\circ$	$Z_\circ$ L $Z_\circ$
Normalized admittance	$\frac{Y(\omega)}{Y_\circ} = j\frac{\omega C}{Y_\circ}$	$-j\frac{Z_\circ}{\omega L}$
Reflectance	$\lvert\Gamma\rvert^2 = \left(\frac{\omega C}{2Y_\circ}\right)^2\left[1+\left(\frac{\omega C}{2Y_\circ}\right)^2\right]^{-1}$	$\left[1+\left(\frac{2\omega L}{Z_\circ}\right)^2\right]^{-1}$
	$\psi_\Gamma = \pi + \arctan\left(\frac{2Y_\circ}{\omega C}\right)$	$\pi - \arctan\left(\frac{2\omega L}{Z_\circ}\right)$
Transmittance	$\lvert\tau\rvert^2 = \left[1+\left(\frac{\omega C}{2Y_\circ}\right)^2\right]^{-1}$	$\left(\frac{2\omega L}{Z_\circ}\right)^2\left[1+\left(\frac{2\omega L}{Z_\circ}\right)^2\right]^{-1}$
	$\psi_\tau = -\arctan\left(\frac{\omega C}{2Y_\circ}\right)$	$\arctan\left(\frac{Z_\circ}{2\omega L}\right)$

TABLE II-A (Continued)

SINGLE-ELEMENT EQUIVALENT CIRCUITS FOR THIN CAPACITIVE AND INDUCTIVE MESHES, AND THEIR OPTICAL PROPERTIES

Type of mesh	Capacitive	Inductive
For complementary meshes	$\frac{C_c}{2Y_\circ} = \frac{2L_i}{Z_\circ}$	
	Low frequency approximations ($\omega \ll 1$)	
Reflectance	$\lvert\Gamma\rvert^2 = \left(\frac{\omega C}{2Y_\circ}\right)^2$	$1 - \left(\frac{2\omega L}{Z_\circ}\right)^2$
	$\psi_\Gamma = \frac{3\pi}{2} - \left(\frac{\omega C}{2Y_\circ}\right)$	$\pi - \left(\frac{2\omega L}{Z_\circ}\right)$
Transmittance	$\lvert\Gamma\rvert^2 = 1 - \left(\frac{\omega C}{2Y_\circ}\right)^2$	$\left(\frac{2\omega L}{Z_\circ}\right)^2$
	$\psi_\tau = -\left(\frac{\omega C}{2Y_\circ}\right)$	$\frac{\pi}{2} - \left(\frac{2\omega L}{Z_\circ}\right)$

(Originally cited from Ulrich (1967a). Above expression was cited from Moller and Rothschild (1971).

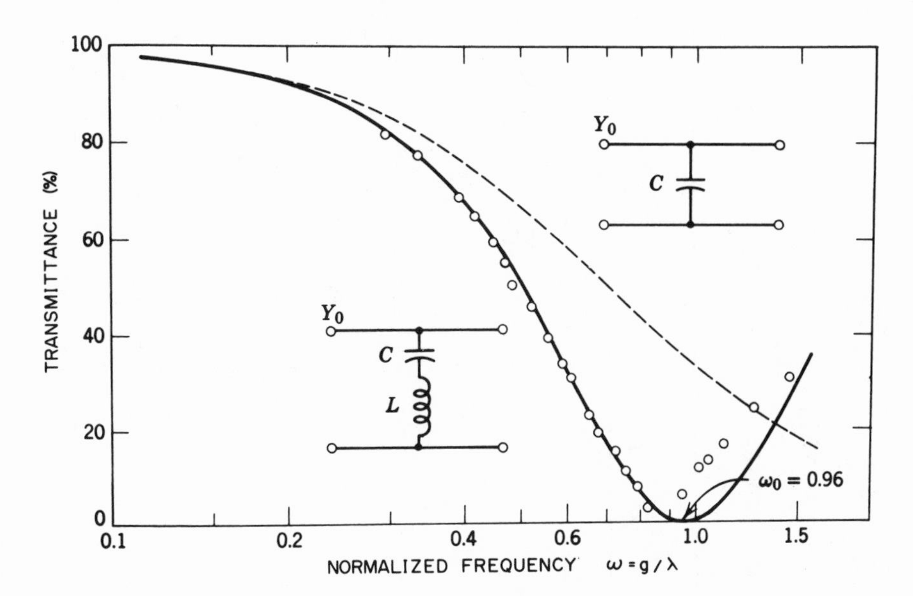

Figure 15. Approximation of the measured transmittance of the capacitive mesh No. 4 by one-element and by two-element equivalent circuit. Measurement ; 0 0 0 . Calculated with one element for Y ; - - - . Calculated with two elements for Y ; ——— . (After Ulrich, 1967a).

for the capacitive meshes represented the experimental points in the region $\omega < 1$ well, but the agreement was poorer around $\omega = 1$. With respect to the inductive meshes, the fit was not so good. This is probably due to the finite thickness of the mesh. The normalized impedance $\hat{Z}$ and the normalized resonant frequency ω_o only depends on the ratio a/g. A plot of $\hat{Z}$ vs a/g for all meshes in Table I is shown in Fig. 16, in which the ordinate of the left side $\hat{Z}_c$ belongs to capacitive meshes and that of the right side $\hat{Z}_i$ to inductive meshes. The relation $\hat{Z}_i \cdot \hat{Z}_c = 1/4$ connects both scales. The solid line in Fig. 16 represents $\hat{Z}_c^{(1)}$ of one-dimensional thin

TABLE II-B

MULTI-ELEMENT EQUIVALENT CIRCUITS FOR THIN CAPACITIVE AND INDUCTIVE MESHES* , AND THEIR OPTICAL PROPERTIES

Type of Mesh	Capacitive	Inductive
Equivalent circuit	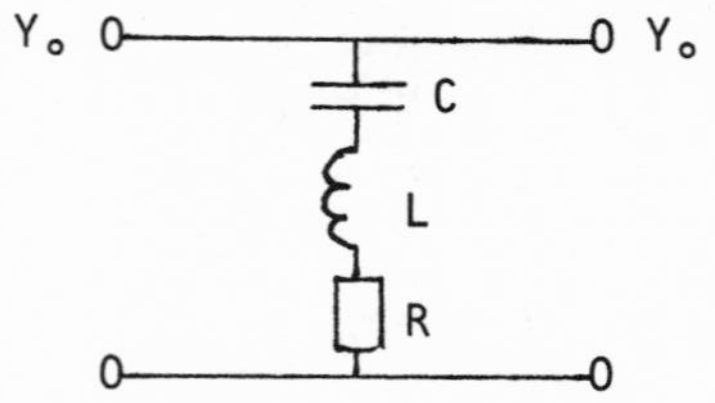 Y_0, C, L, R, Y_0	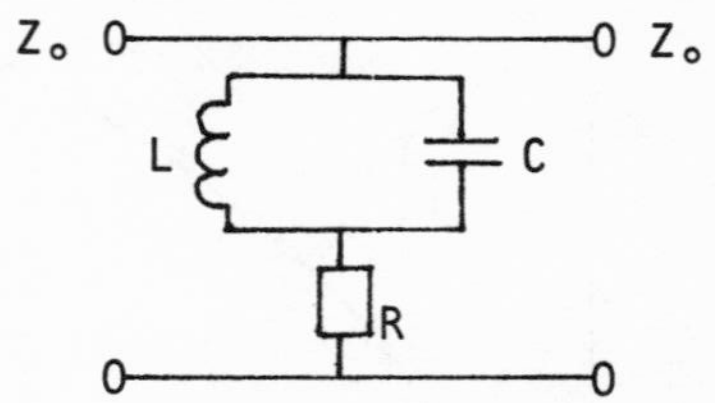 Z_0, L, C, R, Z_0
Resonance frequency	ω_0	
Characteristic impedance of empty space	$Z_0 = 1/Y_0$	
Normalized impedance of L and C at resonance	$\hat{Z} = \frac{\omega_0 L}{Z_0} = \frac{Y_0}{\omega_0 C}$	
Generalized frequency	$\Omega = \frac{\omega}{\omega_0} - \frac{\omega_0}{\omega} = \frac{\lambda_0}{\lambda} - \frac{\lambda}{\lambda_0}$	

TABLE II-B (Continued)

MULTI-ELEMENT EQUIVALENT CIRCUITS FOR THIN CAPACITIVE AND INDUCTIVE MESHES*, AND THEIR OPTICAL PROPERTIES

Type of mesh	Capacitive	Inductive
Normalized admittance	$\frac{Y(\omega)}{Y_\circ} = \frac{1}{\frac{\overline{R}}{Z_\circ} + J\Omega\hat{Z}}$	$\frac{1}{\frac{\overline{R}}{Z_\circ} - j\frac{\hat{Z}}{\Omega}}$
Reflectance	$\lvert\Gamma\rvert^2 = \frac{1}{\left(1 + \frac{2\overline{R}}{Z_\circ}\right)^2 + (2\hat{Z}\Omega)^2}$	$\frac{1}{\left(1 + \frac{2\overline{R}}{Z_\circ}\right)^2 + \left(\frac{2\hat{Z}}{\Omega}\right)^2}$
	$\psi_\Gamma = \pi - \arctan \frac{2\hat{Z}\Omega}{1 + \frac{2\overline{R}}{Z_\circ}}$	$\pi + \arctan \frac{2\hat{Z}/\Omega}{1 + \frac{2\overline{R}}{Z_\circ}}$

(Continued)

TABLE II-B (Continued)

MULTI-ELEMENT EQUIVALENT CIRCUITS FOR THIN CAPACITIVE AND INDUCTIVE MESHES*, AND THEIR OPTICAL PROPERTIES

Type of Mesh	Capacitive	Inductive
Transmittance	$\lvert\tau\rvert^2 = \dfrac{\left(\frac{2\overline{R}}{Z_\circ}\right)^2 + (2\hat{Z}\Omega)^2}{\left(1 + \frac{2\overline{R}}{Z_\circ}\right)^2 + (2\hat{Z}\Omega)^2}$	$\dfrac{\left(\frac{2\overline{R}}{Z_\circ}\right)^2 + \left(\frac{2\hat{Z}}{\Omega}\right)^2}{\left(1 + \frac{2\overline{R}}{Z_\circ}\right)^2 + \left(\frac{2\hat{Z}}{\Omega}\right)^2}$
	$\psi_\tau = \arctan \dfrac{2\hat{Z}\Omega}{\left(\frac{2\overline{R}}{Z_\circ}\right)\left(1 + \frac{2\overline{R}}{Z_\circ}\right) + (2\hat{Z}\Omega)^2}$	$- \arctan \dfrac{\frac{2\hat{Z}}{\Omega}}{\left(\frac{2\overline{R}}{Z_\circ}\right)\left(1 + \frac{2\overline{R}}{Z_\circ}\right) + \left(\frac{2\hat{Z}}{\Omega}\right)^2}$
Absorptance	$A = 4\left(\dfrac{\overline{R}}{Z_\circ}\right)\lvert\Gamma\rvert^2$	

TABLE II-B (Continued)

MULTI-ELEMENT EQUIVALENT CIRCUITS FOR THIN CAPACITIVE AND INDUCTIVE MESHES*, AND THEIR OPTICAL PROPERTIES

Type of Mesh	Capacitive	Inductive
For complementary meshes	$\frac{C_c}{2Y_o} = \frac{2L_i}{Z_o}$	$\frac{C_i}{2Y_o} = \frac{2L_c}{Z_o}$

*Resonance at $\omega = \omega_o$ and the losses are taken into account. Numerical values of $\hat{Z}$ and ω_o can be taken from Fig. 16 and Eq.(28), (29) and (30). $\lambda_o = g/\omega$ is the wavelength corresponding to ω_o.

(Originally cited from Ulrich (1967a). Above expression was cited from Moller and Rothschild (1971).

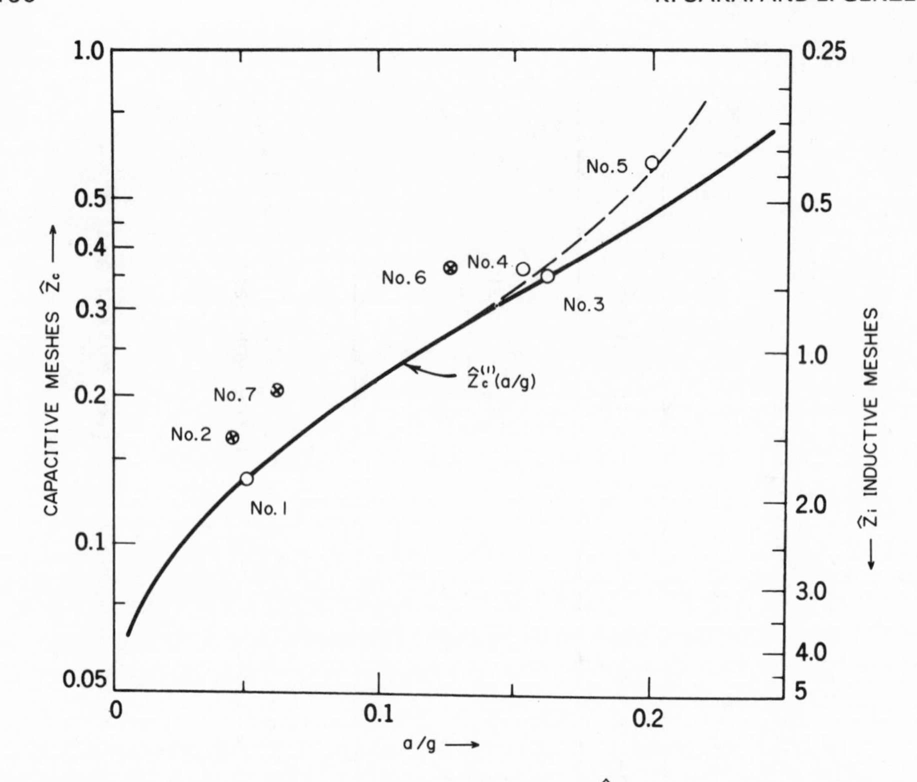

Figure 16. The characteristic impedances ($\hat{Z}$) of the meshes (in Table I) as a function of a/g. Experimental values for the capacitive meshes 0 0 , and for the inductive meshes ⊗ ⊗ . Theoretical characteristic impedance of one-dimensional grating in the limit ω<<1 ——— . (After Ulrich, 1967a).

capacitive strip grating calculated from Marcuvitz's theoretical treatment, i.e.

$$\hat{Z}_c{}^1[(\frac{a}{g}) = 2\ln \operatorname{cosec} (\frac{a\pi}{g})]^{-1*} \tag{28}$$

The impedance $\hat{Z}_i$ of the inductive meshes can be obtained with the help of $\hat{Z}_i \cdot \hat{Z}_c = 1/4$ and Eq. (28). The good agreement between the calculated curve (one-dimensional) and the measured values for

---

* For this formula, Eq. (3.31) in Möller et al.(1971) has been adopted rather than Eq. (13) in Ulrich (1967a) or Eq. (3) in Ulrich (1967b).

capacitive meshes (two-dimensional) at low values of a/g is not surprising, considering that the mesh differs from one-dimensional grating only by the additional gaps of width 2a in the strips of the grating; and since gaps are oriented parallel to the incident electric field, their presence or absence has only little influence on the surface current. The condition is not fulfilled at higher a/g region, and the experimental values deviate from the one-dimensional model. The dashed curve should be used in the range where deviations become apparent. On the other hand, in the case of inductive meshes, the agreement is not good in any range of a/g. This is probably due to the fact that available meshes always have a finite thickness. The dependence of the resonant frequency ω_o on a/g has been obtained as

$$\omega_o\ (a/g) = 0.82 + 0.32\ a/g \tag{29}$$

for the inductive meshes (Sakai et al.) and

$$\omega_o\ (a/g) = 1 - 0.27\ a/g \tag{30}$$

for the capacitive meshes (Ulrich, 1967) by fitting the measured power transmittance of the respective mesh. The equations are, therefore, not absolutely correct. The loss resistance $\bar{R}$ was expressed as

$$\frac{\bar{R}}{Z_o} = \left(\frac{c}{\lambda\sigma}\right)^{\frac{1}{2}} \frac{\eta}{4} \tag{31}$$

by relating it with the power absorptance. Because of the exponent $\frac{1}{2}$, $\bar{R}$ is only slightly dependent on frequency.

3. BEAMSPLITTERS AND FILTERS BASED ON INHERENT TRANSMISSION AND/OR REFLECTION PROPERTIES OF MESHES

The experimental investigations of FIR properties of various types of meshes have suggested the applicability of these meshes for some spectroscopic components. Bell and Russell (1966) used the inductive meshes as beam splitters in a Michelson type FTS. It was found that the metal meshes act as excellent beamsplitters of wavelengths of approximately twice the mesh periodicity. In Fig. 17 the efficiencies of some meshes are comparatively shown together with an ordinary Mylar beamsplitter in the form of spectral background. Because one mesh covers a spectral range less than a Mylar beamsplitter, it has not been used so often for the ordinary Michelson type FTS. Recently, the mesh beamsplitters have begun to be used effectively again for the double-beam FTS, (L.Genzel, 1976) because the inherent absorption of the metal mesh is negligibly small.

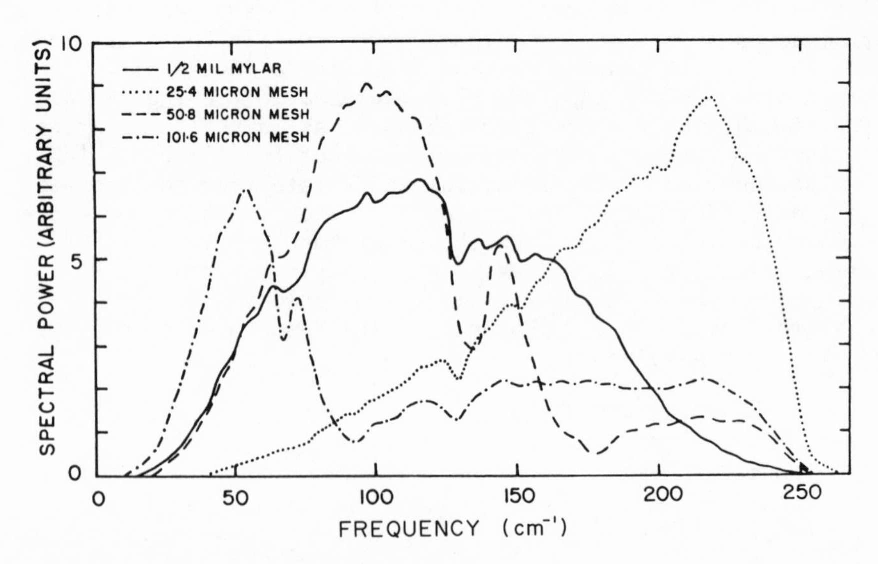

Figure 17. Background spectra obtained with various metal-mesh beam-splitters. A 1/2 mil Mylar beam-splitter is shown for comparison. (After Bell and Russel, 1966a)

Wirecloth metal meshes have been investigated by Mitsuishi et al. (1963), and Ressler and Möller (1967). The power transmittance and power reflectance of one of this type of meshes are shown in Fig.18. Mitsuishi et al. (1963) have pointed out that the reflection

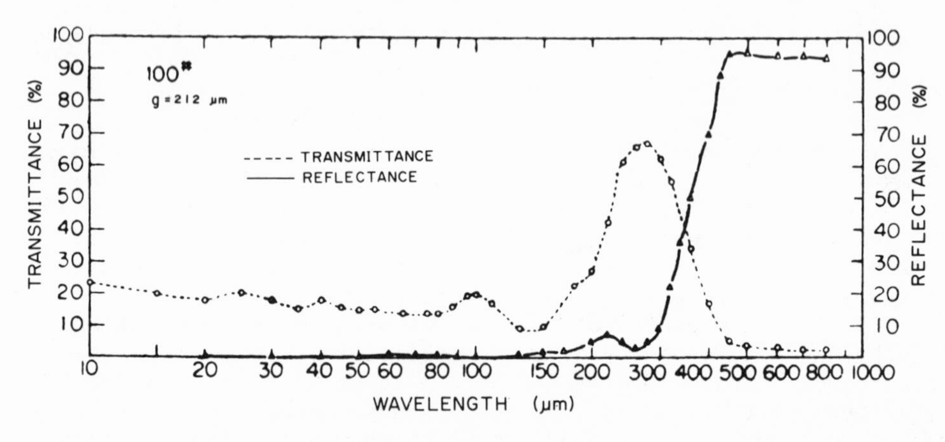

Figure 18. FIR transmittance and reflectance of a 100# wirecloth mesh, measured at 15° incidence angle. (After Mitsuishi et al.,1963)

property of wirecloth metal meshes acts more effectively as a low-pass filter than that of the electroformed metal meshes, since the rejection in the stopband (the diffraction region defined roughly by $\lambda/g<1.3$) is excellent. The measured reflectance of various wire-cloth meshes listed in Table III is shown in Fig. 19. The rejection in the stopband can be improved by two or more reflections.

TABLE III

GEOMETRICAL PARAMETERS OF THE MEASURED WIRECLOTH MESHES

Mesh (#)	g(μm)	$2a^*$(m)	$2a^*/g$
280	95	39	0.41
200	127	46	0.36
145	171	60	0.35
100	212	82	0.39
65	384	177	0.46

*In the original paper a denotes the diameter of a wire instead of the wire radius.
(After Mitsuishi et al.,1963)

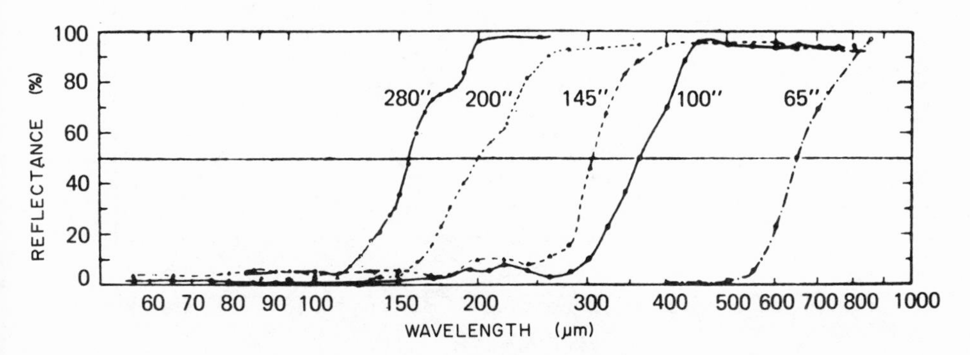

Figure 19. FIR reflectance of five different wirecloth meshes (listed in Table III) useful as lowpass filters. Measured incident angle is 15° . (After Mitsuishi et al., 1963).

The power transmittance of the inductive meshes shows the bandpass property around the wavelength near the mesh periodicity. The electroformed meshes are more suitable for this purpose than the wirecloth meshes. The bandwidth is affected by the strip breadth 2a/g (See Fig. 8 (a)). In order to use the inductive meshes as wide bandpass filters, the meshes with 2a/g less than 0.3 will be suitable. The transmittance of the meshes with different periodicity but with the same 2a/g value of 0.3 has been measured and the results are shown in Fig.20. The transmittance decreases to longer wavelength more rapidly than shorter wavelength. Since the rejection of shorter waves is not sufficient, the filters are used in combination with lowpass filters. Bandpass filters with moderate bandwidth have been reported by various investigations (Ressler et al., 1967; Romero et al., 1973; Varma et al., 1969a; Pradhan et al., 1972) including some which aim toward mesh application in radiometric experiments. The narrowing of bandwidth and the improvement of rejection in the stopband are realized with two or more transmissions. In order to avoid the resonances which could occur at longer wavelengths where the reflectance of meshes is high, two or more inductive meshes were stacked either incoherently or closely (less than the mesh periodicity, g) even if in parallel. Yoshinaga type (sometimes called Yamada type) filters (Yamada et al., 1962; Yoshinaga 1973) were combined to cut off the residual transmission in the shorter wavelength region. Sakai and Yoshida (1978) have tried to narrow the bandwidth of bandpass filters by changing the parameters of meshes. They varied the normalized line breadth (2a/g) to find an optimum condition and pointed out that 2a/g~ 0.5 was best.

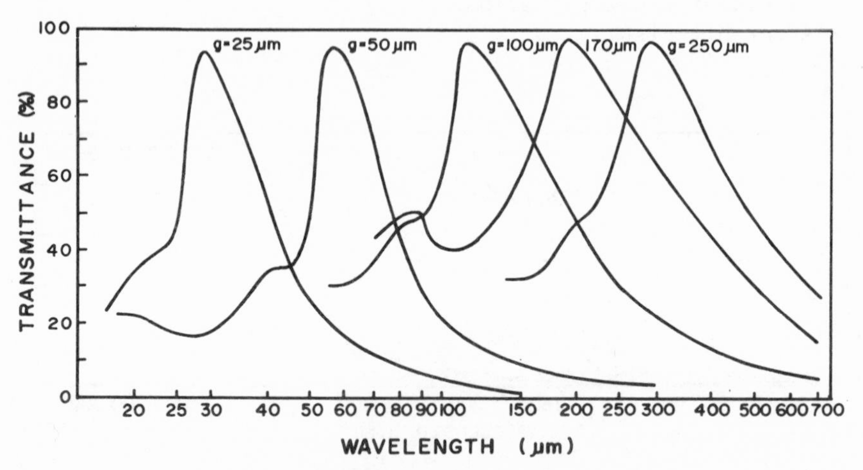

Figure 20. Transmittance of electroformed inductive meshes with 2a/g = 0.3 useful for wide bandpass filters ; mesh periodicity is indicated in the figure. (After Sakai et al., 1969)

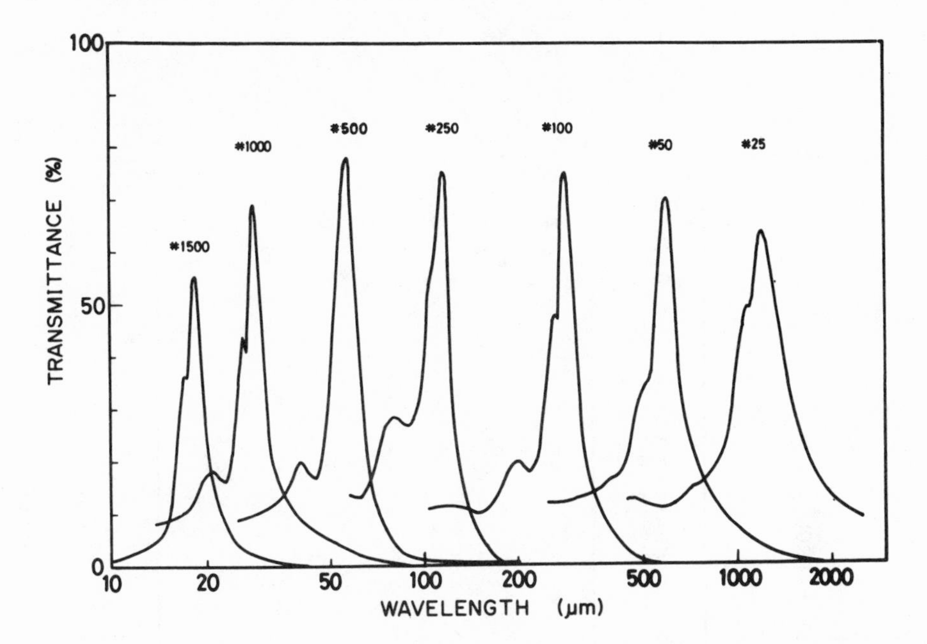

Figure 21. Transmittance of electroformed inductive meshes with 2a/g = 0.5 useful for narrow bandpass filters. Characteristics of these filters are in Table IV. (After Sakai and Yoshida, 1978)

Figure 21 shows the FIR transmission property of narrow bandpass filters of optimum condition. The filters with Q~4 and maximum transmittance ~70% in average have been obtained. The optical properties of these filters are listed in Table IV .

Recently, Timusk and Richards (1981) and Keilmann (1981) have reported highpass transmission filters made of a thick perforated metal plate. An array of regularly perforated holes acts as waveguides. Below the cutoff frequency $\nu_1 = 0.586/d$, for the lowest propagating mode in a circular waveguide with diameter d, the filter has a very low transmittance, and in addition it has a very steep slope (e.g. 1000 db/octave). In figure 22, as an example, the performance of this king of filter is shown. For the calculation of the transmission, in Fig. 22, the equation

$$T = \exp(-2\gamma t) \tag{32}$$

TABLE IV

CHARACTERISTICS OF SINGLE MESH NARROW BANDPASS FILTERS

Mesh (#)	Mesh periodicity g (μm)	Peak wavelength λ_o (μm)	Maximum value T (%)	Q-value
1500	16.9	18.5	55	8.4
1000	25.4	28.5	69	5.0
500	50.8	57.5	78	4.8
250	101.6	114	75	4.2
100	254	285	79	4.0
50	508	599	70	3.7
25	1016	1220	67	2.0

(After Sakai and Yoshida 1978)

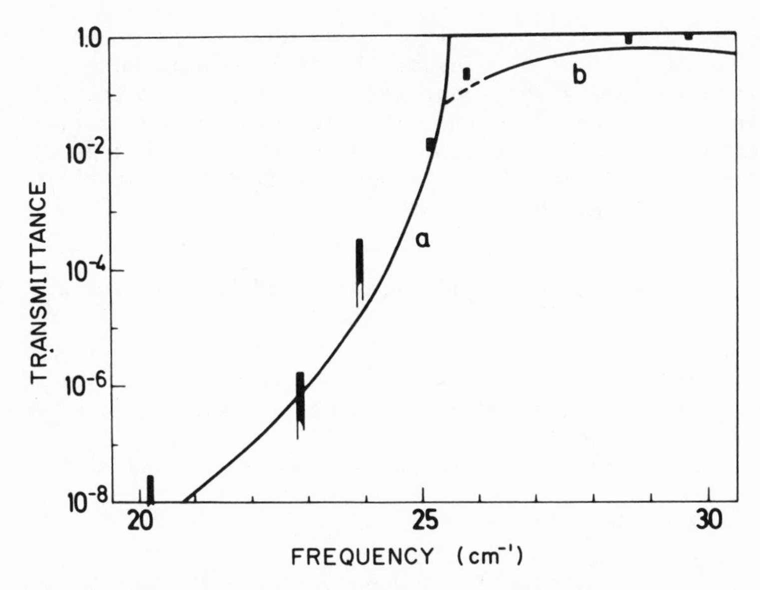

Figure 22. Transmission property of a perforated thick grill. Points, experimental results obtained by use of a laser source. a , theoretical curve obtained from Eq. (32) with $\nu_1 = 25.5$ cm^{-1} and t/d = 4.3. b, Fourier spectrum. (After Keilmann, 1981)

with

$$\begin{aligned} \gamma &= 0 && (\nu > \nu_1) \\ \gamma &= 2\,(\nu_1^2 - \nu^2)^{\frac{1}{2}} && (\nu < \nu_1) \end{aligned} \tag{33}$$

has been used, where ν is the frequency expressed in the unit of (cm^{-1}), t the thickness of the metal plate and γ the propagation constant of the hole.

4. FABRY-PEROT INTERFEROMETERS AND INTERFERENCE FILTERS

4.1. Principles of Fabry-Perot Interferometer

The FPI has been a well established tool for high-resolution spectroscopy for a long time in wavelength regions other than FIR. The general optical properties of FPI are now well known. It is unnecessary to modify the theoretical treatment for the introduction of the interferometer into the FIR region. The FPI utilizes the multiple-beam interference from two highly reflective parallel plates of low loss. The power transmittance $\tilde{T}(\nu)$ of the interferometer is given by the well-known Airy formula:

$$\tilde{T}(\nu) = \left(1 - \frac{A}{1-R}\right)^2 \left[1 + \left(\frac{2\sqrt{R}}{1-R}\right)^2 \sin^2\frac{\phi}{2}\right]^{-1}, \tag{34}$$

where ν is the frequency (or wavenumber) expressed in (cm^{-1}), R and A the power reflectance and absorptance, respectively, of a single reflector plate, and ϕ is the phase difference between two successive beams of the internal multiple reflection, expressed by

$$\phi = 2\pi\nu \cdot 2nd - 2\psi_\Gamma \tag{35}$$

in which ψ_Γ is the phase shift due to the reflection on one plate and nd is the optical path difference between adjacent beams for normal incidence. $\tilde{T}(\nu)$ has maxima of the positions where the phase ϕ becomes $2m\pi$ with $m = 1, 2, 3, \ldots$ The value of maximum transmittance is given by

$$\tilde{T}_{max} = \left(1 - \frac{A}{1-R}\right)^2 = \left(\frac{T}{1-R}\right)^2 \tag{36}$$

At the position where $\phi = 2\pi(m + \frac{1}{2})$ is satisfied, $\tilde{T}(\nu)$ has a minimum value expressed by

$$\tilde{T}_{min} = \left(\frac{T}{1+R}\right)^2 \tag{37}$$

The free spectral range $\Delta\nu$ between adjacent orders is given by

$$\Delta\nu = \frac{1}{2nd} \tag{38}$$

The resolving power Q is defined in terms of the half-width $\delta\nu$ of the peaks centered at ν_m

$$Q = \frac{\nu_m}{\delta\nu} = mF_R \tag{39}$$

with

$$\nu_m = \frac{1}{2nd}\left(m + \frac{\psi_\gamma}{\pi}\right), \tag{40}$$

where F_R, called finesse, is the effective number of multiple reflections. F_F is given the relation

$$F_R = \frac{\Delta\nu}{\delta\nu} = \frac{\pi}{2\arcsin\left[(1-R)\ /2\sqrt{R}\ \right]} \simeq \frac{\pi\sqrt{R}}{1-R} \tag{41}$$

The approximation is valid to better than 3 percent for $R \geq 0.6$. In order to achieve high $\tilde{T}_{max}$, high Q and low $\tilde{T}_{min}$ values, reflectors with high reflectivity and low absorptance are demanded.

The practical resolving power is somewhat smaller than mF_R due to surface defects, nonparallelism, and the finite aperture in use (1976). These factors can all be expressed in terms of the finesse. Besides reflection finesse F_R, a defect finesse $F_D = x/2\sqrt{2}$ for plates good to λ/x, a parallelism finesse $F_p = y/\sqrt{3}$ for plates parallel to λ/y and an aperture finesse $F_A = 8f^2/m$ for a beam of focal ratio f at order m and introduced, and the total finesse F_T is then expressed by

$$\frac{1}{F_T{}^2} = \frac{1}{F_R{}^2} + \frac{1}{F_D{}^2} + \frac{1}{F_P{}^2} + \frac{1}{F_A{}^2} \tag{42}$$

the m-th order peak of a FPI occurs at the position expressed by Eq. (40), so that the interferometer can be tuned by changing nd. But filtering is necessary against overlapping orders. A tunable FPI works as a monochromator and lacks, therefore, the multiplex gain (or Fellgett advantage) (Fellgett, 1967) but it has Jacquinot's light throughput gain (Jacquinot, 1954).

4.2 Realization and Performance of Fabry-Perot Interferometer or Interference Filters

For the realization of FPI's or interference filters various structures have been presented. As meshes are fragile and have no rigidity of their own, the first thing we have to do is to stretch the meshes carefully keeping them flat. At an initial stage of the investigation, the meshes were cemented with thinned collodion onto plane plates of crystal quartz. The existence of

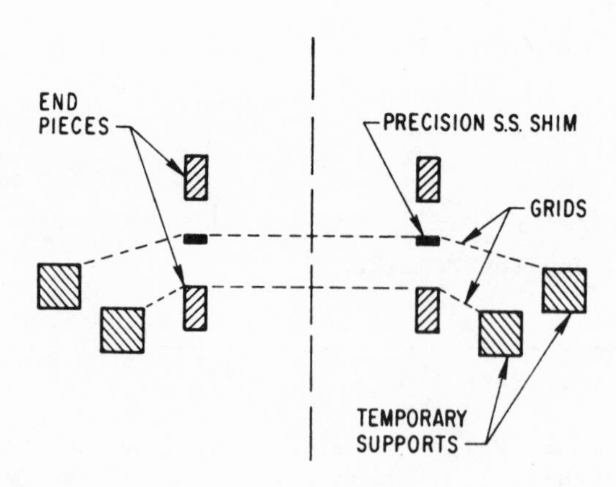

Figure 23(a)Various types of Fabry-Perot interferometers : Interference filter with fixed spacing. After Rawcliffe and Randall (1967).

the substrate, however, not only degrades the performance but is troublesome. Soon the method of making self-supporting meshes was devised. It is analogous to flattening a drum skin. In Fig. 23 we present some representative constructions. Figure 23 (a) (Rawcliffe et al., 1967) can be used only in the static mode (Keeping both reflectors immovable) with fixed spacing; on the other hand, Figure 23 (b) (Ulrich, 1967a) is used in the static mode with variable spacing. Figure 23 (c) (Ulrich et al., 1963; Sakai et al., 1969) can be used both in the static mode and in the scanning mode.

Usually the inductive meshes are used as reflectors when the FPI is used as a monochromator. The measured power transmittance of a FPI is shown in Figure 24. The interferometer consists of self-supporting meshes with #500 ($g \sim 50\ \mu$). The full line represents the Airy curve calculated on the basis of the measured power transmittance of a simgle mesh and the absorptance expressed by Eq. (22). The difference between the measured curve and the calculated curve is due to the slight unparallelism, the beam convergence and the finite spectral slit width. Using refined techniques, Rawcliffe and Randall (1967) investigated an interferometer made of self-supporting meshes with # 750 ($g \sim 34\ \mu$) at a fixed distance of 0.0808 cm. In Figure 25 the finesse calculated from the measured

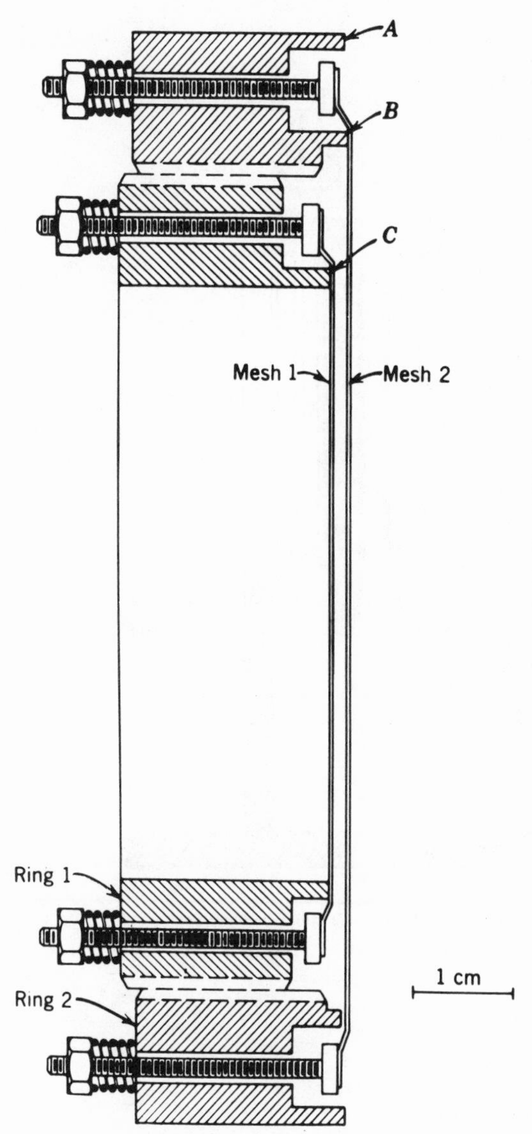

Figure 23(b) Various types of Fabry-Perot interferometers: Interferometer with variable spacing used in a fixed separation. After Ulrich 1969.

half-width (solid dots), is plotted against ν together with the value calculated from the measured power transmittance of a single mesh (open square) showing an ideal case. In the figure, the lowered finesse due to the introduction of the beam angle convergence to the ideal case (open circles) and the further lowered finesse because of the additional correction for the instrumental resolution (open rectangular) are plotted. We see that the measured values are slightly higher than the fully-corrected calculated results but, on the whole, agree satisfactorily with them. In Figure 26, the measured transmittance at the peaks for the same interferometer is plotted for the 4° and 8° beam angles. It can be easily explained from the tendency of the power reflectance and the power transmittance of inductive meshes, and with Eq.(36) and Eq.(41),

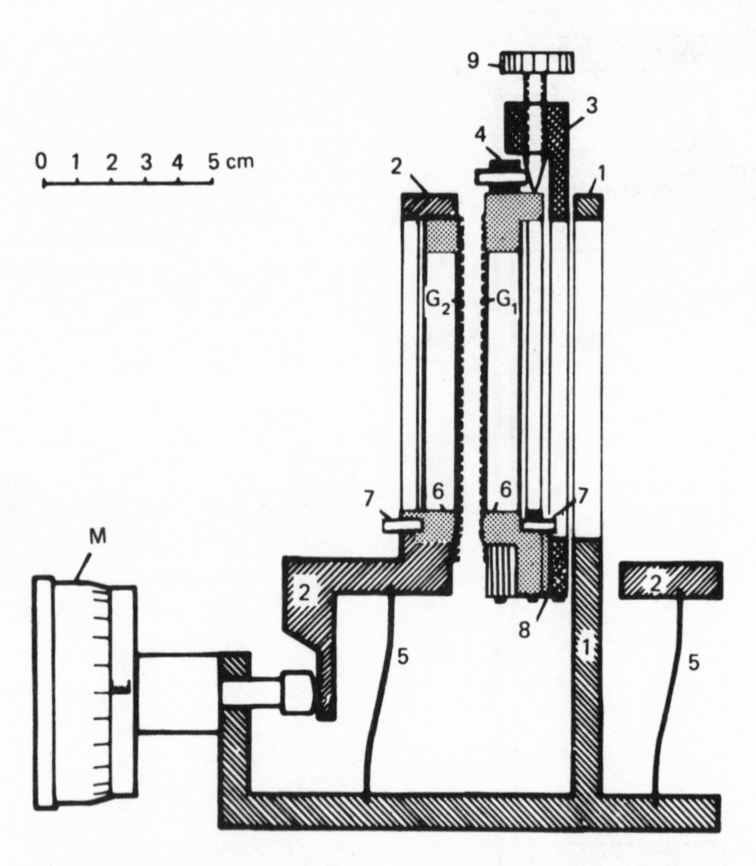

Figure 23(c) Various types of Fabry-Perot interferometers: Tunable Fabry-Perot interferometer. After Ulrich et al., 1963

that the finesse decreases and the peak transmittance increases with increasing wavenumber. The minimum transmittance halfway between peaks is plotted in Figure 27, in which the calculations were based on the transmittance and the reflectance of the mesh and on Eq. (37). Knowing the optical properties of the interferometer, a scanning FPI (Fig. 23 (c)) was practically applied to a monochromator (Ulrich et al., 1963; Sakai et al., 1969). In a simple optical arrangement (Fig. 28 below), the emission spectrum of a

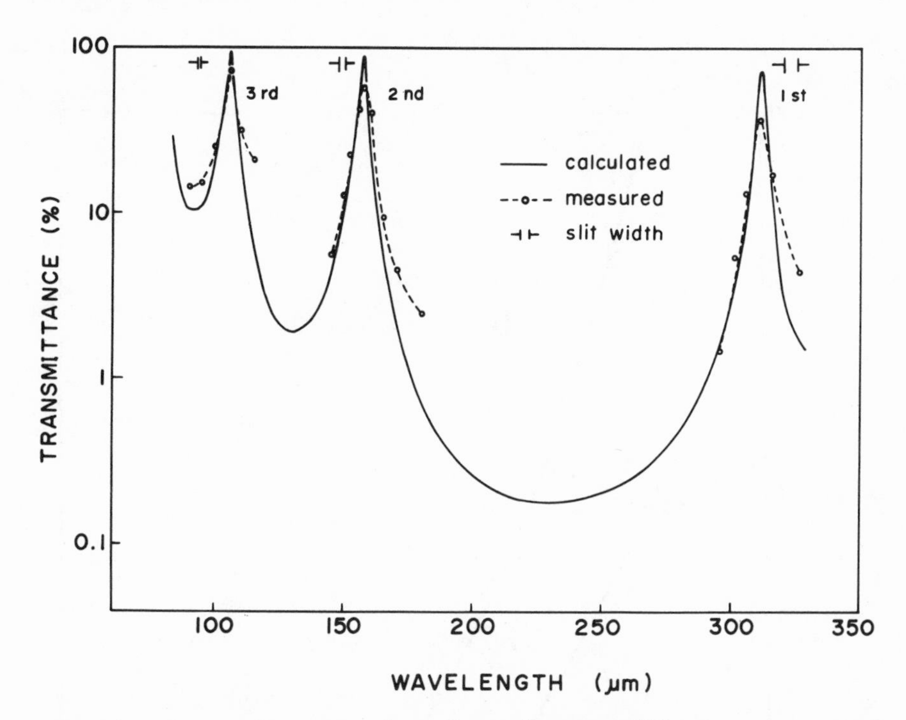

Figure 24. Transmittance of an interference filter constructed from two inductive meshes. Solid line shows the Airy curve calculated from the measured transmittance and Eq.(22). (After Sakai et al., 1969)

mercury arc with absorption lines of water vapor was measured (Fig. 28 above). Grisar et al. constructed a spectrometer (Fig 29)(1967) composed of a FPI in higher orders and a FIR grating. In this spectrometer, the FPI acts as main dispersing element and the grating merely acts as a pre-filter for order separation. With the increase of applications of scanning FPI, more skillfully devised instruments have been presented by Lecullier and Chanin (1976), Holah and Morrison (1977) and many other investigators.

The possibilities of interferometers constructed either with two identical capacitive meshes (symmetrical interferometers) or with one inductive and one capacitive mesh (asymmetrical interferometers) have been pointed out by Ulrich (1967a). The differences between symmetrical capacitive filters and symmetrical inductive ones are that the interference orders do not lie harmonically in the frequency scale, and the finesse increases with frequency.

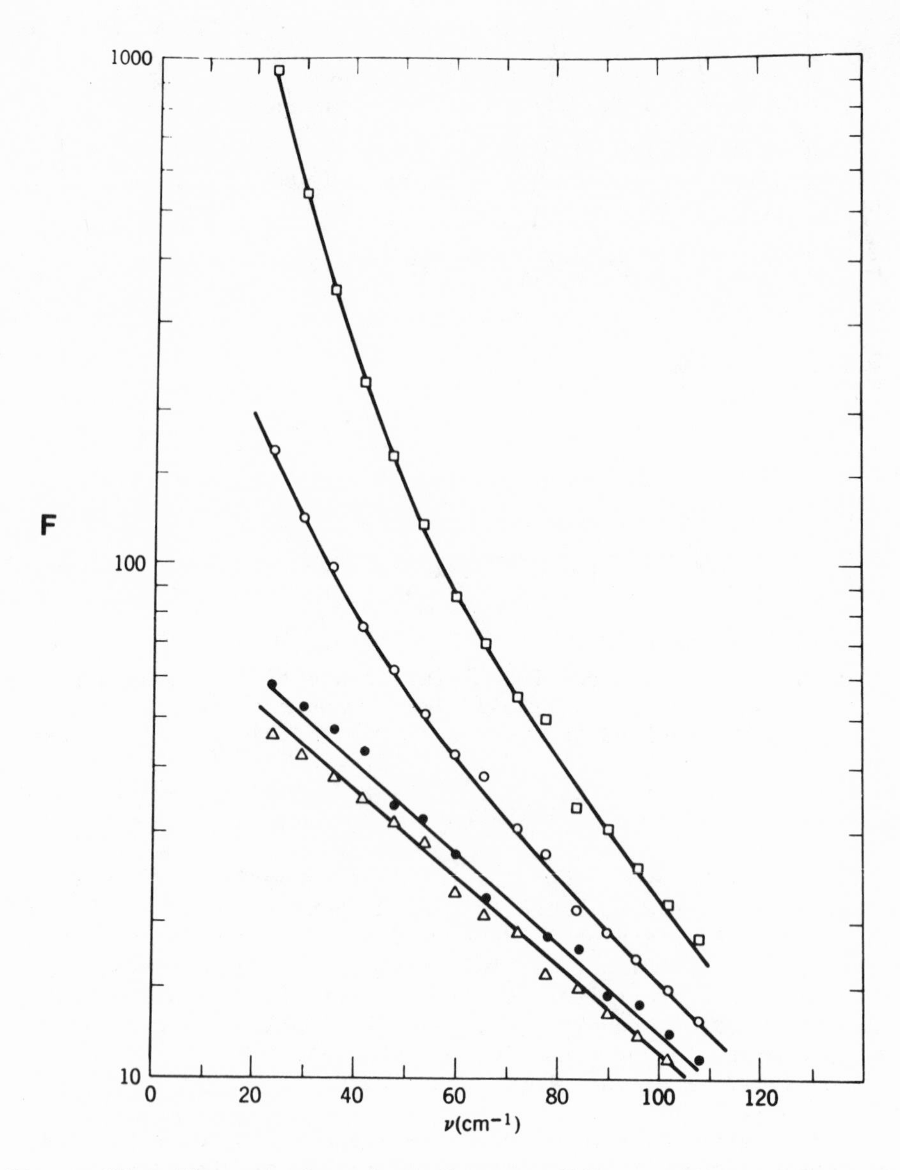

Figure 25. Finesse of an interference filter composed of two #750 (g~34 μm) meshes : experimentally measured finesse ●, finesse calculated from theory using measured transmittance values □, calculated finesse corrected for converging beam o, calculated finesse corrected for converging beam and for finite resolution △. (After Rawcliffe and Randall 1967).

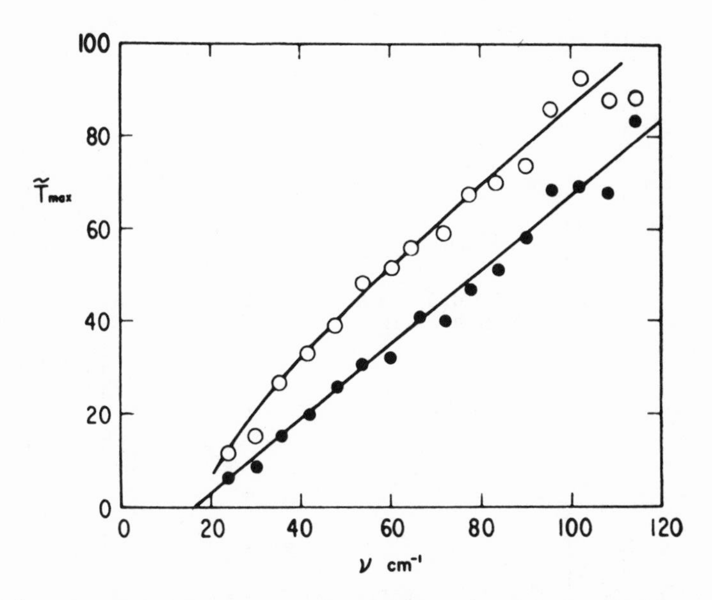

Figure 26. Transmittance at the peaks of an interference filter composed of two #750 meshes and 0.0805 cm spacing, measured for beam convergences of about 8° (●) and about 4° (○). (After Rawcliffe and Randall 1967)

The measured transmittance of this type of interferometer is shown in Figure 30 fabricated using two No. 1 meshes in Fig. 12. The full line represents the transmittance of the interferometer calculated for the equivalent circuit in the figure using the parameters $\hat{Z}$, ω_o and $\bar{R}$ from Table 1. The deviation of the measured values from the calculated curve are attributed to absorption in the meshes, to imperfections of the meshes, to their nonparallelism and to the finite resolution of the spectrometer.

Aside from the symmetrical case, the construction with one inductive and one capacitive mesh gives rise to new features. The Airy formula implies that the interference maxima with highest transmission occurs only when the reflectance of both reflectors is identical. The symmetrical filters satisfy the condition at all frequencies. Because of the optical properties of inductive

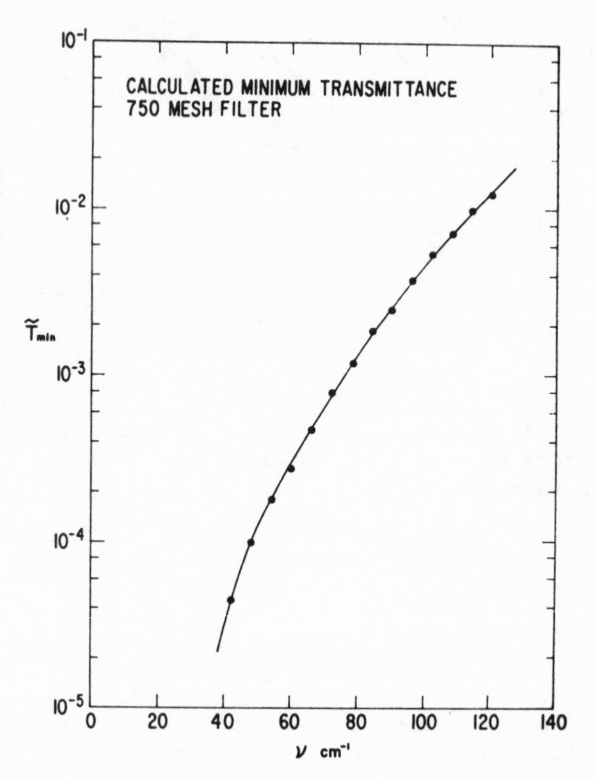

Figure 27. Calculated minimum transmittance of a #750 mesh filter. (After Rawcliffe and Randall 1967)

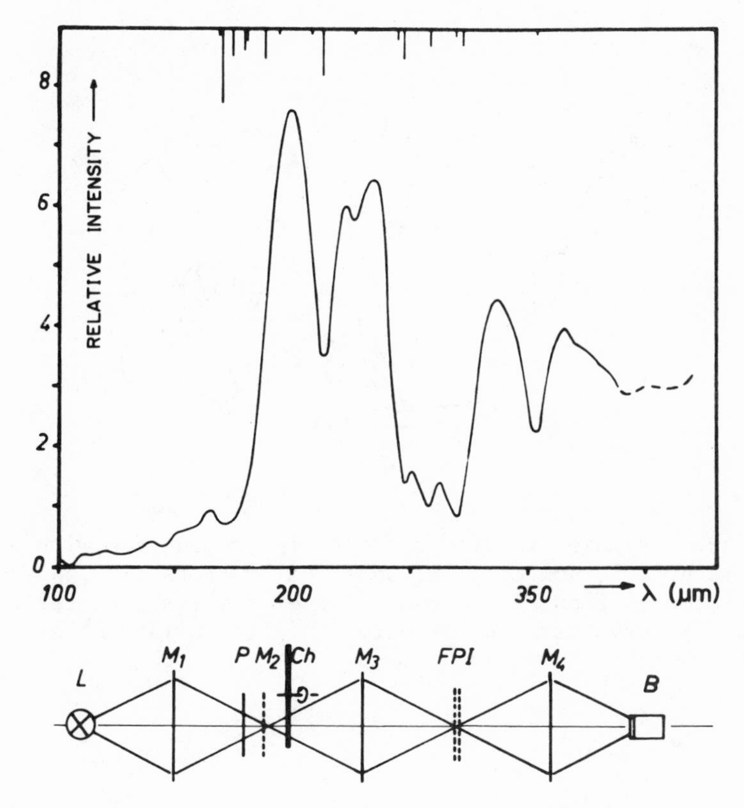

Figure 28. Upper ; absorption spectrum of water vapor obtained with a tunable FPI using a mercury arc lamp as a source. Lower ; simple optical arrangement used for the measurement. (After Ulrich et al., 1963).

and capacitive meshes, that condition is satisfied only at a crossing point of power transmission or power reflection curves of both meshes. If the separation between both meshes is so chosen that the resonance occurs at this crossing point, the interferomenter will show one interference maximum at the frequency with near unity transmittance. The other interference maxima are considerable reduced due to the lack of symmetry of the reflectivity

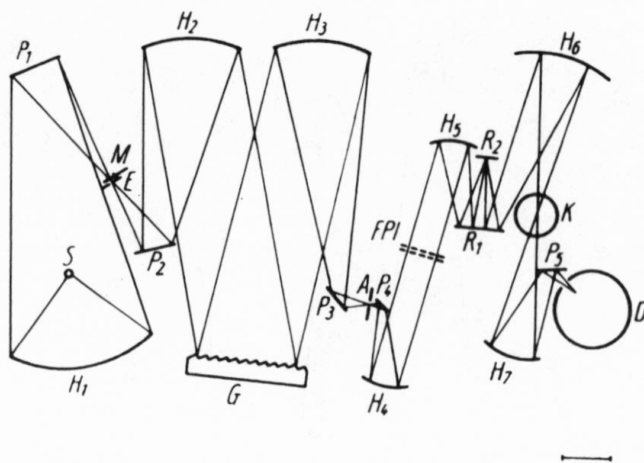

Figure 29. FPI used in higher orders in conjunction with a grating spectrometer : S, source ; H, spherical mirrors ; P, plane mirrors M, modulator ; E and A, entrance and exit slits ; R, reflection filters ; K, cryostat ; D, detector. (After Grisar et al., 1967)

of both reflectors. The peak transmittance of the asymmetrical interferometer at ν_m is given by

$$\tilde{T}_{max} = \frac{|\tau_i(\nu_m)|^2 |\tau_c(\nu_m)|^2}{[1 - |\Gamma_i(\nu_m)||\Gamma_c(\nu_m)|]^2} \tag{43}$$

An example of the asymmetrical interference fringe is represented in Figure 31. It has been combined from the meshes No.2 and No.3 in Table I . The transmittance of both meshes are equal at $\nu = 15\ cm^{-1}$. The separation d= 386 μm was chosen so as to tune the first order interference maximum to the frequency. The full line in Figure 31 is the theoretical transmittance of the interferometer, calculated with the inserted equivalent circuit and the

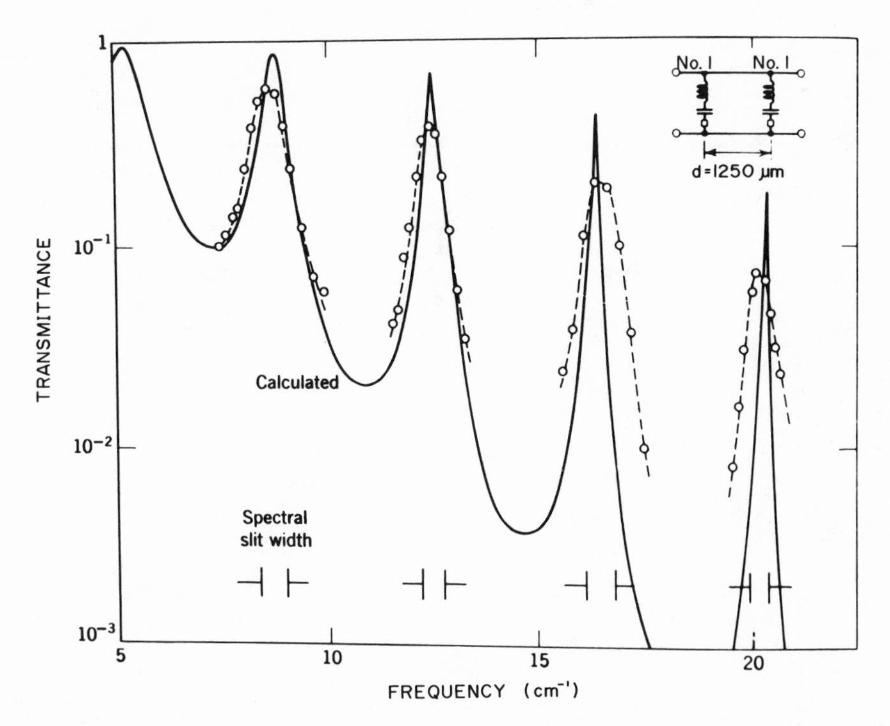

Figure 30. Transmittance of an interference filter constructed from two capacitive meshes No. 1 (in Table I). The full line is the transmittance calculated for the inserted equivalent circuit using the parameters in the table. (After Ulrich, 1967a)

parameters from Table I. The measured points nearly coincide with this curve. The first order maximum, in reality, appears at $\nu = 14.2\ cm^{-1}$ with 92% transmittance, the second order lies at $23\ cm^{-1}$ with 40% transmittance and the third order appears very weakly at $30\ cm^{-1}$. This kind of configuration is useful for bandpass filters rather than as monochromator.

5. MULTIMESH FILTERS

5.1. Principles of Multimesh Interference Filters

The optical properties of coherent multimesh configurations can be investigated theoretically again by use of the transmission line theory. Following the theory, a line of meshes shown in

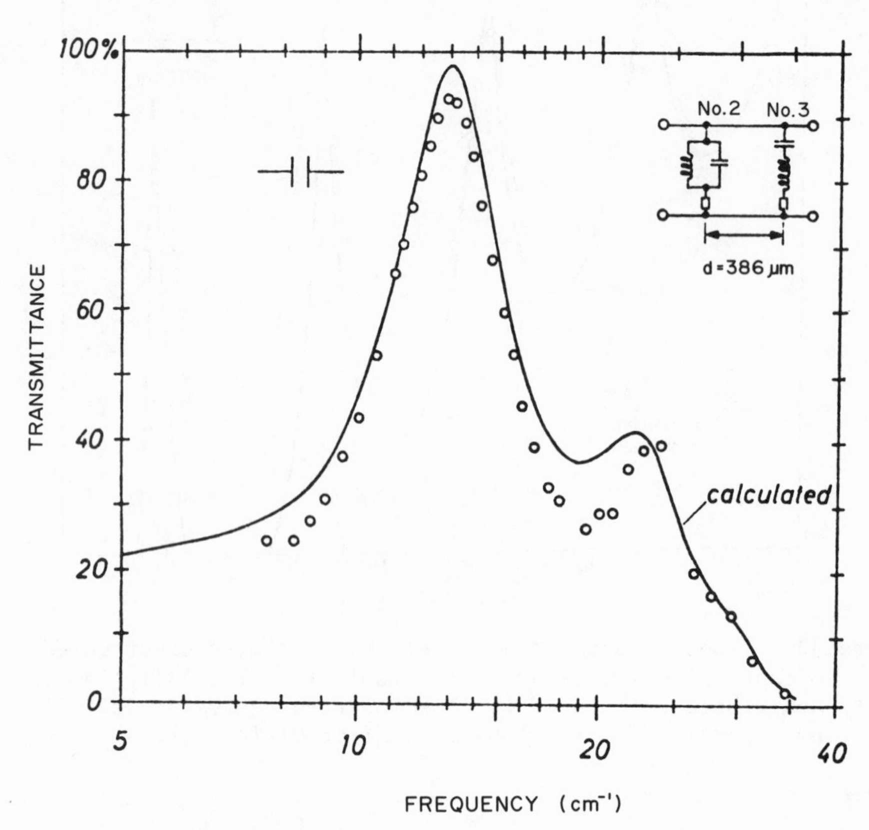

Figure 31. Transmittance of an asymmetric interference filter combined from one inductive and one capacitive mesh. No.2 and No.3 meshes in Table I are used. Distance between the meshes d = 386 μm. (After Ulrich 1967a)

Fig. 32 can be converted to a transmission line expressed in Fig.33, in which Z_o , Z_i (i = 1, 2, 3 ...) and Z denote the impedance of free space, the impedances of meshes and the impedance between meshes, γ and d_i (i = 1, 2 ...) the propagation constant and the distances between meshes. This transmission line is composed of alternative connections of two kinds of four-terminal networks, each of which is expressed conveniently by a matrix having four elements.

In general, for a four-terminal network shown in Fig. 34 and

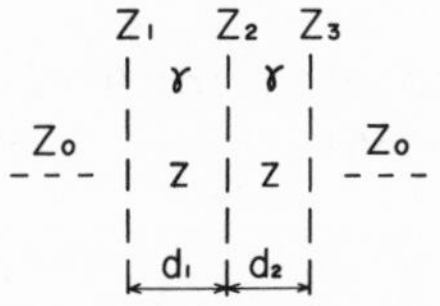

Z_1 Z_2 Z_3

γ γ

Z_o Z_o

z z

d_1 d_2

Figure 32. Schematic of multimesh configuration : Zi (i = 1, 2,3...), impedances of meshes ; Z_o , impedance of free space ; di (i = 1, 2, 3, ...), distances between evanescent meshes ; γ , a propagation constant.

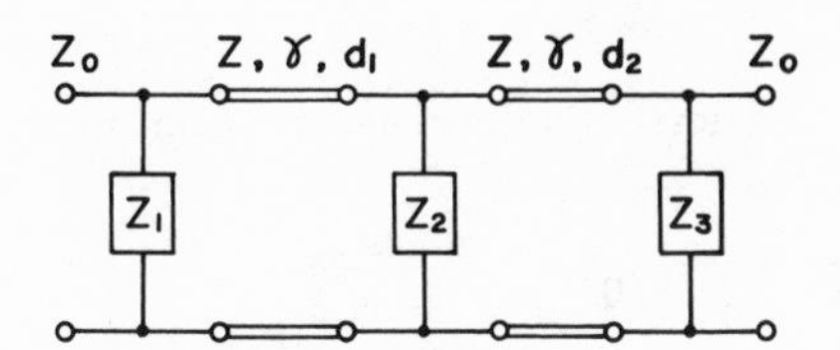

Figure 33. Equivalent circuit representation of the multimesh configuration presented in Fig.32.

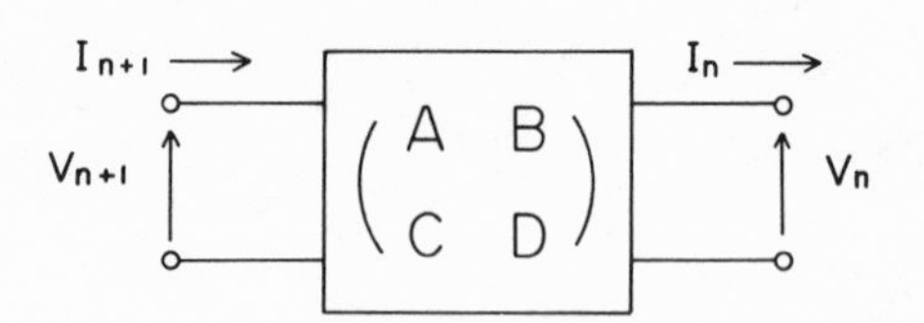

Figure 34. General representation of four-terminal network expressed by a matrix having four elements A, B, C and D.

expressed by matrix elements A, B, C and D, the relation

$$\begin{pmatrix} V_{n+1} \\ I_{n+1} \end{pmatrix} = \begin{pmatrix} A & B \\ C & D \end{pmatrix} \begin{pmatrix} V_n \\ I_n \end{pmatrix} \tag{44}$$

holds. The transmission line theory shows that such networks give rise to a voltage (or amplitude) reflection coefficient

$$\Gamma = \frac{A + B/Z_o - CZ_o - D}{A + B/Z_o + CZ_o + D} \tag{45}$$

and a voltage (or amplitude) transmission coefficient

$$\tau = \frac{2}{A + B/Z_o + CZ_o + D} \tag{46}$$

The matrix representation of a lumped impedance is given in Fig.35 (a), together with that of lumped admittance in Fig.35 (b). Substituting the matrix elements into Eqs. (45) and (46), we obtain an amplitude reflection coefficient and an amplitude transmission coefficient. The results, of course, coincide with Eqs. (24) and (25). The matrix representation of a line between adjacent lumped impedances is given in Fig. 35 (c). The propagation constant γ is a complex number in general and we write it in the form

$$\gamma = \alpha + j\beta \tag{47}$$

where α and β are real and imaginary parts of γ, and $j = \sqrt{-1}$.

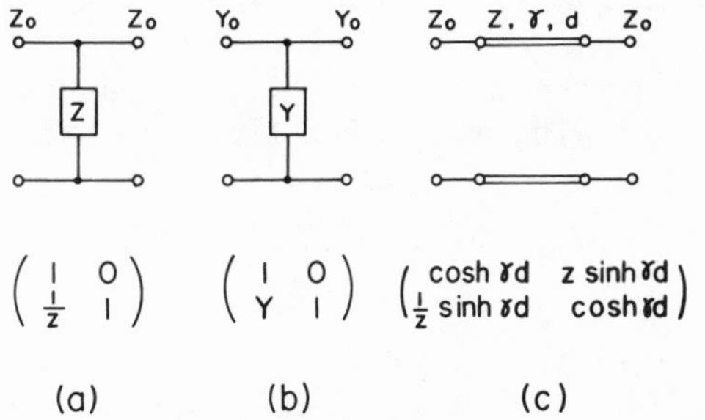

$$\begin{pmatrix} 1 & 0 \\ \frac{1}{Z} & 1 \end{pmatrix} \qquad \begin{pmatrix} 1 & 0 \\ Y & 1 \end{pmatrix} \qquad \begin{pmatrix} \cosh \gamma d & Z \sinh \gamma d \\ \frac{1}{Z} \sinh \gamma d & \cosh \gamma d \end{pmatrix}$$

(a) (b) (c)

Figure 35. Matrix representations of some characteristic four-terminal networks : (a) transmission line shunted by a lumped impedance Z, (b) transmission line shunted by a lumped admittance Y, (c) transmission line for distributed parameter circuit.

For a non-absorbing media, γ is a purely imaginary number and then the matrix is written in the simpler form

$$\begin{pmatrix} \cos \beta d & jZ \sin \beta d \\ j\frac{1}{Z} \sin \beta d & \cos \beta d \end{pmatrix}$$

Since $\beta = \omega\sqrt{\varepsilon\mu}$ and $Z = \sqrt{\mu/\varepsilon}$, we see that $\beta = \omega n/c_o$ and $Z = (Z_o/n)\cdot(\mu/\mu_o)$ which reduces to $Z = Z_o/n$ if the permeability of the medium μ is nearly equal to that of free space μ_o. ω is the frequency of the incident radiation, c_o the light velocity in free space and n the refractive index of the medium with real number. The matrix is, hence, rewritten in the form

$$\begin{pmatrix} \cos 2\pi \dfrac{nd}{\lambda} & j \dfrac{Z_o}{n} \sin 2\pi \dfrac{nd}{\lambda} \\ j \dfrac{n}{Z_o} \sin 2\pi \dfrac{nd}{\lambda} & \cos 2\pi \dfrac{nd}{\lambda} \end{pmatrix}$$

in which λ is the wavelength of radiation in free space.

The properties of the multimesh filters can be investigated by performing the following multiplications of matrices,

$$\begin{pmatrix} A & B \\ C & D \end{pmatrix} = \begin{pmatrix} 1 & 0 \\ \dfrac{1}{Z_1} & 1 \end{pmatrix} \begin{pmatrix} \cos 2\pi \dfrac{nd_1}{\lambda} & j \dfrac{Z_o}{n} \sin 2\pi \dfrac{nd_1}{\lambda} \\ j \dfrac{n}{Z_o} \sin 2\pi \dfrac{nd_1}{\lambda} & \cos 2\pi \dfrac{nd_1}{\lambda} \end{pmatrix} \begin{pmatrix} 1 & 0 \\ \dfrac{1}{Z_2} & 1 \end{pmatrix}$$

$$\cdot \begin{pmatrix} \cos 2\pi \dfrac{nd_2}{\lambda} & j \dfrac{Z_o}{n} \sin \dfrac{nd_2}{\lambda} \\ j \dfrac{n}{Z_o} \sin 2\pi \dfrac{nd_2}{\lambda} & \cos 2\pi \dfrac{nd_2}{\lambda} \end{pmatrix} \cdots \begin{pmatrix} 1 & 0 \\ \dfrac{1}{Z_k} & 1 \end{pmatrix} \tag{48}$$

and by substituting the resultant matrix elements A, B, C and D into Eq. (45) or (46). The Airy formula can be derived, of course, as a special case. The actual design of multimesh filters is then performed with the aid of a computer.

Apart from such an investigation using transmission line theory, Pradhan et al. (1970) have made calculations on the characteristics of multimesh filters on the basis of Airy's method. They presented the expressions for the power transmittance and reflectance of filters as functions of the power transmittance and reflectance of constituent meshes.

5.2. Realization and Performance of Multimesh Filters

The various configurations of meshes give rise to lowpass, highpass, bandpass and bastop features with steep slopes and high attenuation in the stop bands. One of the multimesh constructions in shown in Fig. 36. In order to achieve good performance, the filters must be constructed paying close attention to the parallelism and flatness of the meshes. The most critical parts in the construction are the spacers. Until now, several types have been used; spacers cut from stainless steel shim stock and lapped to remove any burrs (Rawcliffe et al., 1967), spacers cut from brass shim stock and photoetched to remove burrs (Ulrich, 1968) annular copper

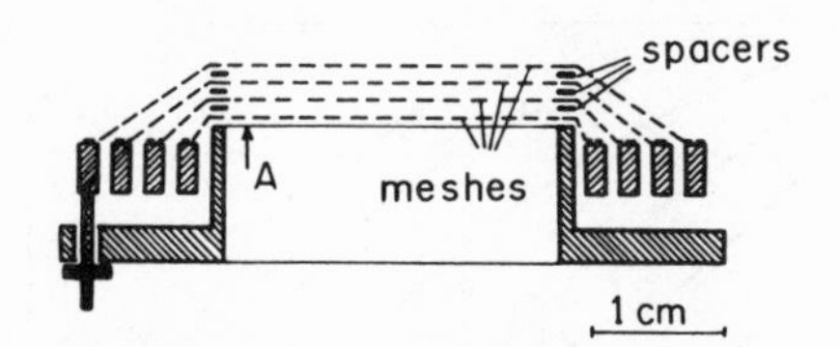

Figure 36. A construction of multimesh interference filters. Each mesh is glued to a ring and stretched like a drum skin by means of three screws and nuts (only one is shown). Surface A is optically flat. (After Ulrich, 1968)

spacers photoetched to the required dimmensions (Holah et al.,1972) spacers formed by pressing a ring of annealed high-purity copper to the desired thickness (Lecullier et al., 1976) and so on. At final assembly, care must be taken so that dust particles do not lie between each spacer and the facing meshes. It is possible to keep two adjacent meshes flat and parallel to within ± 1 μm.

5.2.1. Lowpass Filters

The basic design of a lowpass filter is that of a series of quarter-wave spacers (Holah et al., 1977). However, this construction shows, as in the case of most interference devices, a high transmission peak at the wavelength equal to twice the spacing. A skilful design to reduce this halfwave transmission is to use the spacing $d = \lambda_o/2 = g/2\omega_o$, so that the halfwave transmission appears at λ_o where the transmission of the constituent mesh shows its minimum.

The transmission properties of lowpass filters with two and more meshes have been calculated by Ulrich (1967b) on the basis of the equivalent circuit representation of a single, thin, capacitive mesh in TableIII-B. The results for single, two, three and four meshes are shown in Fig. 37 (a)-(d), as a function of ω/ω_o for various values of $2\hat{Z}$. In the figure, only the range $\omega/\omega_o>1$ is given. For $\omega/\omega_o >1$, the equivalent circuit is no longer applicable

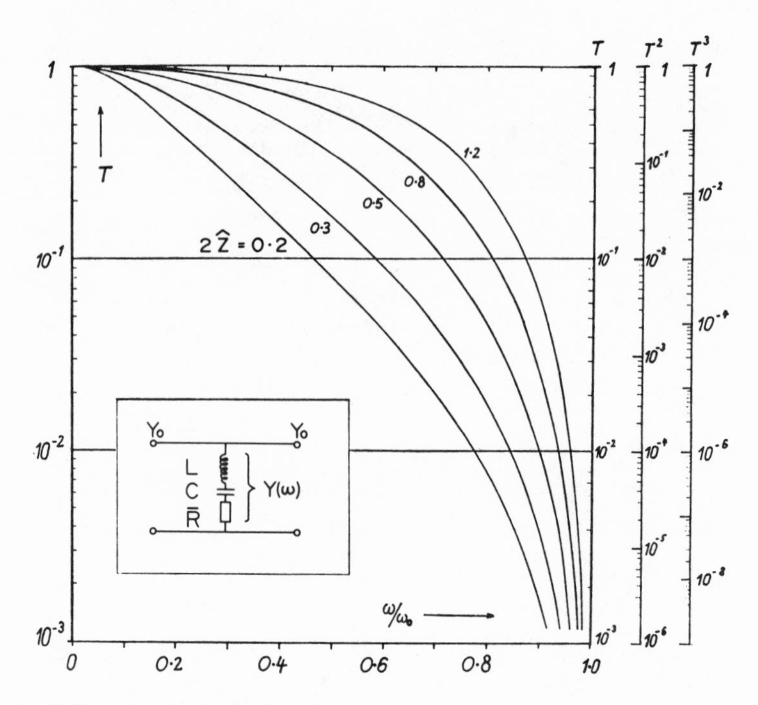

Figure 37(a). Transmittance of single mesh. (After Ulrich, 1967b)

since diffraction appears. For these filters, consisting of more than two meshes, an equal spacing of $d = g/2$ for every separation and a loss resistance of $\bar{R} = 0.005$ are assumed. By increasing the number of meshes, the cutoff slope increases rapidly but, at the same time, the ripple in the passband increases. It will be possible to reduce this ripple by a more sophisticated design using meshes of different impedances and different spacings. Actual performances of two, three and four mesh filters, composed of identical capacitive meshes and with equal spacings, are shown in Fig.38 (a) - (c). The construction of the first two filters belongs to that of Fig.23(b They were measured down to 10^{-3} by use of a periodic FTS of the lamellar grating type (1958). The construction of the filters of Fig. 38 (c) belongs to that of Fig. 36. For three different spacings the cutoff frequency apparently shifts but the slopes are nearly identical. The measurements down to values of the order of 10^{-4} were made using a FTS in asymmetric mode (Bell et al., 1966a,b)

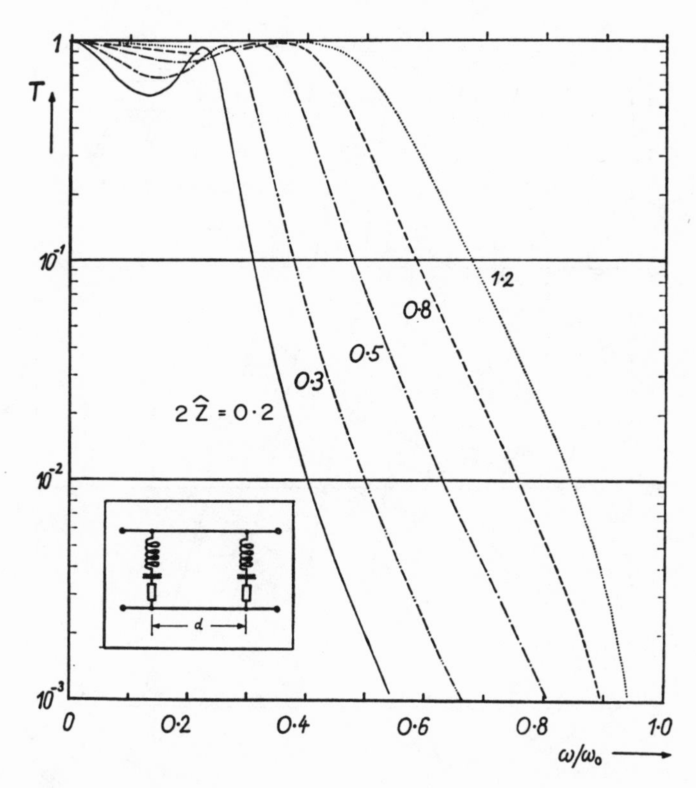

Figure 37 (b). Transmittance of two-mesh. (After Ulrich, 1967b)

which allows the measurement of amplitude transmittance $|\tau|$ with a signal/noise ratio of the order of 100. The filters having spacings $d > g/2\omega_0$ show a spurious maximum in the stopband at $\nu < 1/g$. This arises from the resonance action between adjacent meshes. The peak is actually suppressed due to the influence of slight non-parallelism. In the region $\nu > 1/g$, where the transmission line theory is no more effective, the transmission of filters can be determined only experimentally. The problem of the spurious peaks and the leakage in the

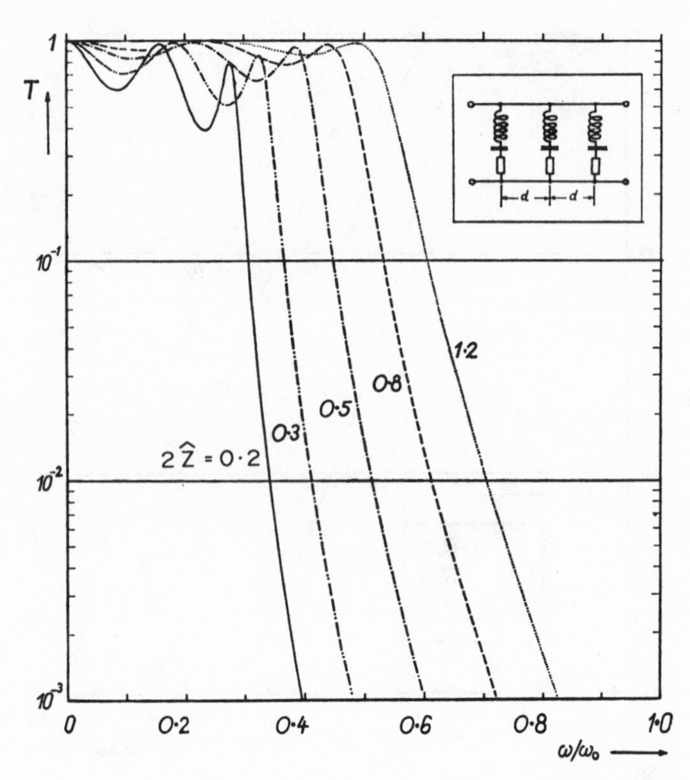

Figure 37 (c). Transmittance of three-mesh . (After Ulrich, 1967b)

stopband can also be considerably reduced by combining capacitive meshes with different periodicity g. The design procedure for the best combination of meshes and the optimal spacing has not yet been developed. One possible guiding principle can be found in the design tehnique for microwave filters (1964). As examples of lowpass filters made from various meshes, the measured transmission characteristics of some four-mesh filters are shown in Fig. 39. The transmittance was again measured by use of a FTS in the asymmetric mode.

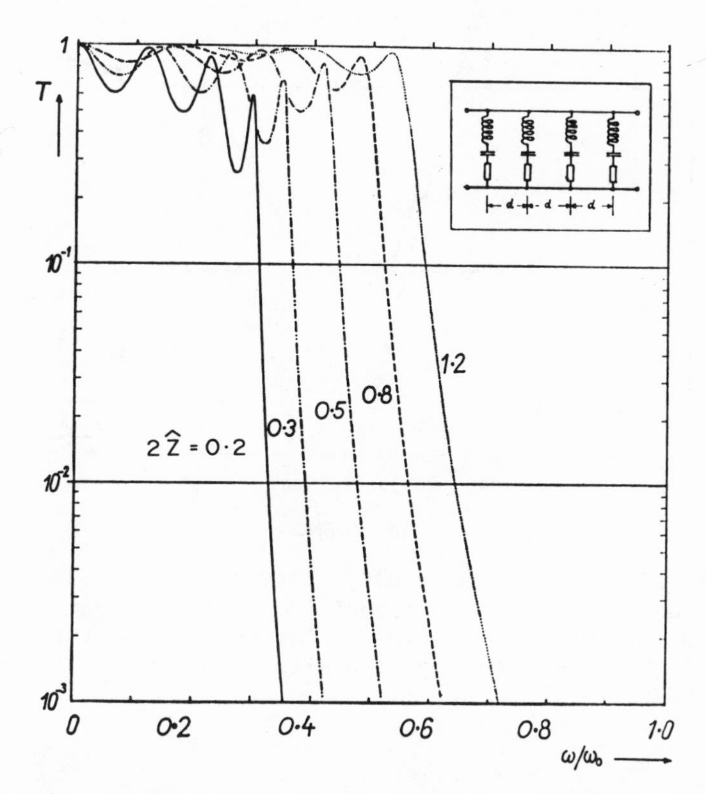

Figure 37 (d). Transmittance of four-mesh interference lowpass filters calculated from the equivalent circuit shown in respective figure. (After Ulrich, 1967b)

The fabrication of lowpass interference filters has been continued subsequently by such investigators as Varma and Moller (1969b) Holah and Smith (1972), Grenier et al.(1973), Holah and Auton (1974) Whitomb and Keene (1980), Holah (1980a)Timusk and Richards (1981) and others. Holah and Auton (1974) and Holah (1980a) showed that it is possible to obtain good filtering characteristics with lowpass filters having

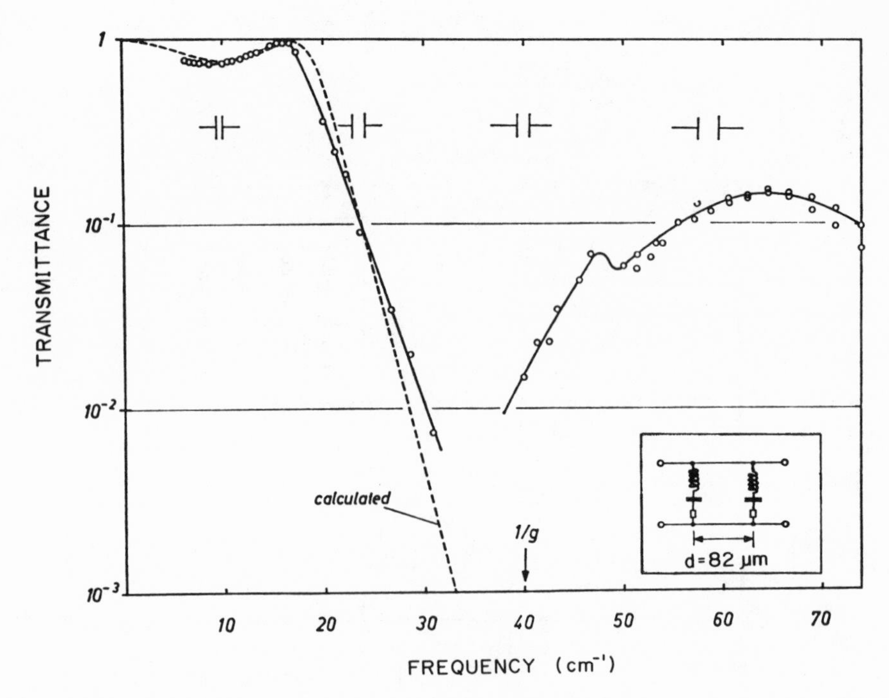

Figure 38 (a). Transmittance of lowpass filters consisting of two identical and equally-spaced capacitive meshes. Parameters of respective filter are listed in Table V. Construction of this belongs to Fig. 23 (b). This was measured with periodic FTS. (After Ulrich, 1967b, 1968).

edges ranging from 70 to more than 2000 μm. According to them, there are two problems preventing the design of a filter as straightforwardly as implied in the theoretical considerations. The first problem is that the optical g value of the capacitive mesh is larger than the physical g value due to the effect of a substrate * (2.5 μm

---

* For capacitive meshes with periodicities greater than 100 μm and with a/g ratios of 0.2, it has been found experimentally that the relationship between physical and effective g values is given by a simple linear relation

$$g_{eff} = 1.1\ g$$

For capacitive meshes having periodicities less than 100 μm the difference between g_{eff} and g is constant and the value is ~8 μm.

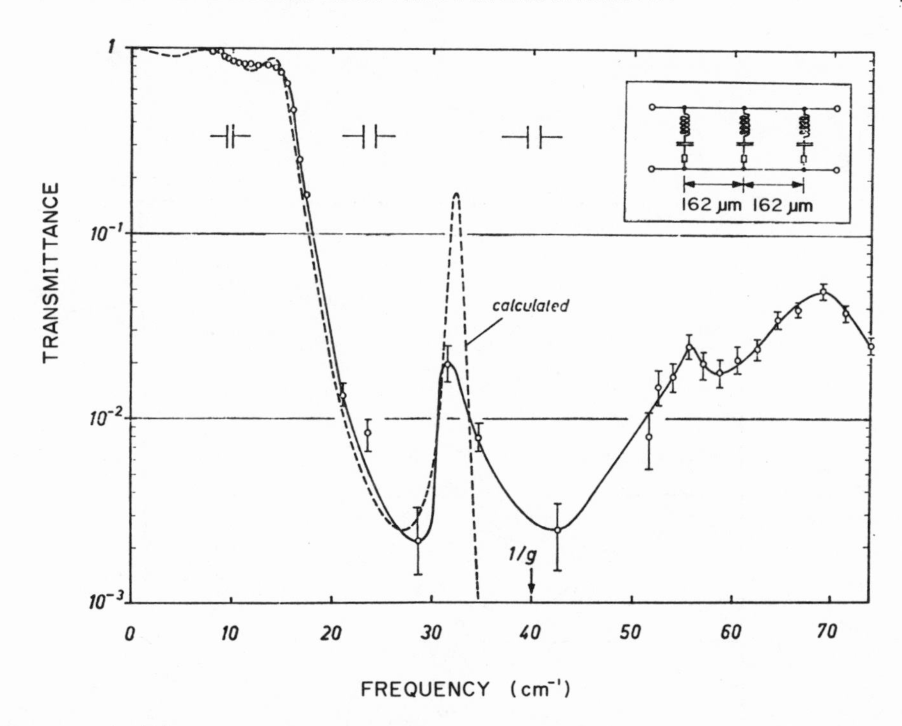

Figure 38 (b). Transmittance of lowpass filters consisting of three identical and equally-spaced capacitive meshes. Parameters of respective filter are listed in Table V. Construction of this belongs to Fig. 23(b). This was measured with periodic FTS. (After Ulrich, 1967b, 1968)

film of Mylar). Secondly, a substrate with 2.5 μm thickness behaves in any optical system as if it were about 4 μm thick because its refractive index is about 1.5. In spite of the these problems they pointed out that they could control the position of the edge within 10 per cent of specifications. The properties of lowpass filters which have been investigated to date are summarized in Table V. Only Timusk and Richards used capacitive meshes on thick (50 to 350 μm) Mylar substrates, while all the other investigators used thin (2.5 of 3 μm) Mylar substrates.

5.2.2. Highpass Filters †

† In this section, only the multimesh type is described. Therefore, the highpass transmission filters of thick perforated metal plate developed by Timusk and Richards (1981) and Keilmann (1981) are not included.

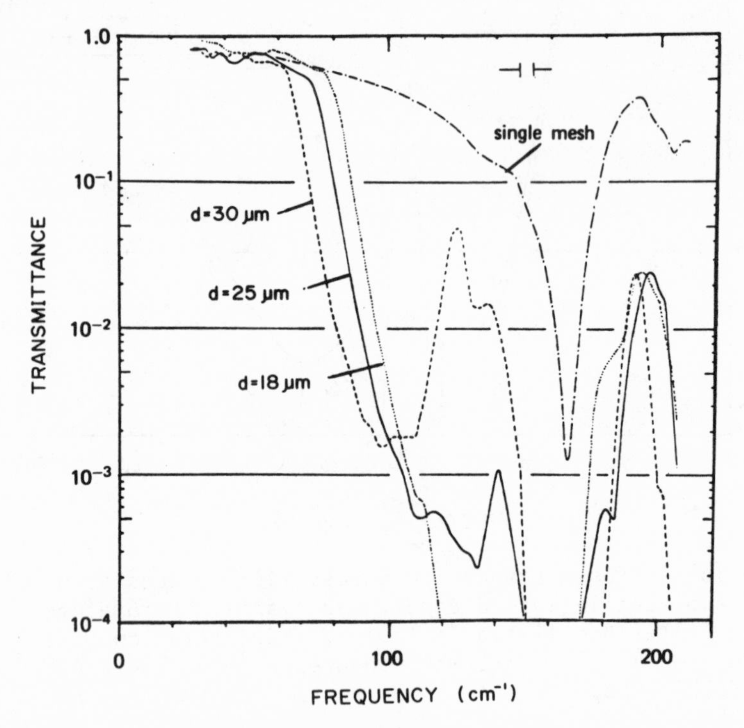

Figure 38 (c). Transmittance of lowpass filters consisting of four identical and equally-spaced capacitive meshes. Parameters of respective filter are listed in Table V. Construction of this was measured with asymmetric FTS. (After Ulrich, 1967b, 1968)

Up to now, not much effort has been put into the development of highpass filters. Highpass filters are best made with inductive meshes. Holah (1980b) reported some highpass filters ranging from 30 to 100 cm $^{-1}$. They were made of three identical meshes and identical spacings. In Fig.40 (a) - (c) typical results are represented. Ulrich (1968) constructed a filter composed of four identical meshes with three equal spacings. The transmission characteristic is shown in Fig.40 (d). According to this work, the highpass filters

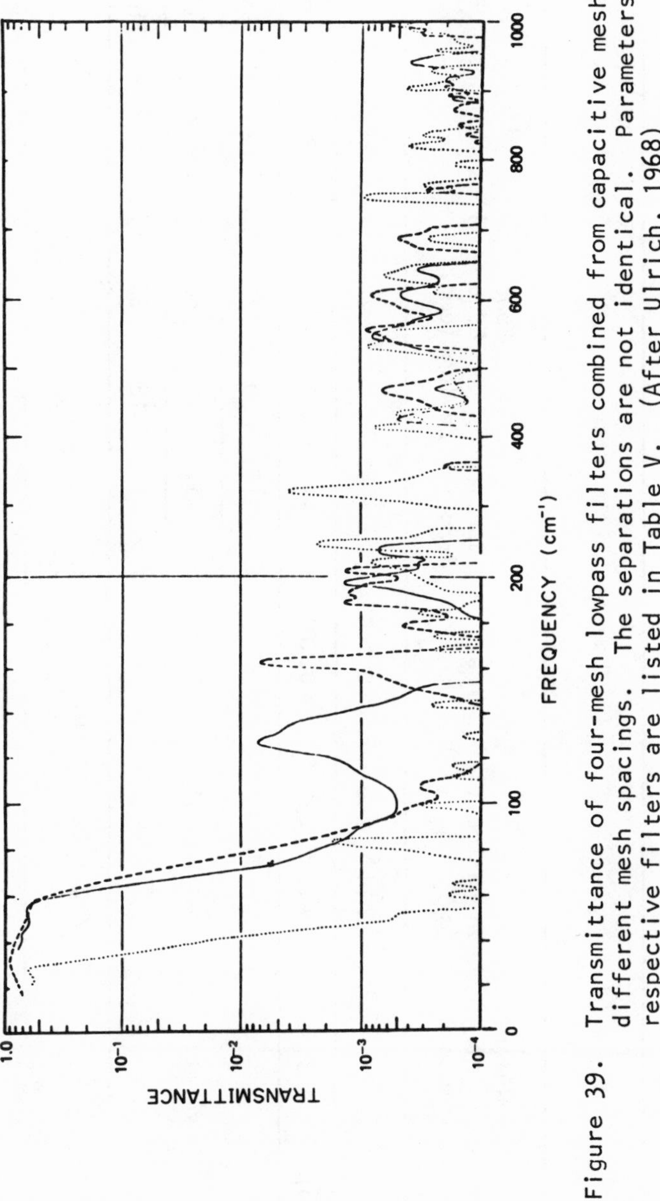

Figure 39. Transmittance of four-mesh lowpass filters combined from capacitive meshes of different mesh spacings. The separations are not identical. Parameters of respective filters are listed in Table V. (After Ulrich, 1968)

TABLE V

SUMMARY OF MULTIMESH LOWPASS, HIGHPASS AND BANDSTOP FILTERS

(1) Lowpass Filters*

Authors [Ref. No.]	Number of Meshes	Mesh Periodicity $g_1/g_2/\ldots g_i$ (μm)	a/g	Spacings $d_1/d_2/\ldots/d_i$ (μm)	Band Edge (cm^{-1})	Illustrations	Remarks
Ulrich (1967b)	2	All 250	0.161	82	24	Fig.38(a)	$Z_\circ=0.70$, $\omega_\circ=0.96$ $\bar{R}=0.02$
	3	All 250	0.161	All 162	18	Fig.38(b)	
Ulrich (1968)	4	All 51	0.18	All 30	71	Fig.38(c)	54dB/Octave
				All 25	80		64dB/Octave
				All 18	86		71dB/octave
	4	102/102/102/51	$0.08(g_1 - g_1)$ $0.18(g_4)$	50/50/40		Fig.39 (....)	
		25/51/51/25	$0.1\ (g_1, g_4)$ $0.18(g_2, g_3)$	28/20/28		(——)	
				All 20		(----)	

(1) Lowpass Filters*

Authors [Ref.No.]	Number of Meshes	Mesh Periodicity $g_1/g_2/\ldots g_i$ (μm)	a/g	Spacings $d_1/d_2/\ldots d_i$ (μm)	Band Edge (cm^{-1})	illustrations	Remarks
Varma and Möller (1969b)	2					(Fig.2,3 in Ref.)	
	4	127/127/78/78	~0.2	37/50/37		(Fig.4 in Ref.)	
		78/127/127/78					
		127/78/78/127					
	4	127/127/78/78	~0.2	37/50/37		(Fig.5 in Ref.)	
		78/78/101/101					
		127/127/181/181					
		181/181/211/211					
	8	211/211/181/181/423/423/317/317	~0.2	37/50/37/100/37/50/37			

(continued)

TABLE V (Continued)

SUMMARY OF MULTIMESH LOWPASS, HIGHPASS AND BANDSTOP FILTERS

(1) Lowpass Filters*

Authors [Ref. No.]	Number of Meshes	Mesh Periodicity $g_1/g_2/\ldots g_i$ (μm)	a/g	Spacings $d_1/d_2/\ldots d_i$ (μm)	Band Edge (cm^{-1})	Illustrations	Remarks
Holah and Smith (1972)	4	25/45/45/33		16/25/25	130	(Fig.11(a)in Ref)	
		/45/52/52/34		All 24		(Fig.12 in Ref.)	
Holah and Auton (1974)	4	18/37/37/25		All 15	133	(Fig.5 in Ref.)	
		42/90/90/37		All 26	67	(Fig.7 in Ref.)	
	5	37/90/180/180/42		All 72	14	(Fig.8 in Ref.)	
	6	180/360/550/550/270/90		All 280	10	(Fig.9 in Ref.)	
Holah (1980a)	4	600/900/900/600		All 375	6.3 4.3	(Fig.3.4 in Ref.)	
Grenier et al.,1973						(Fig.6 in Ref.)	

(1) Lowpass Filters*

Authors [Ref. No.]	Number of Meshes	Mesh Periodicity $g_1/g_2/\ldots g_i$ (μm)	a/g	Spacings $d_1/d_2/\ldots/d_i$ (μm)	Band Edge (cm^{-1})	Illustrations	Remarks
Whitcomb and Keene (1980)						(Fig.1 in Ref.)	
Timusk and Richards (1981)	6	All 508		All 1000	2.8	(Fig.3 in Ref.)	50~350 μm thick Mylar substrate
				All 500	3.8		
				All 495	5		Operation at 1.2K
	4	All 282		All 250	6.7		
	6	All 282		All 178	9		
		All 254		All 236	10.2		

(continued)

TABLE V (Continued)

SUMMARY OF MULTIMESH LOWPASS, HIGHPASS AND BANDSTOP FILTERS

(2) Highpass Filters[†]

Authors [Ref. No.]	Number of Meshes	Mesh Periodicity $g_1/g_2/\ldots/g_i$ (μm)	a/g	Spacings $d_1/d_2/\ldots/d_i$ (μm)	Band Edge (cm^{-1})	Illustrations	Remarks
Ulrich (1968)	4	All 51	0.11	All 20	108	Fig.40(d)	136dB/octave
Holah (1980b)	3	All 59		All 10	∼100	Fig.40(a)	
				All 17	∼ 90		
				All 25	∼ 80		
		All 63.5		All 25	∼ 60	Fig.40(b)	
				All 38	∼ 60		
		All 250		All 50	∼ 30	Fig. 40(c)	

(3) Broad Bandpass Filters

(3) Broad Bandpass Filters

Authors [Ref.No.]	Number of Meshes	Mesh Periodicity $g_1/g_2/\ldots/g_i$ (μm)	a/g	Spacings $d_1/d_2/\ldots/d_i$ (μm)	Band Edge (cm^{-1})	Illustrations	Remarks
Ulrich (1968)	3	51/25/51	0.11(g ,g) 0.16 ( )	All 51		Fig.41(a)	Normal Inductive Meshes
	4	51/25/25/51	0.11(g ,g) 0.16(g ,g)	All 51		Fig.41(a)	
	4	All 102	0.06(a/g) 0.14(b/g)	50/100/50		Fig.41(b)	Inductive Cross Meshes
Holah and Smith (1972)	3	50/25/50		All 42		(Fig. 7 in Ref.)	Double Halfwave Filter
Davis (1980)	4	All 620	0.06(a/g) 0.144(b/g)			Fig.41(c)	Inductive Cross Meshes

(continued)

TABLE V (Continued)

SUMMARY OF MULTIMESH LOWPASS, HIGHPASS AND BANDSTOP FILTERS

(4) Bandstop Filter

Authors [Ref. No.]	Number of Meshes	Mesh Periodicity $g_1/g_2/.../g_i$ (μm)	a/g	Spacings $d_1/d_2/.../d_i$ (μm)	Band Edge (cm^{-1})	Illustrations	Remarks
Ulrich (1968)	2	All 102	0.13(a/g) 0.06(b/g)	38		Fig.42	Capacitive Cross Meshes

* All lowpass filters are composed of normal capacitive meshes.

† All highpass filters are composed of normal inductive meshes.

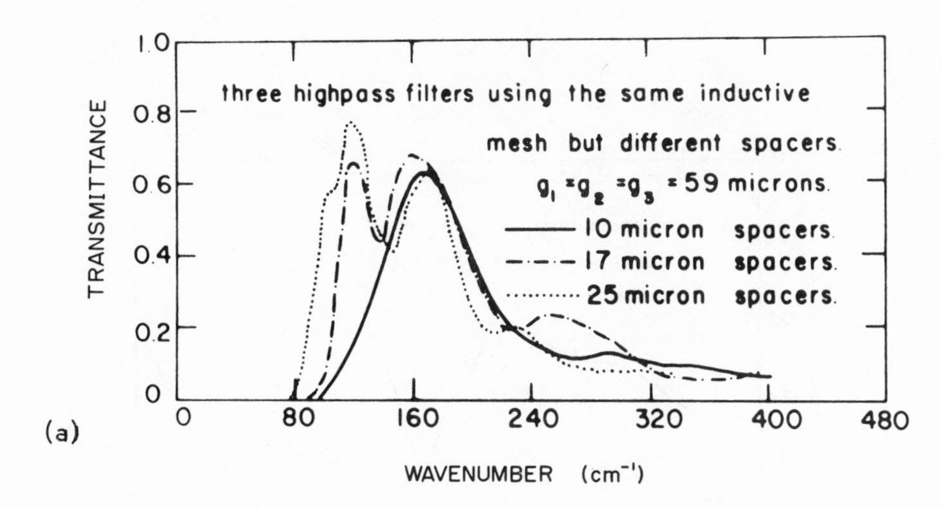

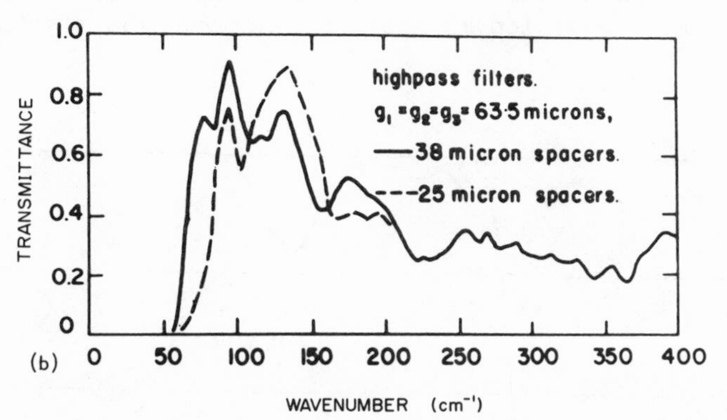

Figure 40 (a) (b). Transmittance of highpass filters. Consisting of three identical and equally-spaced inductive meshes. Parameters of each filter are listed in Table V. After Holah(1980b)

have a rather pseudo-highpass or broad-bandpass characteristic because the transmission falls off gradually above $1/g$ owing to the combined effect of the stopband interference action in the zeroth order and the multiple diffraction. The specifications of this type of filters are also included in Table V.

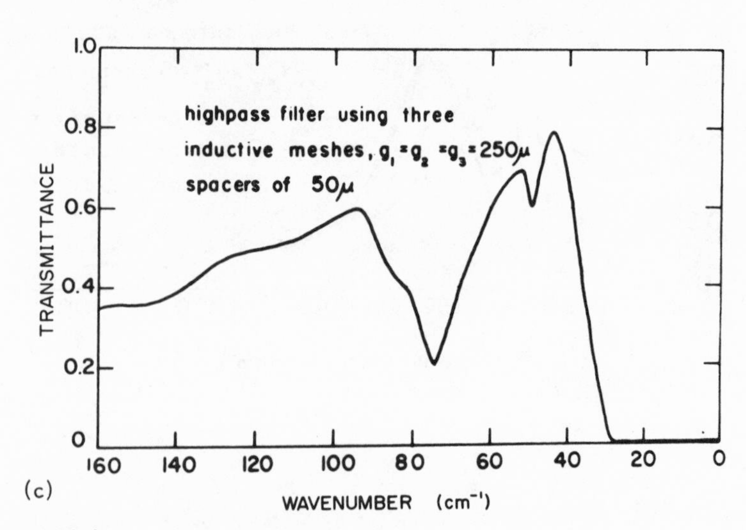

(c)

Figure 40 (c). Transmittance of highpass filters. Consisting of three identical and equally spaced inductive meshes. Parameters of this filter are listed in Table V. After Holah (1980b)

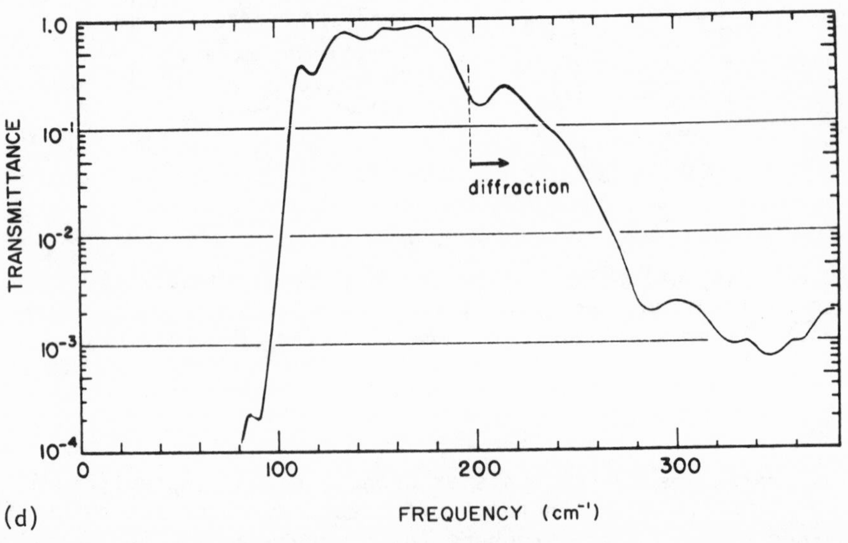

(d)

Figure 40 (d). Transmittance of highpass filters . Consisting of four identical and equally-spaced inductive meshes. Parameters of this filter are listed in Table V. After Ulrich, 1968.

5.2.3. Bandpass Filters

According to the preceding descriptions, narrow-bandpass filters can be fabricated with the simplest two-mesh configuration, and broad-bandpass filters based on the inherent optical property are available. However, broad-bandpass filters with high attenuation in the stop-bands and with a more box-shaped characteristic in the passbands have to be constructed by the multimesh configuration of normal inductive cross-meshes. The performance of three-mesh and four-mesh filters, made of normal inductive meshes, are given in Fig.41 (a) (Ulrich, 1968). The (calculated) transmission of a two-mesh filter with the same half-value (5 cm^{-1}) bandwidth as the three-mesh filter is also shown. In order to form a broad band using only two meshes, inductive meshes with low reflectance must be chosen. This substantially deteriorates the attenuation in the stopbands. Hence, it is necessary to increase the number of meshes. A straightforward design to meet given specifications is possible for this type of filter by the general design procedure for electrical ladder filters. Baldecchi et al. (1974) have made filters having a bandwidth of about 2 cm^{-1} and a rejection factor larger than 10^8, by piling up 20 meshes with the same g value.

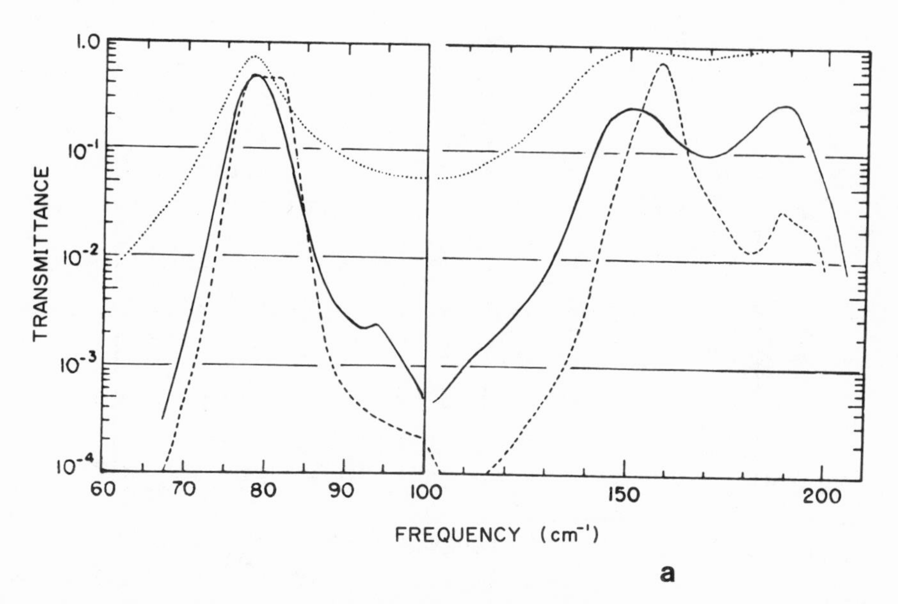

Figure 41 (a). Transmittance of broad bandpass filters: consisting of three (———) or four (----) identical and equally-spaced inductive meshes, or calculated transmittance of a two-mesh filter having bandwidth of 5 cm^{-1} (····)After Ulrich, 1968. Parameters of this filter are listed in Table V.

The bandpass filters made with inductive cross-meshes have been investigated by Ulrich (1968) and by Davis (1980). The dominant optical property of this new structure is that the resonant frequency ω_o, at which the transmission peaks, shifts to 0.5/g instead of 0.85/g for ordinary meshes. Because of the reduction of the resonant frequency to a value far below the onset of diffraction, it is possible to employ this resonance directly for filtering without disturbance by diffraction. The transmittance of a four-mesh construction by Ulrich, together with that of a single inductive cross mesh, is shown in Fig. 41 (b). One main advantage of this type of bandpass filter, in comparison with those of normal inductive meshes, is the complete absence of higher-order responses near two and three times the fundamental passband frequency when the fundamental frequency coincides with the resonant frequency. The performance of the filters constructed by Davis is shown in Fig. 41 (c). The filters are formed of four identical inductive cross-meshes separated by quarter-wave spacings.

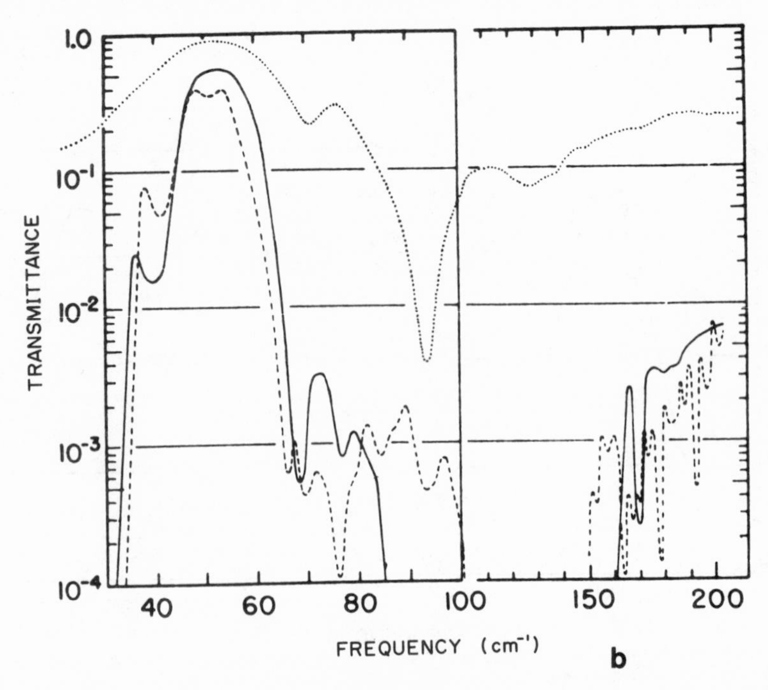

Figure 41 (b). Transmittance of broad bandpass filters:consisting of four identical, equally-spaced inductive cross-meshes, measured at normal incidence (———) or at 27° incidence (- - -), or transmittance of a single inductive cross-mesh (····) Parameters of this filter are listed in Table V. After Ulrich, (1968)

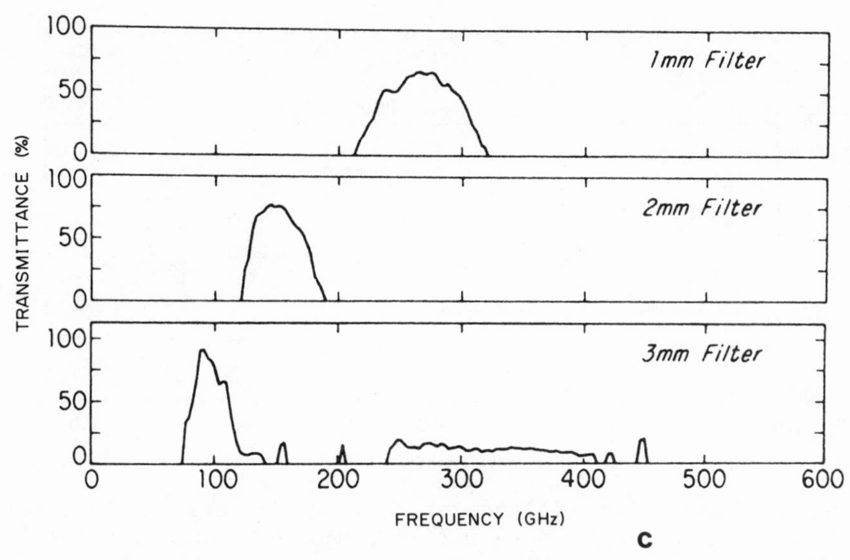

Figure 41 (c). Transmittance of broad bandpass filters: consisting of four identical and equally-spaced inductive cross-meshes. Parameter of this filter are listed in Table V. After Davis (1980).

5.2.4. Bandstop Filters

The capacitive cross-mesh has the possibility to form bandstop filters. As in the case of ordinary meshes, its transmission property is complementary to the inductive cross-mesh. Therefore, the transmittance has a minimum at the resonant frequency. At frequencies sufficiently far away from the resonant frquency, the transmission is fairly transparent on both sides of the resonant frequency. The measured transmittances of first, a single capacitive cross-mesh, and second, filter made from two identical meshes, are shown in Fig. 42 (Ulrich, 1968). Since bandstop filters are rather seldom used, they have not been intensively investigated.

6. APPLICATIONS OF METAL MESH FILTERS AND FABRY-PEROT INTERFEROMETERS

6.1. Applications for Laser Work

The history of the FIR laser (Kneubühl et al., 1980) began in 1962-1964 with electric discharge laser studies of the He-Ne and noble gas-Ne systems by McFarlane, Patel and others, and investigations of the H_2O and HCN systems by Crocker, Gebbie and others. Electric-discharge molecular laser research was actively pursued by

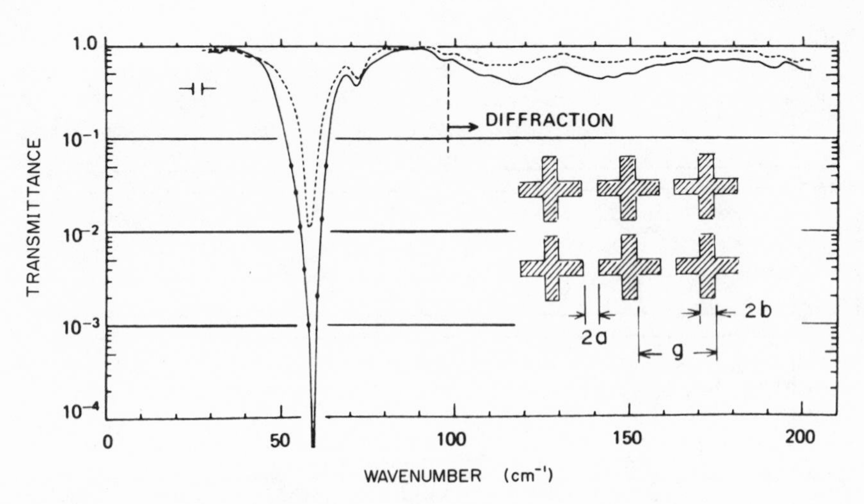

Figure 42. Power transmittance of bandstop filters: - - - single capacitive cross-mesh, ——— filter composed of two identical capacitive cross-meshes. Parameters of the filter are listed in Table V. (After Ulrich, 1968)

some investigators, during the period 1964-1970, aiming at development of a powerful source in the FIR region. In 1970, a new era was opened by Chang and Bridges (Chang et al., 1970) with the invention of the optically pumped FIR lasers. Since then, numerous papers have been reported by a large number of workers, and more than 1000 lines are now reported over the FIR region. However, some workers are still continuing to investigate electric discharge lasers (Kneubühl, 1980; Veran, 1979) such as the HCN laser, the DCN laser, the H_2O laser and so on, for the purpose of obtaining powerful output and for applying them to electron density measurement in plasmas.

The output power of the lasers depends, among other parameters, on the optical properties of the output coupler. For the electric discharge lasers or for the transversely optically pumped lasers, a single inductive mesh is conveniently used (Prettl et al., 1966; Bruneau et al., 1978) instead of an output mirror with a hole pierced in its center. In contrast to hole coupling, the metal-mesh coupler provides uniform coupling over the entire cross-section of the resonator. This reduces the diffraction losses and improves the angular distribution of the output beam. Usually the mesh is effectively used optimizing the power transmission and reflection by selecting the mesh periodicity and the line breadth. In addition to the single metal mesh couplers, Fabry-Perot type couplers (Ulrich et al., 1970; Brown et al.,1974) are also used. By adjusting the spacing between the two parallel meshes, the reflectance and the transmittance can be continuously varied, so that this type is convenient to optimize the coupling conditions. Recently this type of coupler has also been used for FIR free-electron lasers (Shaw et al., 1981). In case of longitudinal optical pumping the simple metal mesh is not applicable because of its low reflectivity in the middle infrared (IR) region. In order to overcome this disadvantage, Danielewicz et al. (1975,1976) developed a hybrid metal-mesh dielectric mirror, the constitution of which is represented in Fig. 43(a). This mirror is highly reflective for the IR pump laser because of the multilayer dielectric coating and the FIR output signal is controlled by the metal mesh. They used Si as a substrate, Ge and CaF_2 as high index and low index materials.

Recently, a similar hybrid mirror has been presented by Durschlag and DeTemple (1981). The structure is shown in Fig. 43 (b). They used ZnS instead of CaF_2 because the reststrahlen absorption of ZnS is considerably lower than in CaF_2 , and because ZnS is not hygroscopic and deposits in a hard smooth layer. In addition, the dielectrics are now evaporated directly onto the substrate. The mesh, either capacitive or inductive, is deposited on top to remove residual FIR loss in the dielectrics. Weitz et al.(1978) reported simpler couplers fabricated by evaporating thin Al on a crystal quartz substrate in either form shown in Fig. 44. Though the mesh periodicity g is comparable to the wavelength of the FIR radiation, it is much greater than the wavelength of pumping radiation and, therefore, the reflectance is reasonably high for the 9~10 μm pumping radiation. This permits the use of a crystal quartz substrate, which

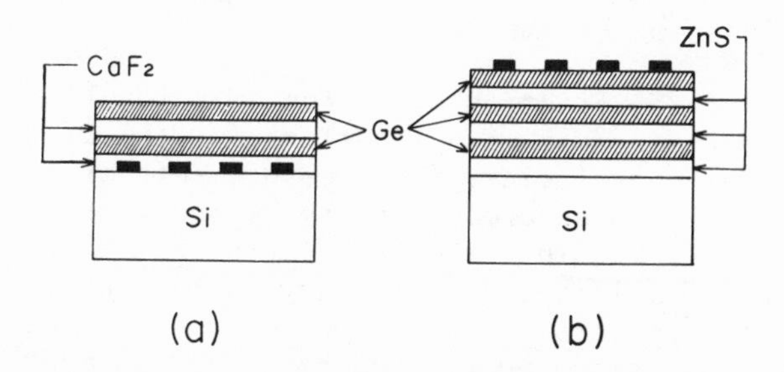

Figure 43. Construction of hybrid metal-mesh multilayer dielectric film mirror. ((a), After Danielewicz et al., 1975. (b) After Durschalag and DeTemple, 1981)

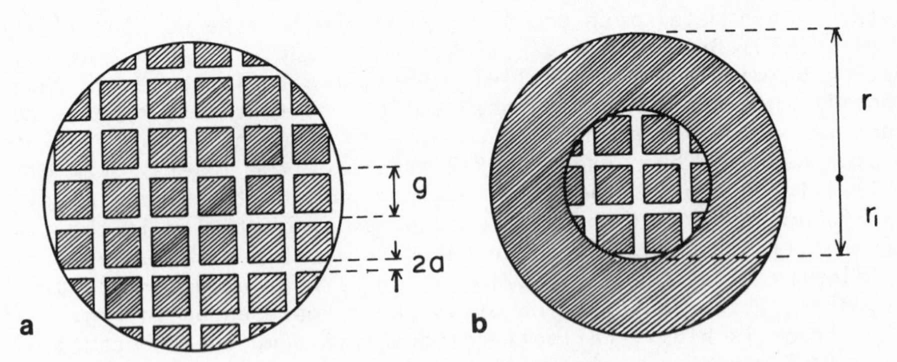

Figure 44. Construction of capacitive-mesh evaporated crystal quartz mirrors: (a) capacitive-mesh output coupler, (b) hybrid capacitive-mesh hole coupler. (After Weitz et al., 1978)

is strongly absorbing at 10 μm, eliminating the necessity of filtering any remaining pump power from the output. It also removes the need for dielectric coating which would cause severe absorption lossess for the wavelengths less than 100 μm. According to them, the structure in Fig. 44 (b) is more effective for lasing at wavelengths less than 100 μm.

The wavelength measurements of FIR lasers are usually performed with an accuracy of about one percent by using the laser resonator itself as a scanning interferometer. The additional use of an external FPI (1974) can increase the accuracy to about one part in 10^4 This FPI works in a very high order, i.e. in an order of several hundreds (1976) to one thousand (Brown et al., 1974), because of the increase of spectral resolution and because there is no need to have a larger free spectral range.

6.2. Applications for Plasma Diagnostics

The study of the electron cyclotron emission (ECE) from present-day tokamak experiments is very important for the diagnostics of their plasmas. In particular, from measurements of the emission it is usually possible to determine the space and time dependence of the electron temperature. Since the ECE occurs on a time scale of the order of a hundred milliseconds, the device must be able to repetitively obtain the spectral information in times of the order of ten milliseconds. Fast scanning FTS has been used on many tokamaks since 1974 when Costley et al. (1974) applied their Martin-Puplett type

FTS to the CLEO tokamak. The FTS offers the whole spectral distribution of the fundamental and higher harmonics of ECE lying in the longer submillimeter and millimeter range. For most tokamaks, the extraordinary mode at the second harmonic of ECE (the electric vector E of the emission is perpendicular to the toroidal field B) is optically thick and the emission from this mode is considered to be that of a blackbody. Hence, this mode is usually used for the determination of the electron temperature and its profile. Using the fast scanning FTS, typical time and frequency resolution are 15 ms and 12 GHz respectively. These limitations essentially come from an electro-mechanical scanning device.

More recently, because of the need to measure some fast and potentially important phenomena, fastscanning FPI's have begun to be used which scan only the second harmonic. The instrument which was originally devised by Blanken, Brossier and Komm (1974; Komm et al., 1975) offers a number of significant advantages for this type of investigation. The apparatus developed by Walker et al.(1981) is shown in Fig. 45. The scanning is carried out repetitively with an electro-mechanical scanning device. It is capable of scanning over several tens of gigaherz in a few milliseconds with a finesse greater than 30. Typically, the frequency resolution of 2.5 ms is obtained (1980). The cyclotron radiation enters the FPI via a hole in the side of the cylinder and a stationary 45° mirror located inside the cylinder. Generally the FPI needs a lowpass filter to reject high orders. The filter is placed behind the apparatus.

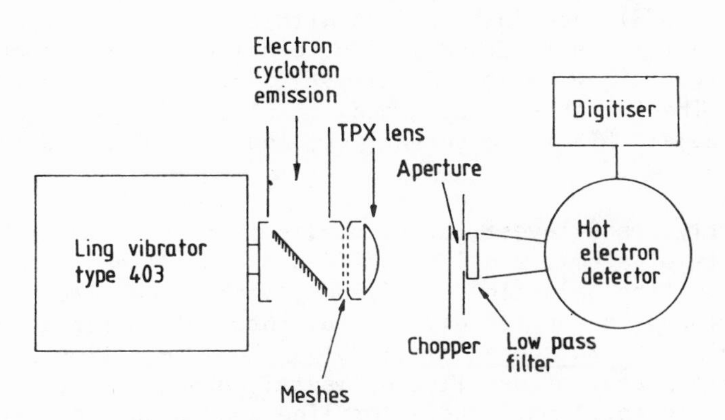

Figure 45. Schematic diagram of fast-scanning FPI for plasma diagnostics. (After Walker et al., 1981)

As a filter, the multimesh lowpass type is currently in use.

When a FPI is used in a static mode, it is possible to trace the time behavior of ECE more precisely at the sacrifice of spectral information. The time resolution is increased up to the value determined by the detector response time, which is ~0.2 μs for the InSb hot electron detector. Aside from the FPI, the bandpass filter also has been applied (1980) for such a purpose.

6.3. Applications for Far Infrared Astronomy

The investigation of FIR astronomy has begun with the discovery of FIR sources in the direction of the galactic center by Hoffmann and Frederic (1971) in 1971. From the survey of a portion of the galactic plane at a wavelength of about 100 μm, they found as many as 72 FIR sources. These sources seem to originate in the thermal emission from interstellar dust heated by the light of stars or other energy sources. The observations of these sources were made in a broad band or board a balloon, to avoid the background radiation using a filter system including a broad bandpass filter based on the inherent transmission property of an inductive mesh. Spectroscopic observations of the broad band sources were made by some investigators who used a Michelson FTS (Erikson et al.,1973) or a grating spectrometer (Ward et al., 1974; Harper, 1974) in 1973 and 1974. Brandshaft et al.(1975) followed them in 1975, with a scanning FPI having an instrumental resolution of 6 μm at wavelengths from 80 to 135 μm.

On the other hand, at the same time, the measurements of cosmic background radiation has become an interesting subject for FIR astronomy (Muehlner et al., 1970; Mather et al., 1971) Muehlner and Weiss (1970.1973) made observations with a system including lowpass interference filters immersed in liquid Helium. Their apparatus is shown in Fig. 46. With this instrument, they obtained the value of 2.7 K. The Richards group (Woody et al., 1975, 1979) newly used a Martin-Puplett FTS for observations, and they obtained the value of 2.96 K.

Recently, many investigators became interested in high resolution measurements of atomic and ionic fine structure lines from galactic H II regions, the galactic center and planetary nebulae. These fine structure lines in the FIR give useful information for diagnosing physical plasmas (Watson et al.,1980). The line ratios are sensitive and accurate probes of density, elemental abundances, and ionization structure. In addition, studies of line shapes and Doppler shifts are valuable in studies of the gross dynamics of H II regions and galactic center. Among several investigating groups (Watson et al., 1980) who make use of Michelson FTS, grating spectrometers or heterodyne systems, the groups at Berkeley (Storey et al.,1980) and at CNRS (Lecullier et al.,1976; Belland et al., 1980; Chanin et al., 1978) utilize tandem FPI systems, similar with each other, because

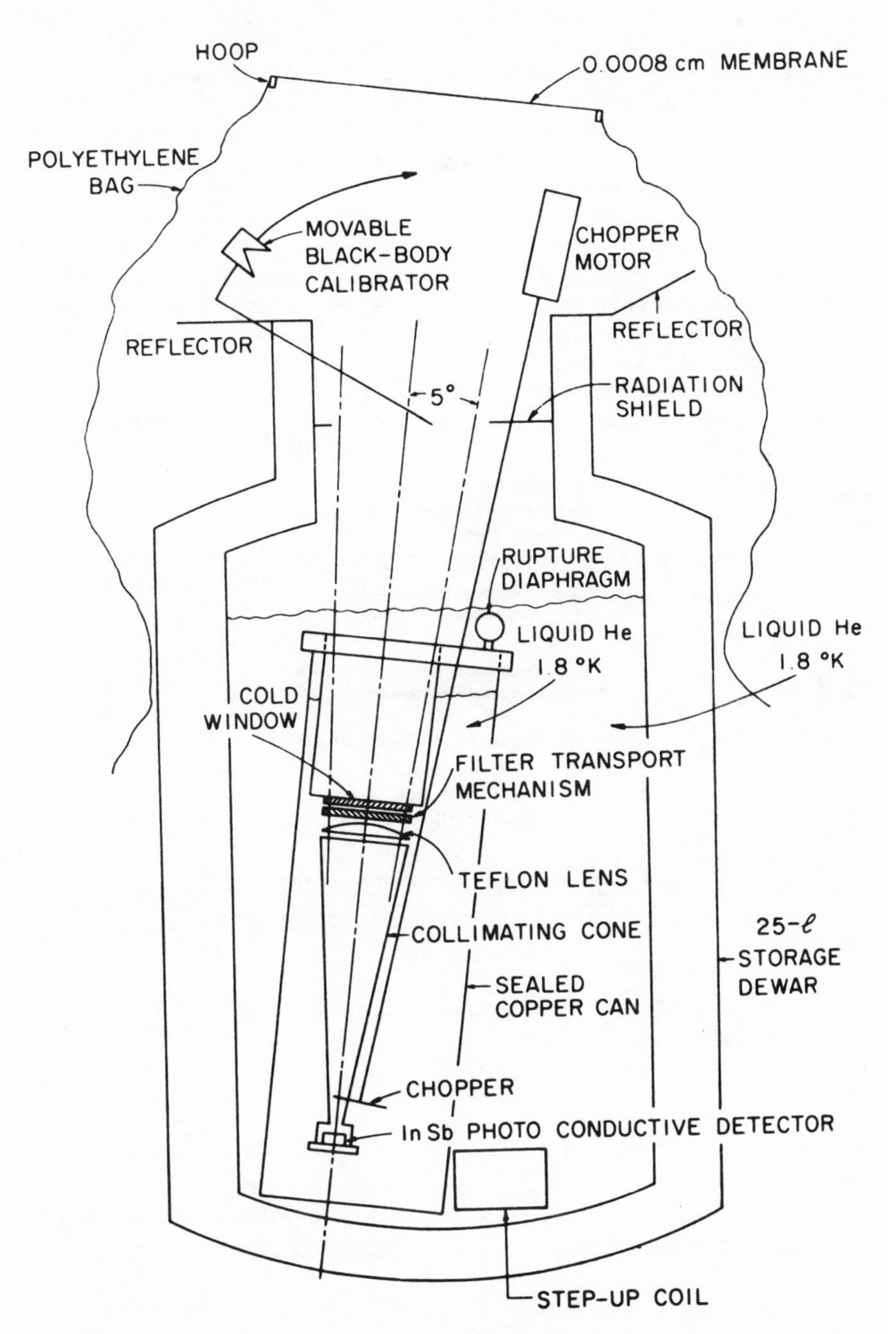

Figure 46. Schematic diagram of balloon-borne radiometer for cosmic background measurements. (After Muehlner and Weiss, 1970, 1973)

of the Jacquinot advantage without Fellgett advantage (Fellgett, 1967) (in the present case, a narrow band effectively reduces the thermal background radiation) and because of its compactness. The instrument used by the group at Berkeley (Storey et al., 1980) is shown in Fig. 47, in which a scanning FPI in high order is used together with a fixed Fabry-Perot interference filter to isolate a single order of the scanning FPI. This spectrometer covers the spectral range between 50 and 200 μm with a resolution above 10^3. Both for the scanning FPI and for the fixed interference filter, the inductive mesh is used and the scanning of the FPI is performed by use of a piezoelectric tube. From the measurement of the transmission profile of a methanol laser at 118.8 μm, it was found that the scanning FPI achieved a finesse as high as 133. The fixed interference filters are mounted on a wheel to allow a choice of several operating wavelength regions during the course of a single flight with this airborne instrument. Though the scanning FPI is operated at room

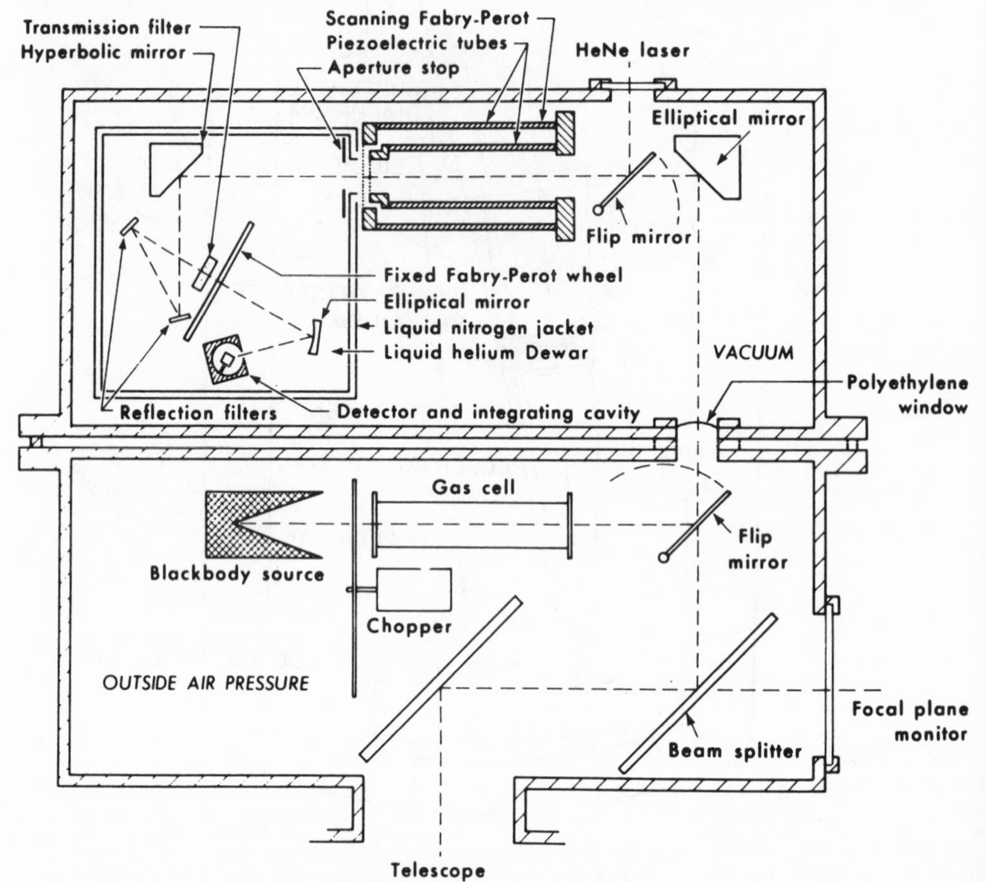

Figure 47. Schematic diagram of airborne FPI for FIR astronomy. (After Storey et al.,1980)

temperature, the fixed filters and the whole optics up to the detector operate at liquid helium temperature by placing them inside a helium cooled chamber. They use this spectrometer for airborne observations, and detected the [O III] 88.35 μm and the [O I] 63.17 μm lines at the first flight (1979) (Storey et al., 1979). In succession, the [O III] 51.8 μm and the [N III] 57.30 μm lines (Watson et al., 1979) strong emissions from the rotational transitions of CO at 124.2 μm and 118.6 μm (Watson et al.,1980) and strong absorption lines of OH at 119.23 μm and 119.44 μm (Storey et al., 1981) have been observed with a resolving power ranging from 200 to 4000. A summary of observations made to date by all of the groups can be seen in Ref. 94.

In astrophysical investigations, almost all cases need He-cooled filters of 4.2 K or below to reduce the instrumental radiation loading the sensitive detectors. This demands, therefore, temperature-stable bandpass (Holah et al., 1979) and lowpass (Timusk et al.,1981; Holah et al., 1974) filters.

7. CONCLUSIONS

Metal-mesh filters and corresponding FPI's act as powerful tools for many applications in the FIR region today. The advent of the capacitive mesh, in addition to the use of the inductive mesh, has further extended the usefulness of meshes. Plenty of experimental studies on the FIR properties of meshes have been made since the initial stage of the investigations. Theoretical approaches to describe the optical properties also have been carried out. The equivalent circuit and transmission line model can be successfully and conveniently used in the diffraction-free region but fails in the region where the diffraction appears. A recent approach based on the method of harmonic expansion and field-matching by Chen (1973) and Lee (1971) or McPhedran and Maystre (1977) describes the experimental results fairly well for both the diffraction-free and the diffraction regions. With the increase of applications, some refined or sophisticated designs for FPI's and for interference filters have been presented. Although the applications sometimes require that the filters are operated at liquid helium temperature, filters satisfying such requirements are now available without deterioration of the performance. The FPI, as a monochromator, has the optical throughput gain but has not the multiplex gain. Because of this property the FPI seems disadvantageous, in general, compared to FTS which has both advantages. However, the special property of FPI is more effectively used when background radiation disturbs the measurements or in cases where a rather small wavelength region is to be investigated. Further progress in this field may be achieved with the development of meshes having new structures, and therefore, having new optical properties which can be matched to the special problem.

ACKNOWLEDGEMENT

The authors are grateful to Prof. H. Okuda at the Institute of Space and Astronautical Science for useful suggestions in reviewing the applications of metal mesh filters and FPI to the astronomy.

REFERENCES

Baker, E.A.M., and Walker, B., 1980, Measurements of electron cyclotron emission with Fabry-Perot interferometers, Proceedings of Joint Workshop on ECE and ECRH, Oxford.

Balakhanov, V. Ya., 1966, Properties of a Fabry-Perot interferometer with mirrors in the form of a backed metal grid, Sov. Phys. Doklady, 10:788-790.

Baldecchi, M.G., Dall'Oglio, G., Melchiorri, B., Melchiorri, F., and Natale, V., 1974, Narrow band filters for astronomical applications in the far infrared, Infrared Phys., 14:343-345.

Bell, E.E., and Russell, E.E., 1966a, Measurement of the far infrared optical properties of solids with a Michelson interferometer used in the asymmetric mode: Part II, the vacuum interferometer, Infrared Phys., 6:75-84.

Bell, E.E.1966b, Measurement of the far infrared optical properties of solids with a Michelson interferometer used in the asymmetric mode: part I, mathematical formulation, Infrared Phys., 6:57-74.

Belland, P., and Lecullier, J.C., 1980, Scanning Fabry-Perot interferometer: performance and optimum use in the far infrared range Appl. Opt., 19:1946-1952.

Beunen, J.A., Costley, A.E., Neill, G.F., Mok, C.L., Parker, T.J., and Tait, G., 1981, Performance of free-standing grids wound from 10- μm diameter tungsten wire at submillimeter wavelengths: computation and measurement, J.Opt.Soc. Am., 71:184-188.

Blanken, R.A., Brossier, P., and Komm, D.S., 1974, A Technique to measure submillimeter tokamak synchrotron radiation spectra, IEEE Trans. Microwave Theory Tech. MTT-22: 1057-1060.

Brandshaft, D., McLaren, R.A., and Werner, M.W., 1975, Spectroscopy of the Orion nebula from 80 to 135 microns, Astrophys.J.,199: L115-L117.

Brown, F., Kronheim, S., and Silver, E., 1974, Tunable far infrared methyl fluoride laser using transverse optical pumping, Appl. Phys. Lett., 25:394-396

Bruneau, J.L., Belland, P., and Veron, D., 1978, A cw DCN waveguide laser of high volumetric efficiency, Opt. Commun., 24:259-264

Casey, Jr. J.P., and Lewis, E.A., 1952, Interferometer action of a parallel pair of wire gratings, J.Opt. Soc. Am., 42:971-977.

Chang, T.Y., 1974, Optically pumped submillimeter-wave sources, IEEE Trans. Microwave Theory Tech., MTT-22:983-988.

Chang, T.Y., and Bridges, T.J., 1970, Laser action at 452, 496, and 541 μm in optically pumped CH_3F, Opt. Commun., 1:423-426

Chanin, G., and Lecullier, J.C., 1978, A scanning Fabry-Perot interferometer, Infrared Phys., 18:589-594.

Chen, C.C., 1970, Transmission through a conducting screen perforated periodically with apertures, IEEE Trans. Microwave Theory Tech., MTT-18:627-632.

Chen, C.C., 1971, Diffraction of electromagnetic waves by a conducting screen perforated periodically with circular holes, MTT-19:475-481.

Chen, C.C., 1973, Transmission of microwave through perforated flat plates of finite thickness, MTT-21:1-6.

Costley, A. E., Hastie, R., Paul, J.W.M., and Chamberlain, J., 1975, Electron Cyclotron emission from a tokamak plasma: experiment and theory, Phys. Rev. Lett., 33:758-761.

Culshaw, W., 1959, Reflectors for a microwave Fabry-Perot interferometer, IRE Trans. Microwave Theory Tech., MTT-7:221-228.

Culshaw, W., 1960, High resolution millimeter wave Fabry-Perot interferometer, IRE Trans. Microwave Theory Tech.,MTT-8:182-189.

Danielewicz, E.J., and Coleman, P.D., 1976, Hybrid metal mesh-dielectric mirrors for optically pumped far infrared lasers, Appl. Opt., 15:761-766.

Danielewicz, E.J., Plant, T.K., and DeTemple, T.A., 1975, Hybrid output mirror for optically pumped far infrared lasers, Opt. Commun., 13:366-369.

Davies, J.B., 1973, A least-squares boundary residual method for the numerical solutions of scattering problems, IEEE Trans. Microwave Theory Tech., MTT-21:99-104.

Davis, J.E., 1980, Bandpass interference filters for very far infrared astronomy, Infrared Phys., 20:287-290.

Durschlag, M.S., and DeTemple, T.A., 1981, Far-IR optical properties of freestanding and dielectrically backed metal meshes, Appl. Opt., 20:1245-1253.

Erickson, E.F., Swift, C.D., Witteborn, F.C., Mord, A.J., Augason, G. Caroff, L.J., Kunz, L.W., and Giver, L.P., 1973, Infrared spectrum of the Orion nebula between 55 and 200 microns, Astrophys.J 183:535-539.

Evans, D.E., James, B.W., Peebles, W.A.,and Sharp, L.E., 1976, Spectral composition of far-infrared laser radiation optically excited in methyl fluoride, Infrared Phys., 16:193-195.

Fellgett, P., 1967, Conclusions on multiplex methods, J. Phys., 28: C2 165-171.

Garg, R.K., and Pradhan, M.M., 1978, Far-Infrared characteristics of multi-element interference filters using different grids, Infrared Phys., 18:292-298.

Genzel, L., and Sakai, K., 1977, Interferometry from 1950 to the present, J. Opt. Soc. Am., 67:871-879.

Genzel, L., and Weber, R., 1958, Spektroskipie im fernen Ultrarot durch Interferenz-Modulation, Z. angew. Phys., 10:195-199.

Genzel, L., Chandrasekhar, H.R., and Kuhl, J., 1976, Double-beam Fourier spectroscopy with two inputs and two outputs, Opt. Comm. 18:381-386.

Gornik, E., Müller, W., and Gaderen, F., 1976, Optimization of a tunable far-infrared InSb source, Infrared Phys., 16:109-115

Grenier, P., Langlet, A., Talureau B., and Coron, N., 1973, Photomete for Submillimeter measurements, Appl. Opt., 12:2863-2868.

Grisar, R.G.J., Reiners, K.P., Renk, K.F., and Genzel, L., 1967, Impurity-induced far-infrared absorption in the phonon gap regions of KI and Kbr, Phys. stat. sol., 23:613-620.

Harper, D.H., 1974, Far-infrared emission from H II regions. Multicolor photometry of selected sources and 2.2 resolution maps of M42 and NGC 2024, Astrophys. J., 192:557-571.

Hoffmann, W.F., Frederick, C.L., and Emery R.J., 1971, 100-micron map of the galactic-center region, Astrophys. J., 164:L23-L28.

Hoffmann, W.F., Frederick, C.L., and Emery R.J., 1971, 100-micron survey of the galactic plane, Astrophys. J., 170:L89-L97.

Holah, G.D., 1980a, Very longwavelength lowpass interference filters, Int. J. IR and MM Waves, 1:225-234.

Holah, G.D., 1980b, High-frequency-pass metallic mesh interference filters, Int. J. IR and MM Waves, 1:235-245.

Holah, G.D., and Auton, J.P., 1974, Interference filters for the far infrared, Infrared Phys., 14:217-229.

Holah, G.D., and Morrison, N., 1977, Narrow-bandpass interference filters for the far infrared, J. Opt. Soc. Am., 67:971-974.

Holah, G.D., and Smith, S.D., 1972, J. Phys. D :Appl. Phys., 5:496-509.

Holah, G.D., and Smith S.D., 1977, Far infrared interference filters, J. Phys. E., 10:101-111.

Holah, G.D., Morrison, N.D., and Goebel, J.G., 1979, Low temperature narrowbandpass mesh interference filters, Appl. Opt., 18:3526-3532.

Jacquinot, P., 1954, The luminosity of spectrometers with prisms, grating or Fabry-Perot etalons, J.Opt. Soc. Am., 44:761-765.

Kawahata, K., Sato, M., Yoshida, T., and Sakai, K., 1980, Measurement of electron cyclotron emission by mesh filter method, Jpn. J. Appl. Phys., 19:1003-1004.

Keilmann, F., 1981, Infrared high-pass filter with high contrast, Int. J. IR and MM Waves, 2:259-272.

Kneubühl, K.F., and Sturzenegger, Ch., 1980, Electrically Excited Submillimeter-Wave Lasers, in "Infrared and Millimeter Waves" (K.J.Button, ed.) Vol.3 (Academic Press, New York)

Komm, D.S., Blanken, R.A., and Brossier, P., 1975, Fast-scanning far-infrared Fabry-Perot interferometer, Appl.Opt., 14:460-464.

Lamarre, J.M., Coron, N., Courtin, R., Dambier, G., and Charra, M., 1981, Metallic mesh properties and design of submillimeter filters, Int. J. IR and MM Waves, 2:273-292.

Larsen, T., 1962, A survey of the theory of wire grids, IRE Trans. Microwave Theory Tech., MTT-10:191-201.

Lecullier, J.C., and Chanin, G., 1976, A scanning Fabry-Perot interferometer for the 50-1000 μm range, Infrared Phys., 16:273-278

Lee, S.W., 1971, Scattering by dielectric-loaded screen, IEEE Trans. Antennas Propagat., AP-19:656-665.

Lewis, E.A., and Casey, J.P., 1951, Metal grid interference filter, J.Opt. Soc. Am., 41:360.

Lewis, E.A., Casey, Jr. J.P., 1952, Electromagnetic reflection and transmission by gratings of resistive wires, J. Appl. Phys. 23:605-608.

Marcuvitz, N., 1951, "Waveguide Handbook", McGraw-Hill, New York.

Mather, J.C., Werner, M.W., and Richards, P.L., 1971, A search for spectral features in the submillimeter background radiation, Astrophys. J., 170:L59-L65.

Matthaei, G.L., Young, L., and Jones, E.M.T., 1964, "Microwave Filters, Impedance-Matching Networks and Coupling Structures" McGraw-Hill, New York.

McPhedran, R.C., and Maystre, D., 1977, On the theory and solar application of inductive grids, Appl. Phys., 14:1-20.

Mitsuishi, A., Ohtsuka, Y., Fujita S., and Yoshinaga, H., 1963, Metal mesh filters in the far infrared region, Jpn. J. Appl. Phys., 2: 574-577.

Möller, K.D., and Rothschild, W.G., 1971, "Far-Infrared Spectroscopy 90-127 Wiley-Interscience , New York.

Muehlner, D., and Weiss, R., 1970, Measurement of the isotropic background radiation in the far infrared, Phys. Rev. Lett., 24:742-746.

Muehlner, D., and Weiss, R., 1973, Balloon measurements of the far-infrared background radiation, Phys. Rev. D, 7:326-344.

Pradhan, M.M., 1970, Reciprocal grids as low pass transmission filters for the far infrared region, Infrared Phys., 10:199-206.

Pradhan, M.M., 1971, Multigrid interference filters for the far infrared region, Infrared Phys., 11: 241-245.

Pradhan, M.M., and Garg, R.K., 1976, Five-and six-grid interference filters for far infrared, Infrared Phys., 16:449-452.

Pradhan, M.M., and Garg, R.K., 1977, Far infrared transmission characteristics of multielement reciprocal grid interference filters, Infrared Phys., 17:253-256

Pradhan, M.M., Gerbaux, X., Hadni, A., and Chanin, G., 1972, Narrow band metallic grid filters for the very far infrared, Infrared Phys., 12:263-266.

Prettl, W., and Genzel, L., 1966, Notes on the submillimeter laser emission from cyanic compounds, Phys. Lett., 23:443-444.

Rawcliffe, R.D., and Randall, C.M., 1967, Metal mesh interference filters for the far infrared, Appl. Opt., 6:1353-1358.

Renk, K.F., and Genzel, L., 1962, Interference filters and Fabry-Perot interferometers for the far infrared, Appl. Opt., 1:643-648

Ressler, G.M., and Möller, K.D., 1967, Far infrared bandpass filters and measurements on a reciprocal grid, Appl. Opt., 6:893-896.

Romero, H.V., and Blair, A.G., 1973, Bandpass filters for the 100-cm^{-1} region using metallic mesh grids, Appl. Opt., 12: 84-86.

Romero, H.V., Gursky, J., and Blair, A.G., 1972, Far infrared filters for a rocket-borne radiometer, Appl. Opt., 11:873-880.

Sakai, K., Fukui, T., Tsunawaki, Y., and Yoshinaga, H., 1969, Metallic mesh bandpass filters and Fabry-Perot interferometer for the far infrared, Jpn. J. Appl. Phys., 8:1046-1055.

Sakai, K., and Yoshida, T., 1978, Single mesh narrow bandpass filters from the infrared to the submillimeter region, Infrared Phys., 18:137-140.

Sakai, K., and Yoshida, T., an analysis of the data at hand including Fig. 8(a).

Saksena, B.D., Pahwa, D.R., Pradhan, M.M., and Lal, K., 1969, Reflection and transmission characteristics of wire gratings in the far infrared, Infrared Phys., 9:43-52.

Selby, M.J., Jorden, P.R., and MacGregor, A.D., 1976, A helium cooled Fabry-Perot interferometer for infrared astronomical spectroscopy, Infrared Phys., 16:317-323.

Shaw, E.D., and Patel, C.K.N., 1981, Use of intracavity filters for optimization of far-infrared free-electron lasers, Phys. Rev. Lett., 46:332-335.

Storey, J.W.V., Watson, D.M., and Townes, C.H., 1979, Observations of far-infrared fine structure lines: [O III] 88.35 microns and [O I] 63.2 microns, Astrophys. J., 233:109-118.

Storey, J.W.V., Watson, D.M., and Townes, C.H., 1980, An airborne far-infrared spectrometer for astronomical observations, Int. J. IR and MM Waves, 1:15-25.

Storey, J.W.V., Watson, D.M., and Townes, C.H., 1981, Detection of interstellar OH in the far-infrared, Astrophys. J., 244:L27-L30.

Timusk, T., and Richards, P.L., 1981, Near millimeter wave bandpass filters, Appl. Opt., 20:1355-1360.

Tomaselli, V.P., Edewaard, D.C., Gillan, P., and Möller, K.D., 1981, Far-infrared bandpass filters from cross-shaped grids, Appl. Opt. 20:1361-1366.

Ulrich, R., 1967a, Far-infrared properties of metallic mesh and its complementary structure, Infrared Phys., 7:37-55.

Ulrich, R., 1967b, Effective low-pass filters for far infrared frequencies, Infrared Phys., 7:65-74.

Ulrich, R., 1968, Interference filters for the far infrared, Appl.Opt. 7:1987-1996.

Ulrich, R., 1969, Preparation of grids for far infrared filters, Appl. Opt., 8:319-322.

Ulrich, R., Renk, K.F., and Genzel, L., 1963, Tunable submillimeter interferometers of the Fabry-Perot type, IEEE Trans. Microwave Theory Tech., MTT-11:363-371.

Ulrich, R., Bridges, T.J., and Pollack, M.A., 1970, Variable metal mesh coupler for far infrared lasers, Appl. Opt., 9:2511-2516.

Varma, S.P., and Möller, K.D., 1969a, Far infrared band-pass filters in the 400 - 16 cm^{-1} spectral region, Appl. Opt., 8:2151-2152.

Varma, S.P., and Möller, K.D., 1969b, Far infrared interference filters, Appl. Opt., 8:1663-1666.

Veron, D., 1979, Submillimeter Interferometry of High-Density Plasmas, in "Infrared and Millimeter Waves" (K.J.Button, ed.), Vol.2, (Academic Press, New York.

Vogel, P., and Genzel, L., 1964, Transmission and reflection of metallic mesh in the far infrared, Infrared Phys., 4:257-262.

Walker, B., Baker, E.A.M., and Costley, A.E., 1981, A Fabry-Perot interferometer for plasma diagnostics, J. Phys. E; Sci. Instrum., 14:832-837.

Ward, D.B., and Harwit, M., 1974, Observations of the Orion nebula at 100 μm, Nature, 252:27.

Watson, D.M., and Storey, J.W.V., 1980, Far infrared fine structure lines in the interstellar medium, Int. J. IR and MM Waves, 1:609-629.

Watson, D.M., Storey, J.W.V., and Townes, C.H., 1979, Observations of far-infrared fine structure lines in the interstellar medium, Digest of 4th International Conference of IR and MM Waves and their Applications (Miami, Dec.1979), 175

Watson, D.M., Storey, J.W.V., and Townes, C.H., 1980, Detection of CO J=21 →20 (124.2 μm) and J=22→ 21 (118.6 μm) emission from the Orion Nebula, Astrophys. J., 239:L129-L132.

Weitz, D.A., Skocpol, W.J., and Tinkham, M., 1978, Capacitive-mesh output couplers for optically pumped far-infrared lasers, Opt. Lett., 3:13-15.

Whitcomb, S.E., and Keene, J., 1980, Low-pass interference filters for submillimeter astronomy, Appl. Opt., 19:197-198.

Woody, D.P., and Richards, P.L., 1979, Spectrum of the cosmic background radiation, Phys. Rev. Lett., 42:925-929.

Woody, D.P., Mather, J.C., Nishioka, N.S.,and Richards, P.L., 1975, Measurement of the spectrum of the submillimeter cosmic background, Phys. Rev. Lett., 34:1036-1039.

Yamada, Y., Mitsuishi, A., and Yoshinaga, H., 1962, Transmission filters in the far-infrared region, J. Opt. Soc. Am., 52:17-19.

Yoshinaga, H., 1973, in "Progress in Optics" (E. Wolf ed.) XI:77-122. North-Holland Publishing Co., Amsterdam.

MAGNETIC RESONANCES IN PEROVSKITE-TYPE LAYER STRUCTURES

Reinhart Geick and Karlheinz Strobel

Physikalisches Institut der Universität Würzburg
Röntgenring 8, (8700) Würzburg
Fed. Rep. of Germany

1. INTRODUCTION

Halide perovskite-type layer structures have obtained their name from the fact that they have structural units, namely transition metal halogen octahedra, in common with the cubic perovskite structure. There, the octahedra form a three-dimensional corner-sharing network. In the layer structure, the octahedra are linked together to corner-sharing two-dimensional arrays (cf. Fig. 1). Materials in this class with paramagnetic divalent ions in the center of the octahedra are good models for quasi-two-dimensional magnetism (de Jongh et al., 1974). The two-dimensional character of these compounds arises mainly from the fact that the intra-layer exchange interaction is the dominant one, in particular the interaction between nearest neighbors. The exchange interaction between further neighbors in the layers or between adjacent layers (cf.Fig.1) is usually smaller by orders of magnitude (Skalyo et al., 1969; Birgeneau et al., 1970a; Hirakawa et al., 1973; Birgeneau et al., 1973; de Wijn et al., 1973a; Hutchings et al., 1976; Day, 1979; Schröder et al., 1980). In many cases, the anisotropy energy is also rather small, and the predominant part in the Hamiltonian is the isotropic Heisenberg term. These materials are often called quasi-two-dimensional ferro- or antiferromagnets, although the three-dimensional long range order can only originate from the anisotropy or the interlayer exhange interaction (Mermin et al.,1966). As already mentioned, there may be found ferro- and antiferromagnetic order in the perovskite-type layer structures. In Table 1, we have compiled examples of ferromagnetic, antiferromagnetic and nonmagnetic materials of the class under consideration. From the examples in Table 1, it will be noted that the inorganic alkali ion K^+ and Rb^+ may be replaced by molecular ions like $(C_nH_{2n+1}NH_3)^+$

and $(NH_3(CH_2)_mNH_3)^{2+}$ which leads to a larger separation of the layers. In Fig. 2, we show some perovskite-type layer structures with organic ions together with Rb_2MnCl_4, for comparison. In most cases, the organic ions invoke structural phase transitions on these materials due to orientation or reorientation of the molecular ions. The actual structures at low temperatures are therefore of orthorhombic or monoclinic symmetry. The octahedra remain regular, but

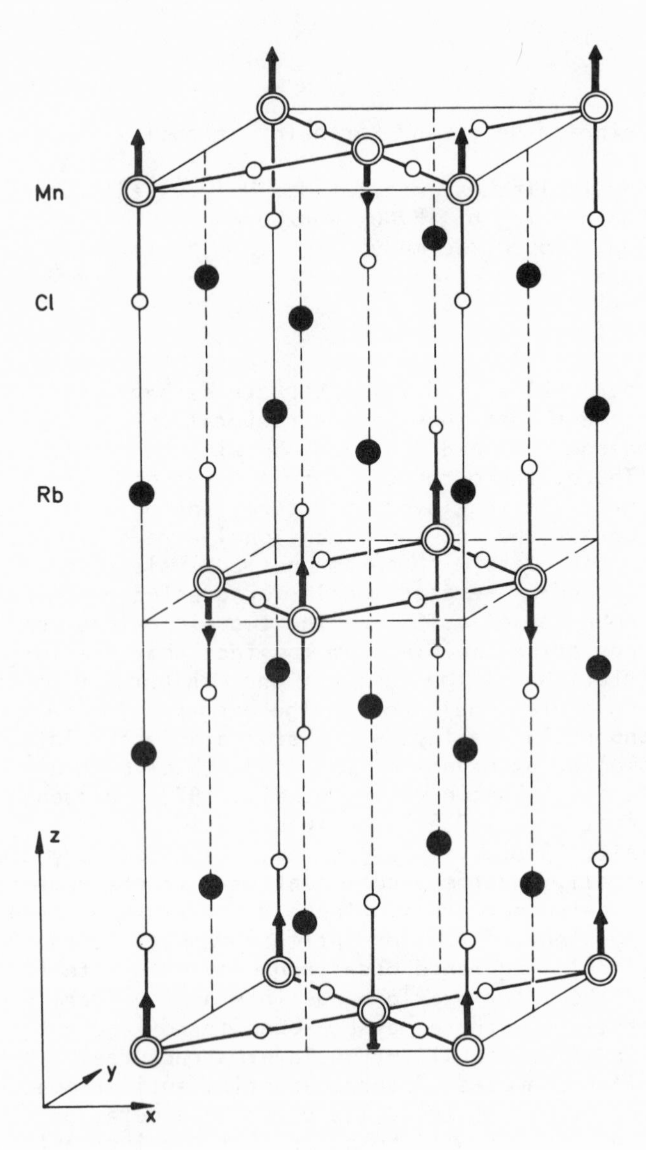

Figure 1. Crystallographic and magnetic structure of Rb_2MnCl_4.

TABLE 1. Examples of perovskite-type layer structures

Ferromagnets	K_2CuF_4, Rb_2CrCl_4, $(CH_3NH_3)_2CuCl_4$ $(C_3H_7NH_3)_2CuCl_4$
Antiferromagnets	K_2NiF_4, K_2MnF_4, Rb_2MnF_4, Rb_2MnCl_4, $(C_nH_{2n+1}NH_3)_2MnCl_4$, $NH_3(CH_2)_mNH_3MnCl4$, $(C_2H_5NH_3)_2CuCl_4$
Nonmagnetic	K_2ZnF_4, Rb_2CdCl_4

are canted and rotated. For reasons of more clarity, we have not shown these details of the structures in Fig. 2. The structures are given there in a simplified and approximate way with the tetragonal symmetry of the parent K_2NiF_4 structure (cf. Fig. 1). The details of the structures under consideration are given in Table 4 in section 4.3. In our context of magnetic resonances, we are interested in the magnetic properties and not so much in the structural phase transitions and in lattice modes which eventually soften near one of the transitions. And we need not consider the phase transitions in detail since the magnetic ions show only very small distortions from a tetragonal lattice in all cases.

The basic aim of our experiments is to obtain information and experimental data about the $\vec{q} = 0$ magnetic excitations. In ordered materials, these magnons or spin waves are the elementary excitations of the spin system above its ground state which depends on the magnetic order. For the measurement of $\vec{q} = 0$ magnetic excitations it is most suitable to perform spectroscopic experiments. For K_2NiF_4, these have been performed by means of Fourier transform spectroscopy (Birgeneau et al., 1970b), and for manganese compounds by means of magnetic resonance techniques in the microwave range (de Wijn et al., 1973b). These optical methods yield generally more accurate information about the anisotropy field and are thus suited for

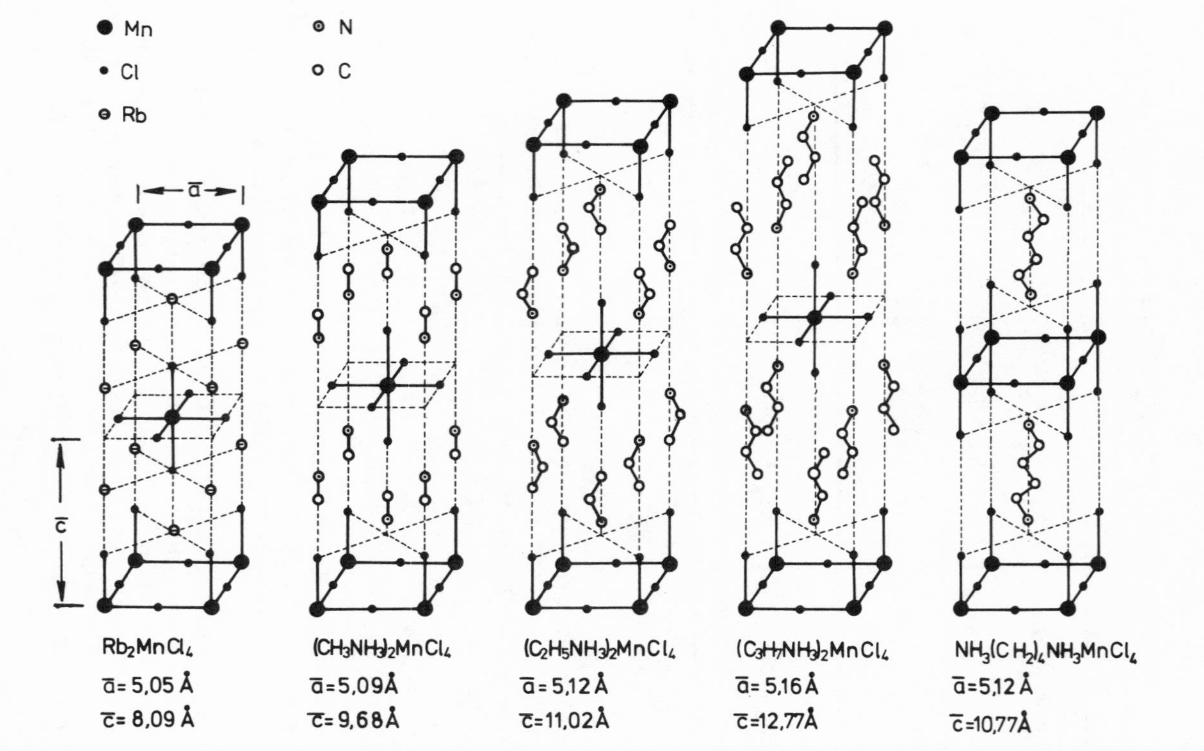

Figure 2. Comparison of the perovskite-type layer structures Rb_2MnCl_4, $(C_nH_{2n+1}NH_3)_2MnCl_4$ with n=1,2,3 and $NH_3C_2H_4NH_3MnCl_4$. $\vec{a}$ and $\vec{c}$ are the average Mn-Mn distance within the layers and the layer separation, respectively. For further details see text, in particular Table 4.

investigations of the magnetic structure, in addition to neutron diffraction experiments. Neutron inelastic scattering, on the other hand, can be used to study the elementary excitations, i.e. the spin waves, over the whole Brillouin zone and thus to provide information about the exchange interaction. We have performed magnetic resonance experiments in the millimeter-wave range where the experimental technique is a mixture of optical and microwave techniques. For the analysis of our experimental data, we have to include electromagnetic propagation of polariton effects as will be explained in detail below. Therefore, our experimental data provide for the first time, to the best of our knowledge, clear experimental evidence for these effects which have been predicted theoretically for ferro- as well as for antiferromagnets(Auld, 1960; Pincus, 1962; Akhiezer et al., 1968; Manohar et al., 1972; Bose et al., 1975; Sarmento et al., 1976a,b,c). The effects of the interaction of a photon with elementary excitations are well known from the study of infrared active phonons and from the "reststrahlenband" in the reflection spectra of such crystals (Born et al., 1954). The excitation of a phonon occurs by means of an electical dipole transition, the excitation of a magnon by means of magnetic dipole transition. Thus the propagation or polariton effects will be smaller in the latter case. Finding them experimentally is of principal importance for the understanding of the optical properties of magnetic materials. With respect to the magnetic properties of perovskite-type layer structures, the magnetic resonance in the millimeter wave range is mostly performed at higher magnetic fields due to the higher frequencies in comparison to the conventional microwave range and due to the use of superconducting solenoids with fields up to 10 T. For example, we have been able by means of this technique to study uniaxial easy-axis antiferromagnets not only in the antiferromagnetic phase but also in the spin-flop regime.

Our review will be organized as follows. In section 2, the experimental equipment which we have used for our investigations is explained. It is outlined in detail that broad-band optical components can be used with advantage in the millimeter-wave range. Moreover, they are usually much cheaper then the equivalent microwave components. In section 3, a brief presentation of the theory for the electromagnetic propagation effects is given in order to introduce our notation and the extensions necessary for obtaining theoretical expressions that can be compared to the experimental data. In section 4.1, the results are compiled which were obtained for the paramagnetic resonance of $(CH_3NH_3)_2MnCl_4$ and Rb_2MnCl_4 at temperatures above their Néel temperatures. On the one hand, these data are a good example for electromagnetic propagation effects. On the other hand, the broadening of the paramagnetic line width is of interest for temperatures approaching the transition temperature. In section 4.2, we present our results for the ferromagnet K_2CuF_4 which are interesting with respect to the pronounced features of the propagation effects in the transmission spectra and to the high-field magnetic properties of a ferromagnet. In section 4.3, a number of magnetic resonance results are discussed for manganese compounds with antiferromagnetic order. For Rb_2MnCl_4, a great variety of data has been obtained in the antiferromagnetic phase and the spin-flop regime at various temperatures. These data enable us to make some predictions about the magnetic phase diagram of Rb_2MnCl_4. By studying the antiferromagnetic resonance of the compounds $(C_nH_{2n+1}NH_3)_2$ with n = 1, 2, 3 and of the compounds $(NH_3C_mH_{2m}NH_3)MnCl_4$ with m = 2, 4, 5, we have obtained information about the dependence of the magnetic properties on the layer separation.

2. EXPERIMENTAL APPARATUS

As already mentioned in the introduction, most of our magnetic resonance experiments were performed in the millimeter-wave range (30-300 GHz). Magnetic resonance in this range differs from that in the conventional microwave range (10-30 GHz, usual ESR equipment) in several aspects. First of all, increase of the frequency means in most cases an increase of the magnetic field in order to obtain resonance. And in fact, a great number of our data were obtained at relatively high fields between 1 and 9 T which were provided by a superconducting solenoid. Secondly, the millimeter-wave range is located between the far infrared spectral region and the microwave range, just at the boundary between different techniques. The microwave range has been developed from the low-frequency end of the electromagnetic spectrum, i.e. from the radio frequency range. It is characterized by technical components like transmitters (IMPATT oscillators, klystrons etc.), resonators (cavities), receivers (diode detectors) and a circuit network (waveguides) The far infrared, on the other hand, has been made usable for spectroscopy starting from the visible and near-infrared range with typically optical methods. In other words, an experimental setup consists of plane and spherical (or even aspherical) mirrors, of interferometers (Fabry-Perot or Michelson), gratings etc. In the millimeter-wave range at the boundary between these techniques, it is advantageous to use a mixture of both, that means to employ optical as well as microwave components. In this range for example, the losses in waveguides become already uncomfortably large. The losses (Hughes) are 22-32 dB/m for 300-220 GHz (waveguide J WR-3) in comparison to 2.5-4 dB/m for 40-25 GHz (waveguide KA WR-28). On the other hand, there are no longer the difficulties which are encountered with free-space radiation at 10 GHz in a laboratory. Therefore, it can be advantageous to send the radiation via a horn antenna into free space to focus it into the next component by means of a mirror.

Our experimental equipment for magnetic resonance in the millimeter-wave range as described first by Schröder et al.(1978) is shown schematically in Fig. 3. As usual in magnetic resonance experiments, the sample is irradiated by monochromatic radiation generated by the source. The resonance condition is fulfilled by varying the magnetic field of the superconducting solenoid. In practice, the transmission of the sample is recorded as a function of magnetic field by means of an X-Y recorder. A voltage proportional to the magnetic field is obtained from the current of the solenoid to drive the X-deflection of the recorder. The radiation is modulated by a chopper in order to employ the usual lock-in technique for processing the signal. The equipment shown in Fig. 3 indeed consists of the above-mentioned mixture of optical and microwave parts. The radiation source is an example for the latter. It is an IMPATT oscillator (Kuno, 1979) with a cavity as resonator and with an isolator in front of it in order to protect the IMPATT diode from radiation reflected back to the source (cf. Fig. 4). We also have some oscillators with Gunn diodes. The millimeter-wave radiation leaves the source as free-space radiation from the horn antenna. All these components are connected by standard rectangular waveguides. Frequencies higher than 100 GHz are obtained by generating harmonics in a frequency multiplier (cf. Fig.5) with a Schottky barrier diode as the nonlinear element (Schneider et al., 1981). Apart from the semiconductor devices (IMPATT, Gunn, and Schottky barrier diodes) and some millimeter wave components as the isolator, the oscillator and the horn antenna have been fabricated in our machine-shop for reasons of limited funds.

Within the free-space radiation path (cf. Fig. 3), the first spherical mirror with a focal length of 20 cm forms a parallel beam which is sent through a Fabry-Perot interferometer for frequency measurements and selection. In contrast to equivalent microwave components, i.e. resonant cavities, the Fabry-Perot interferometer (Renk et al., 1962; Ulrich et al., 1963; Storey et al., 1980; Holah 1980; Ulrich, 1968) can easily be tuned over the whole frequency range 30-300 GHz by varying the distance between the metal mesh reflectors (cf. Fig. 6), the efficiency of which does not change significantly over this range (Mitsuishi et al., 1963; Vogel et al., 1964; Ulrich, 1967; Lamarre et al., 1981). With a grating constant $g = 100$ μm of the metal mesh, the ratio of the wavelength and the grating constant is $\lambda/g = 30$ for $\nu = 100$ GHz, and the transmission of the interferometer is a few percent. The resolution is then sufficient that we are able to resolve the modes of the cavity in the IMPATT source, which are about 0.025 GHz apart. Fig. 7 shows a typical spectrum of the IMPATT oscillator in connection with a frequency multiplier (cf. Fig. 5). The spectrum was recorded by variation of the distance between the reflectors of the Fabry-Perot (cf. Fig. 6). The basic frequency of the IMPATT oscillator was 82.5 GHz. Then, the frequencies of the first, second, third and fourth harmonic are 164.9 GHz, 247.4 GHz, 329.9 GHz, and 412.4 GHz,

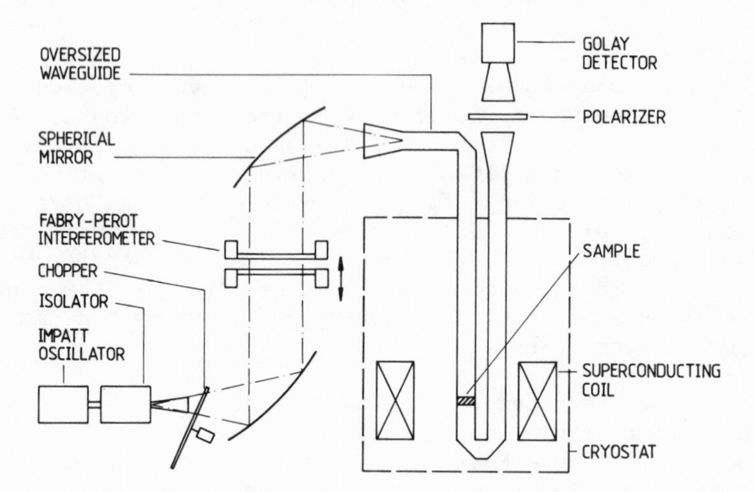

Figure 3. Schematic diagram of the apparatus used for magnetic resonance experiments in the millimeter wave range.

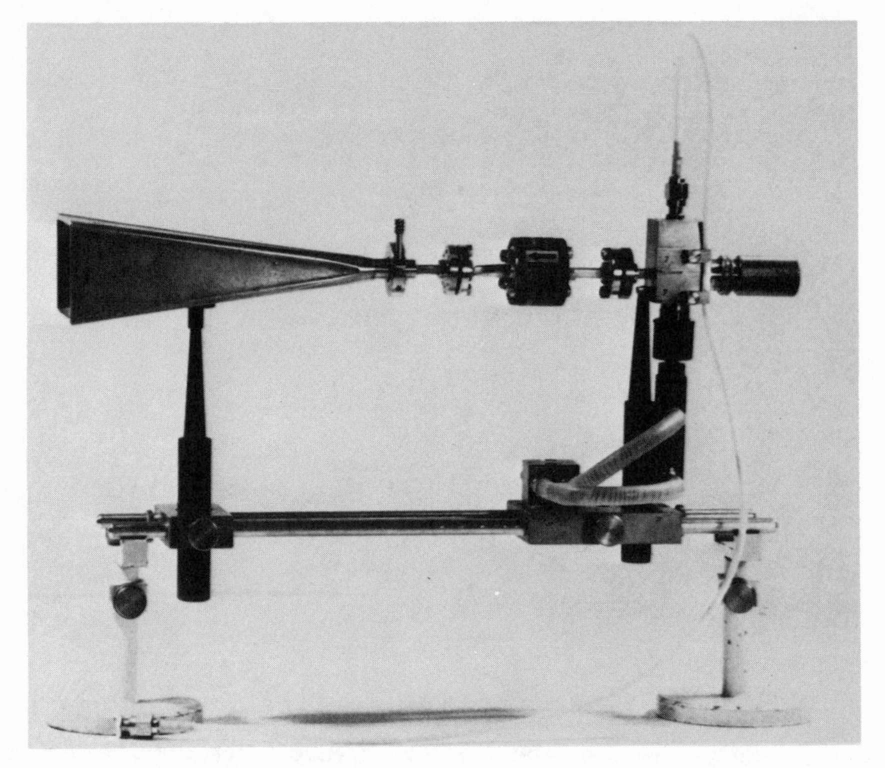

Figure 4. IMPATT oscillator (to the right, has electrical connection for power supply and hoses for water cooling) with isolator (center) and horn antenna (to the left). Except for the isolator and the IMPATT diode, all parts have been fabricated in our machine-shop.

respectively. For this experiment, the high frequency waveguide at the mixer was chosen in such a way that its cutoff frequency was about 200 GHz. Therefore, the basic frequency (82.5 GHz) and the first harmonic (164.9 GHz) are completely suppressed, and the second harmonic (247.4 GHz) is still close to the cutoff frequency and, therefore, strongly attenuated. For these reasons, the third harmonic (329.9 GHz) exhibits the largest intensity while the intensity of the next harmonic is smaller according to the usual decrease of intensity with increasing order of frequency multiplication. The diameter of the reflectors of the Fabry-Perot (cf. Fig. 5) is about 7 cm, which is somewhat smaller than the diameter of the mirrors. For such values of the mirror radius, the losses due to diffraction are still tolerable for millimeter wave radiation ($1 \text{ mm} \leqslant \lambda \leqslant 1 \text{ cm}$). Within the Kirchhoff approximation (Hönl et al., 1966) we obtain for $R = 5$ cm and $\lambda = 3$ mm that the first zero of the diffraction

Figure 5. Multiplier for the millimeter wave range with horn antenna for transmission of the harmonics. Except for the Schottky-barrier diode, all parts have been fabricated in our machine-shop.

function $J_1(q\cdot R\cdot\sin\delta)/q\cdot R\cdot\sin\delta$ occurs at an angle of about 2°. For comparison, the corresponding angle would be about 21.5° for λ = 3 cm (X-band) which indicates intolerably large diffraction losses.

The second spherical mirror focusses the radiation on the entrance of the oversized waveguide. This is a cylindrical tube through which the radiation is led into the cryostat with the superconducting solenoid and to the sample which is placed in the most homogeneous part of the field of the solenoid. The sample has typically a diameter of 5 mm and a thickness of several mm (thicker samples for investigation of antiferromagnets and thinner ones for ferromagnets), and the field homogeneity over the sample volume is about 0.3 %. The diameter of the oversized waveguide is 1 cm, and it can be used broad-band from its cut-off frequency of about 20 GHz to higher frequencies over the whole millimeter wave range. The total length of this waveguide system is about 4.50 m. The system

Figure 6. Fabry-Perot interferometer with metal mesh reflectors, one of them adjustable for parallelicity of the reflectors, and micrometer screw for tuning.

includes six right angle reflectors which have a much better transmissivity than bended waveguides (Kneubühl et al., 1979). Our most surprising experience with the oversized waveguide was that linearly polarized light sent into it comes out linearly polarized within the limits of experimental error. Therefore, we have placed a polarizer in front of the detector which is employed here as an analyzer for the radiation transmitted by the sample. As will be discussed in detail below there is a transmitted nonzero signal at the resonance field under crossed analyzer conditions where the linearly polarized light from the source cannot pass the analyzer. In simple cases of magnetic resonance this signal for crossed

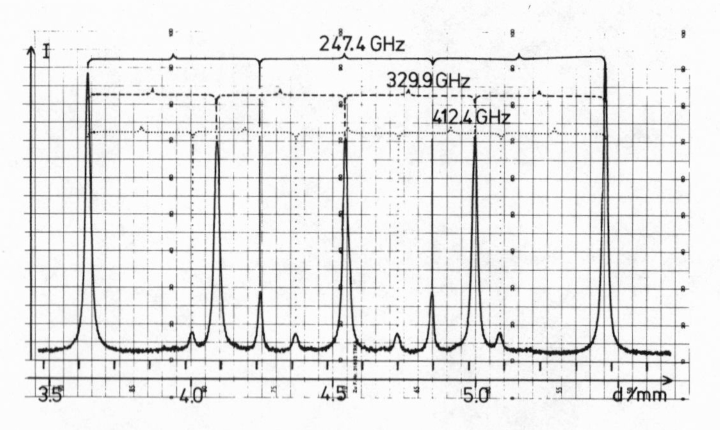

Figure 7. Spectrum of the harmonics of an IMPATT oscillator: at the detector recorded by means of variation of the distance d between the reflectors of the Fabry-Perot. The identification of the peaks is indicated.

analyzer results from one circularly polarized component being absorbed by the sample while the other nonresonant circular component is transmitted and from rotation of the plane of polarization due to the difference of the refractive indices of the two circularly polarized modes. The analyzer in our experimental equipment (cf. Fig. 3) is a wire grid polarizer (gold wire on mylar, 1000 lines per inch). The grating constant of the analyzer (25.4 μm is much smaller than the wavelength. Therefore, a good efficiency is to be expected. As detector we use a Golay cell, a well-known far-infrared, broad-band thermal detector of high sensitivity (see for example:Houghton et al., 1966). Its only disadvantage is its low speed; for high speed experiments a detector with a semiconductor diode, as is usually used in microwave experiments, would be more suited.

Finally, it should be noted that the whole apparatus for magnetic resonance as shown in Fig.3 could also be built up using only microwave components. Apart from the huge amount of money necessary to cover the whole range under consideration and apart from the fact that millimeter-wave components are much more expensive than usual microwave components (X- or K-band), our optical setup with mirrors, Fabry-Perot and oversized waveguide has the great advantage of offering the possibility of broad-band operation. In other words, we have to change the source and to tune the Fabry-Perot correspondingly when performing measurements at different frequencies over the whole millimeter-wave range. The sample stays in the waveguide in the cryostat at a certain temperature during this procedure. The sample temperature can be set in the range $2 \leqslant T \leqslant 300$ K. It is controlled by capacitive temperature controllers which are not influenced by a magnetic field. Another detail worth of mentioning is the Z-like arrangement of the spherical mirrors in the free-space radiation path (cf. Fig. 3) in order to reduce the optical aberrations (Czerny et al., 1930).

In addition to those in the millimeter range, we have carried out some experiments in the conventional X-band microwave range, the results of which will also be reported below. In this case, the equipment consisted of standard microwave components with a Gunn oscillator as the source and with the sample placed as usual in a cavity. Such apparatus is very similar in principle to what is found in a usual ESR spectrometer. The main difference is that we use a superconducting solenoid for providing magnetic fields up to 10 T. Such high fields are generally not achievable with a commercial spectrometer which, on the other hand, yields a much larger homogeneity of the magnetic field in comparison to our solenoid. For our measurements, the cavity with the sample has to be placed in the cryostat at low temperatures.

3. ELECTROMAGNETIC PROPAGATION EFFECTS

As will be demostrated with experimental results below,

electromagnetic propagation effects have to be taken into account for magnetic resonance experiments in the millimeter wave range. In the language of particles or, better, quasiparticles, these effects can also be called magnon-photon coupling or polaritons. For ferromagnets, they have been treated theoretically two decades ago (Auld, 1960; Pincus, 1962; Akhiezer et al., 1968) while similar treatments for antiferromagnets have appeared more recently (Manohar et al., 1972; Bose et al., 1975). Even though the Raman scattering from magnetic polaritons has also been studied theoretically (Sarmento et al., 1976a,b,c), there have been reported no experimental results so far, which are to be considered an evidence for electromagnetic propagation effects or for magnon-photon coupling. We believe our data reported below to be the first of this kind. Before explaining our results in detail and introducing our notation and definitions, we shall briefly outline the electromagnetic theory of magnetic resonance including some extensions which will be needed in the discussions below.

For magnetic resonance, i.e. for the treatment of the spin system near $\vec{q} = 0$, we are allowed to treat the system classically and to consider the macroscopic magnetization $\vec{M}$ only (Keffer, 1966). For our purposes, it is not necessary to take into account a spatial dependence of $\vec{M}$ in order to include exhange effects (proportional to $\vec{q}^2$) in a ferromagnet. As the equation of motion for $\vec{M}$, we take the Bloembergen-Bloch equation (Keffer, 1966)

$$\dot{\vec{M}} = \gamma(\vec{M} \times \vec{H}_{eff}) - \Lambda \vec{m} \tag{1}$$

where $\dot{\vec{M}}$ denotes the time derivative of $\vec{M}$. It will be shown below that we are far from being able to discuss line forms. Therefore, the actual form of the damping is not of interest and we may use the mathematically simpler form of Bloembergen-Bloch instead of the Landau-Lifschitz equation with a more complex form of the damping (For further details, see F.Keffer, 1966). In Eq.1, $\gamma = g\mu_B/\hbar$ is the gyromagnetic ratio, $\vec{H}_{eff}$ is the effective field which in general can be written as follows

$$\vec{H}_{eff} = \vec{H}_{intern} + \vec{H}_o + \vec{h} \tag{2}$$

where $\vec{H}_{intern}$ is the internal field acting on the magnetization (only in ordered materials like ferro- or antiferromagnets), $\vec{H}_o$ is the static external field and $\vec{h}$ the oscillating magnetic field ($\sim e^{i\omega t}$) of the electomagnetic radiation. The magnetization $\vec{M}$

$$\vec{M} = \vec{M}_o + \vec{m} \tag{3}$$

consists of a static part $\vec{M}_o$ which originates from the external field $\vec{H}_o$ or from the internal field $\vec{H}_{intern}$ (in ordered materials) and of an oscillatory or precessing part $\vec{m}$ ($\sim e^{i\omega t}$) which constitutes the response of the system to the oscillating field $\vec{h}$. In Eq.1, Λ is a measure for the damping or for the relaxation of the system which have included for realistic calculations to be compared with experimental data. It should be noted that antiferromagnetically ordered systems the equation of motion (Eq.1) has to be written separately for the two sublattices a and b

$$\dot{\vec{M}}_a = \gamma(\vec{M}_a \times \vec{H}_{a_{eff}}) - \Lambda\, \vec{m}_a$$
$$\dot{\vec{M}}_b = \gamma(\vec{M}_b \times \vec{H}_{b_{eff}}) - \Lambda\, \vec{m}_b \qquad (1')$$

Since the magnetization $\vec{M}_o$ is usually known as a function of temperature and applied field $\vec{H}_o$, the equations of motion (Eq. 1 or Eq. 1') form an inhomogeneous system of linear equations for the transverse oscillating or, better, precessing components of the magnetization, when the differential operator d/dt is replaced by the factor $i\omega$ as usual. Solution of this system of equations yields the dynamic susceptibility $\overleftrightarrow{\chi}$ (a tensor quantity)

$$\vec{m} = (\overleftrightarrow{\chi})\, \vec{h} \qquad (4)$$

Examples of the solution of the equations of motion and for the dynamic susceptibility will be given below.

The next step in our theoretical outline is the inclusion of Maxwell's equations. We assume plane waves, which means

$$\vec{e},\ \vec{h},\ \vec{d},\ \vec{b}\ \sim e^{i(\omega t - \vec{q}\vec{r})} \qquad (5)$$

where $\vec{e}$, $\vec{d}$, $\vec{b}$ are the electric field, the electric displacement and magnetic induction, respectively. As our samples are the perovskite-type layer structure materials, we have to take anisotropic dielectric properties (orthorhombic symmetry) into account in our considerations

$$\vec{d} = (\overleftrightarrow{\varepsilon})\, \varepsilon_o\, \vec{e} = \begin{pmatrix} \varepsilon_{xx} & & \\ & \varepsilon_{yy} & \\ & & \varepsilon_{zz} \end{pmatrix} \cdot \varepsilon_o\, \vec{e} \qquad (6)$$

The magnetic properties of our samples, on the other hand, are described by Eq. 4:

$$\vec{b} = \mu_o (\vec{h} + \vec{m}) - \mu_o(\overleftrightarrow{1} + \overleftrightarrow{\chi})\,\vec{h} \tag{4'}$$

Since for magnetic resonance experiments the magnetic field $\vec{h}$ is the important and basic quantity, we eliminate the electric field $\vec{e}$ by means of curl $\vec{e} = -\dot{\vec{b}}$ and obtain from curl $\vec{h} = \dot{\vec{d}}$ the following equation for $\vec{h}$

$$\frac{c^2}{\omega^2}\,\vec{q} \times \begin{pmatrix} 1/\varepsilon_{xx} & & \\ & 1/\varepsilon_{yy} & \\ & & 1/\varepsilon_{zz} \end{pmatrix} \cdot (\vec{q} \times \vec{h}) = -(\overleftrightarrow{1} + \overleftrightarrow{\chi})\,\vec{h} \tag{7}$$

where the identity $\varepsilon_o\mu_o = c^{-2}$ has been used. Once the direction of $\vec{q}$ has been specified, Eq. 7 is simply a homogeneous system of linear equation for the components of $\vec{h}$. There are two solutions $c^2q^2/\omega^2 = (n - ik)^2$ of the secular equation, which correspond to two electromagnetic waves with different polarizations (eigenvectors of Eq.7) and with different refractive index n and absorption coefficient k. If a particular solution

$$(n - ik)^2_\alpha = \left(\frac{c^2q^2}{\omega^2}\right)_\alpha = f_\alpha(\omega) \qquad \alpha = 1,2 \tag{8}$$

is brought to the form $\omega = f(q)$ one is able to discuss polariton effects (Auld, 1960; Pincus, 1962; Akhiezer et al., 1968; Manohar et al., 1972; Bose et al., 1975; Sarmento et al., 1976a, b, c) which are naturally very similar to the case where an infrared active phonon interacts with electromagnetic waves (Born et al., 1954). Further details will be discussed with the examples below.

So far we have derived the propagation constant cq/ω of electromagnetic waves including the dielectric and magnetic properties of the materials under consideration. Naturally, the latter properties will depend on the applied static field $\vec{H}_o$ and, thus, n and k of at least one of the electromagnetic modes will vary with $\vec{H}_o$. It is this dependence upon which the magnetic resonance experiments are based. On the other hand, n and k cannot be measured directly with the experimental setup explained above. But the actually measured transmissivity of the sample will depend on n and k. Therefore, the last step in our theoretical outline of the electromagnetic propagation effects is to calculate the transmissivity of the sample. It will be shown (see also K.Strobel et al., 1980) that this can be done assuming a plane wave at normal incidence to a planparallel plate of thickness d (cf.Fig.8) without any discrepancy between experimental and calculated data. The incident linearly polarized wave $\vec{h}_o$ is divided into two waves, one of the mode 1 ($\vec{h}_1$) and of mode 2 ($\vec{h}_2$) which propagate according to $(n - ik)_1$ and $(n - ik)_2$, respectively, inside the sample. The amplitudes $\vec{h}_1$ and $\vec{h}_2$ are determined from the conditions that their sum is to be equal to $\vec{h}_o$ and that they have to be eigenvectors of Eq. 7.

For each mode, the amplitude, i.e. the magnetic field of the transmitted wave, is

$$\vec{h}_{\alpha t} = \tilde{t}_\alpha \vec{h}_\alpha \qquad \alpha = 1,2 \tag{9}$$

$$\tilde{t}_\alpha = \left(\frac{t_{01} t_{10} \; e^{-2\pi i (n-ik) d\tilde{\nu}}}{1 - (r_{10})^2 \; e^{-4\pi i (n-ik) d\tilde{\nu}}} \right)_\alpha$$

where we have included the effects due to multiple reflections in

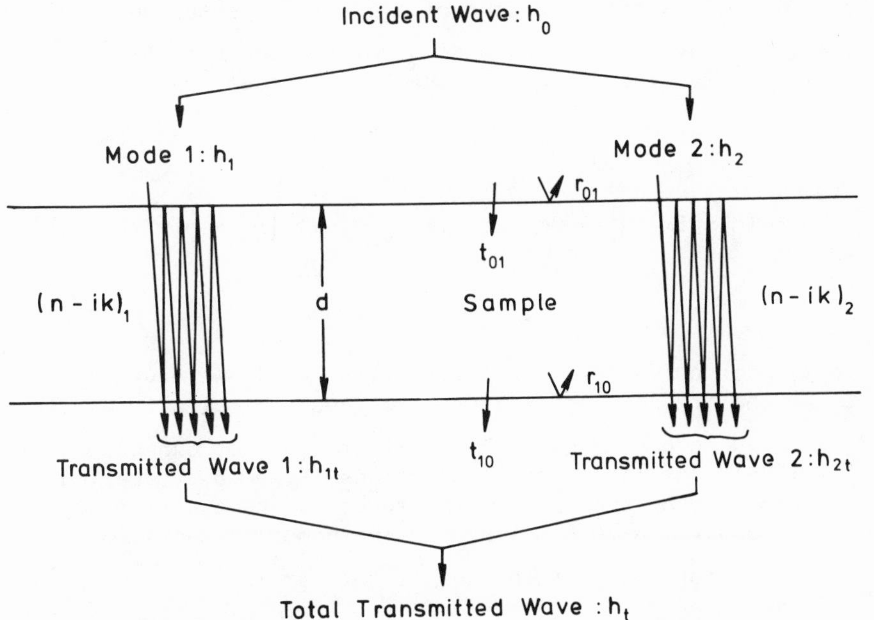

Figure 8. Illustration of the interference effects at a plane-parallel sample in the oversized waveguided:interference due to multiple reflections and due to interaction of the two modes. For further explanations see text.

the sample. As is well known, the denominator in Eq. 9 arises from the summation of all partial waves having the sample passed after one, two, three, etc. zig-zag reflections in it. In Eq. 9 d and $\tilde{\nu}$ are the thickness of the sample (in cm) and the frequency of the incident wave (in cm^{-1}) while t_{01}, t_{10} and r_{10} are the Fresnel coefficients for transmission and reflection as indicated in Fig.8, respectively. The subscript α ($\alpha = 1,2$) at the brackets means that all quantities inside the brackets have to be evaluated with $(n - ik)_\alpha$ for mode α, which means for mode 1 or for mode 2. It should be noted that these coefficients have to be used in our context of magnetic resonance in a somewhat modified form (here given for normal incidence)

$$r_{01} = - r_{10} = \frac{n - ik - \varepsilon_{eff}}{n - ik + \varepsilon_{eff}} \tag{10}$$

$$t_{01} = \frac{2\,\varepsilon_{eff}}{n - ik + \varepsilon_{eff}} \qquad t_{10} = \frac{2(n - ik)}{n - ik + \varepsilon_{eff}}$$

In Eq. 10, ε_{eff} is the component of dielectric tensor appropriate for the component of $\vec{e}$ corresponding to the component of $\vec{h}$ under consideration in accordance with the Maxwell equation

$$\vec{e} = -\,(n - ik)\sqrt{\frac{\mu_o}{\varepsilon_o}}\begin{pmatrix} 1/\varepsilon_{xx} & & \\ & 1/\varepsilon_{yy} & \\ & & 1/\varepsilon_{zz} \end{pmatrix}\left(\frac{\vec{q}}{q} \times \vec{h}\right) \tag{11}$$

Naturally, for nonmagnetic materials with $\mu = 1$, we have $(n - ik)^2 = \varepsilon_{eff}$, and the Fresnel coefficients are reduced to their familiar form as given in every textbook on optics (see for example : Born et al., 1975). At last, the total transmitted wave received by the detector is the sum of the amplitudes of the transmitted waves in modes 1 and 2 ($\vec{h}_t = \vec{h}_{1t} + \vec{h}_{2t}$). The power transmission resulting from the detector signal with and without sample is (without using an analyser)

$$T = \frac{|\tilde{t}_1\,\vec{h}_1 + \tilde{t}_2\,\vec{h}_2\,|^2}{|\,\vec{h}_o|^2} \tag{12}$$

with (cf. Eq. 9)

$$\tilde{t}_\alpha = \left(\frac{t_{01} t_{10}\ e^{-2\pi i (n-ik) d\tilde{\nu}}}{1 - (r_{10})^2\ e^{-4\pi i (n-ik) d\tilde{\nu}}} \right)_\alpha \qquad \alpha = 1,2$$

In Eq. 12, the "absolute value squared" of the amplitude of the wave refers to its vector character as well as to the fact that it is a complex quantity. If an analyzer is used in a position parallel to the polarization of the incident radiation, we have to replace $\vec{h}_1$ and $\vec{h}_2$ by $\vec{h}_1 \cdot \vec{u}_o$ and $\vec{h}_2 \cdot \vec{u}_o$, respectively, in Eq. 12 where $\vec{u}_o$ is a unit vector parallel to $\vec{h}_o$. If the analyzer is set to a position perpendicular to the polarization of the incident radiation, a similar projection is to be used with a unit vector perpendicular to $\vec{h}_o$ and to $\vec{q}$, the wave vector. This calculated power transmission (Eq. 12) is finally the quantity which may be compared with experimental data in magnetic resonance.

From this comparison, we will be able to conclude whether electromagnetic propagation effects (or magnon-phonon coupling or polariton effects) play an important role in the millimeter wave range or not.

4. RESULTS AND DISCUSSIONS

4.1. Paramagnetic Resonance

We have studied the paramagnetic resonance of $(CH_3NH_3)_2MnCl_4$ (= MAMC) and of Rb_2MnCl_4 as a function of temperature above their Néel temperatures. Below them, these perovskite-type layer structure materials exhibit an antiferromagnetic long range order (cf. section 1). The main purpose of these investigations is to obtain experimental data for the width of the resonance line at various temperatures and at various magnetic fields (various frequencies). Generally, the paramagnetic resonance line width increases with decreasing temperatures and seems to diverge at the Néel temperature (Strobel et al., 1980; Yokazawa, 1971; de Wijn et al., 1972). Another aspect of our investigation of the paramagnetic resonance at temperatures above the critical temperature is that the experimental results in the millimeter range represent a good example for electromagnetic propagation effects as introduced in the last section. The decrease of the line width with increasing temperatures reaches such small values of the damping parameter Λ (cf. Eq.1) at 200 K that χ (cf. Eq.4) approaches unity in the neighborhood of the resonance.

This causes drastic changes of (n - ik) (cf. Eq. 7) near the resonance, and, therefore, propagation effects become visible in the transmission spectra.

For paramagnetic materials, the internal field $\vec{H}_{intern}$ vanishes. Assuming the external static field $\vec{H}_o = (0,0,H_o)$ to be applied along the z-axis, the equations of motion (cf. Eq. 1) read in this particular case:

$$\begin{aligned} \dot{m}^x &= \gamma(m^y H_o - M_o h^y) - \Lambda m^x \\ \dot{m}^y &= -\gamma(m^x H_o - M_o h^x) - \Lambda m^y \end{aligned} \tag{13}$$

where m^x, m^y, h^x, h^y are the x and y components of $\vec{m}$ and $\vec{h}$, respectively, $M_o = \chi_o H_o$ is the static magnetization of the sample parallel to the applied field H_o, and is also proportional to the static susceptibility χ_o which is a function of temperature. As mentioned already in the introduction, perovskite-type layer materials like Rb_2MnCl_4 and MAMC are good examples for quasi-two-dimensional antiferromagnets (Schröder et al., 1980). Typically in such materials, the static susceptibility increases with increasing temperature above the Néel temperature, passes through a maximum and then varies according to a Curie-Weiß-Law (Van Amstel et al., 1972; Breed, 1967). The value of the static susceptibility is of the order 10^{-3}. We remember that m^x, m^y, h^x, and h^y are the oscillating components of the magnetization and the magnetic field ($\sim e^{i\omega t}$), respectively, and we realize that the two equations of motion (Eq. 13) become decoupled for $m^{\pm} = m^x \pm im^y$ and $h^{\pm} = h^x \pm ih^y$ (circularly polarized waves). Then the solution of Eq. 13 (cf. Eq. 4) is easily obtained

$$m^{\pm} = \chi^{\pm} h^{\pm} = \chi_o \frac{\gamma H_o}{\gamma H_o \pm (\omega - i\Lambda)} \tag{14}$$

Next we have to include Maxwell's equations into our considerations (cf. Eq.7). Rb_2MnCl_4 and MAMC may be assumed to have uniaxial symmetry (cf. Figs. 1 and 2) with respect to their dielectric properties in the temperature range under consideration (Schröder et al., 1978):

$$\varepsilon_{xx} = \varepsilon_{yy} = \varepsilon_{\perp} \qquad\qquad \varepsilon_{zz} = \varepsilon_{\parallel} \tag{15}$$

These layer materials easily form platelets with the major faces perpendicular to the crystallographic c-axis, which direction is chosen to be the z-axis. Therefore, the easiest way of investigating samples of these materials by means of spectroscopic methods is to send the radiation on these faces, i.e. the wave vector $\vec{q}$ is parallel to the z-axis:

$$\vec{q} = (0,0,q) \tag{16}$$

Under these assumptions, the system of equations for the components of $\vec{h}$ (cf. Eq.7) reads

$$\begin{aligned} [(n - ik)^2 - \varepsilon_\perp (1 + \chi^\pm)] \; h^\pm &= 0 \\ h^z &= 0 \end{aligned} \tag{17}$$

The solutions (eigenvalues) of this system are

$$[(n - ik)_\pm]^2 = \varepsilon_\perp [1 + \chi_o \frac{\gamma H_o}{\gamma H_o \pm (\omega - i\Lambda)}] \tag{18}$$

and the two modes in the crystal are circularly polarized waves. One of them (-) is the resonant mode where the denominator becomes purely imaginary at the resonance ($\omega = \gamma H_o$) while the other mode is nonresonant. For the calculation of the power transmission T (cf. Eq. 12) we assume the incident radiation to be polarized along the x-direction:

$$\vec{h}_o = (h_o, 0,0) \tag{19}$$

Then, we obtain for the two modes (cf. Fig.8)

$$\begin{aligned} \vec{h}_1 = \vec{h}_+ &= \frac{1}{2} (h_o, -ih_o, 0) \\ \vec{h}_2 = \vec{h}_- &= \frac{1}{2} (h_o, +ih_o, 0) \end{aligned} \tag{20}$$

Under the above mentioned assumptions, the Fresnel coefficients (cf. Eq. 10) have to be calculated with $\varepsilon_{eff} = \varepsilon_\perp$ for both modes. Finally, if the analyzer is set parallel to the direction of polarization of the incident radiation (see also Strobel et al., 1980), we obtain for the power transmission

$$T_\parallel = \frac{1}{4} |\tilde{t}_+ + \tilde{t}_-|^2 \tag{21}$$

If, on the other hand, the analyzer is set perpendicular to the direction of polarization of the incident radiation, we obtain

$$T_{\perp} = \frac{1}{4}\left|\tilde{t}_{+} + \tilde{t}_{-}\right|^2 \tag{22}$$

Similar to Eq. 9, the amplitude transmission including multiple reflections is given by

$$\tilde{t}_{\pm} = \left(\frac{t_{01} t_{10}\; e^{-2\pi i(n-ik)d\tilde{\nu}}}{1 - (r_{10})^2\; e^{-4\pi i(n-ik)d\tilde{\nu}}} \right)_{\pm} \tag{9'}$$

where the subscript $\pm$ refers to the two modes (cf. Eq.18). Far from resonance, we may assume $\chi^{\pm} \approx 0$. Then we have $n - ik = \sqrt{\varepsilon_{\perp}}$, and $T_{\parallel}$ (cf. Eq. 21) reduces to the form familiar to almost everyone working in the field of spectroscopy (see for example: Abelés, 1972)

$$T_{\parallel} = |\tilde{t}_{\pm}|^2 = \left| \frac{t_{01} t_{10} e^{-2\pi i(n-ik)d\tilde{\nu}}}{1 - (r_{10})^2\; e^{-4\pi i(n-ik)d\tilde{\nu}}} \right|^2 \tag{23}$$

while $T_{\perp}$ (cf. Eq.22) vanishes as to be expected.

Figures 9 and 10 show the experimental results for the paramagnetic resonance of MAMC at 80 K and 220 K, respectively. The upper trace [denoted "$H_{\parallel} + H_{\perp}$" in Figs. 9 and 10] represents the intensity recorded as a function of applied magnetic field without using an analyzer. The applied field is denoted B instead of H in the figures since it is given there in Tesla, the units of the magnetic flux. The intensity is normalized with its value for $B = 0$ [$I(B)/I(B=0)$]. The middle trace [$H_{\perp}$] and the lower trace [$H_{\parallel}$] show the corresponding intensity when the analyzer is set perpendicular and parallel to the direction of polarization of the incident radiation, respectively. The intensity is normalized to that obtained for $H_o = 0$, i.e. $B = 0$ in Figs. 9 and 10. At a first glance, the results obtained without analyzer at 220 K exhibit a rather odd line shape. It looks as if we were observing a mixture of what is usually called the "absorption mode" and "dispersion mode" in magnetic resonance spectroscopy (see for example: Slichter, 1978). But looking more carefully into the matter, we were led to the conclusion that our line shape is the result of electromagnetic propagation effects, in particular the result of interference between the resonant and the nonresonant circularly polarized mode. As already mentioned, the paramagnetic line width of quasi-two-

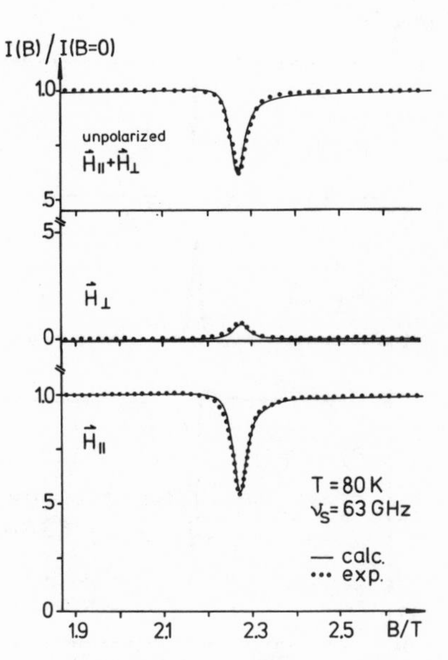

Figure 9. Paramagnetic resonance of MAMC at 80 K: Normalized transmitted intensity I(B)/I(B=0), experimental and calculated values, without analyzer and with the analyzer set perpendicular and parallel to the direction of polarization of the incident radiation, respectively.

dimensional antiferromagnets like MAMC and Rb_2MnCl_4 is strongly dependent on temperature (Strobel et al., 1980; Yokazawa, 1971; de Wijn et al., 1972). Decreasing with increasing temperature,

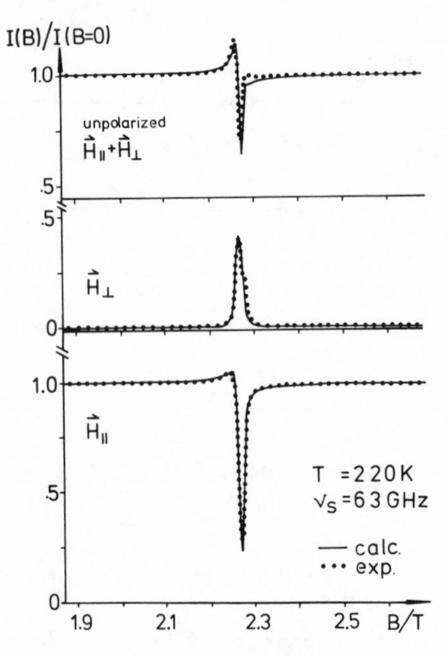

Figure 10. Paramagnetic resonance of MAMC at 220K: Normalized transmitted intensity I(B)f/I(B=0), experimental and calculated values, without analyzer and with the analyzer set perpendicular and parallel to the direction of polarization of the incident radiation, respectively.

it is sufficiently small at 220 K that the absolute value of χ of the resonant mode (cf, Eq. 14) will be close to one near the resonance despite the relatively small value of the numerator in the expression for (Eq. 14). In order to verify these ideas, we have calculated $T_{\parallel} + T_{\perp}$, $T_{\perp}$ and $T_{\parallel}$ by means of Eqs. 21 and 22, respectively, as a function of applied magnetic field H_o (B in Figs. 9 and 10), and normalized to their value for zero magnetic field. In this way, we are able to compare the experimental data [I(B)/I(B=0)] with the corresponding calculated power transmission. The parameter values for MAMC used in the calculation are compiled in Table 2. For a realistic approach, we have used a finite width of the oscillator frequency in our calculations, modelling the actual properties of the IMPATT oscillator source. Under these circumstances, an excellent agreement is achieved between experimental and calculated data for both temperatures. We conclude from these results that electromagnetic propagation effects cannot be neglected even in the case of paramagnetic resonance where it is, of course, questionable to call them polariton effects, as in the case of ordered materials. In our case, there does not exist a magnon branch over the whole Brillouin zone. This may be the reason why theoretical considerations like those in the preceding paragraphs have not been made before. Furthermore, our results clearly indicate that the damping Λ (cf. Eq. 1) has to be included for a realistic theoretical treatment, and the transmission of the sample has to be considered in order to take interference effects into account. These may prove an important indication of the effects under consideration in many cases of spectroscopy on magnetic materials, even though the gap in the $\omega(q)$ dispersion relations (Auld, 1960; Pincus, 1962; Akhiezer et al., 1968; Manohar et al., 1972; Bose et al., 1975; Sarmento et al., 1976a,b,c) is extremely small. In our example MAMC, the resonance field is B = 2.25 T for an oscillator frequency of ν= 63 GHz. At 60 K, the value of the gap is $\gamma\chi_o H_o$ = 85 MHz (= 0.003 T) which indeed is very small. Neglecting the damping ($\Lambda = 0$, $\varepsilon''_{\perp} = 0$) as usual for these considerations (Auld, 1960; Pincus, 1962; Akhiezer et al., 1968; Manohar et al., 1972; Bose et al., 1975; Sarmento et al., 1976 a,b,c), we may solve the equation

$$\frac{c^2 q^2}{\omega^2} = \varepsilon_{\perp} \left(1 + \frac{\chi_o\ \gamma H_o}{\gamma H_o - \omega}\right) \tag{24}$$

for the dispersion relation $\omega = f(q)$ which has two branches with the following asymptotic values of the frequencies

$$\begin{array}{llll} \text{lower branch :} & \text{small } q{:}\ \omega \approx cq/\sqrt{\varepsilon_{\perp}(1+\chi_o)} & \text{(EM-wave)} & \\ & \text{large } q{:}\ \omega = \gamma H_o & \text{(resonance)} & (25) \\ \text{upper branch :} & \text{small } q{:}\ \omega = \gamma(1+\chi_o)H_o & \text{("longitudinal")} & \\ & \text{large } q{:}\ \omega \approx cq/\sqrt{\varepsilon_{\perp}} & \text{(EM-wave)} & \end{array}$$

TABLE 2: Values of various parameters of MAMC

gyromagnetic ratio	$\gamma = 27.99$ GHz/T (g = 2.000)
dielectric constant	$\varepsilon_\perp = \varepsilon'_\perp - i\varepsilon''_\perp$ $\varepsilon'_\perp = 7.0$ $\varepsilon''_\perp = 0.3$ [a]
static susceptibility	$\chi_o = 1.356.10^{-3}$ (T = 80K) [b] $\chi_o = 1.000.10^{-3}$ (T = 220 K) [b]
damping constant	$\Lambda \triangleq 0.4$ GHz = 0.014 T (T = 80 K) $\Lambda \hat{=} 0.2$ GHz = 0.007 T (T = 220K)
sample thickness	d = 1.9 mm

(a) average values, assumed to be independent of temperature for the calculation of the sample transmission. Data taken from B. Schroder et al., 1978.

(b) W.D. van Amstel and L.J. de Jongh (1972)

In Eq. (25), small and large q mean $q \ll \gamma H_o \sqrt{\varepsilon_\perp}/c$ and $1/a \gg q \gg \gamma H_o \sqrt{\varepsilon_\perp}/c$ (a = lattice constant), respectively. It should be noted that the lower branch which starts as an electromagnetic wave (EM-wave) at small q does not pass into a branch of elementary excitations in this case of a paramagnet, in contrast to magnetically ordered materials. The second branch starts at a frequency which corresponds to that of the longitudinal optical mode in lattice dynamics (Born et al., 1954), and it passes into an electromagnetic wave or photon-like branch (EM-wave) at large q. Between the two branches, there is the above-mentioned gap. It has been widely discussed in recent theoretical considerations for ferro- and antiferromagnets (Auld, 1960; Pincus, 1962; Akhiezer et al., 1968; Manohar et al., 1972; Bose et al., 1975)where the damping has been neglected. In our case of a paramagnet, the width of the gap $\Delta\omega = \gamma\chi_o H_o$ (cf.Eq.25) is too small to be seen directly in the spectra. At 220 K, it is mainly the line width of the source which prevents these effects from producing signatures in the spectra. But the dispersion curves (Eq. 25) and the gap between them determine the optical properties and become visible in this indirect way in the transmission spectra. Therefore our experimental results are an evidence for electromagnetic propagation effects in mm wave magnetic resonance.

The basic results towards a better understanding of the magnetic ordering and other properties of antiferromagnetic perovskite-type layer structures are the line widths of the paramagnetic resonance (cf. Fig.23 in Section 4.3.2) as already mentioned. Our results are similar to those of de Wijn et. al. for K_2MnF_4 and Rb_2MnF_4 (de Wijn, 1973b). In detail however, we observe a line width that becomes smaller and smaller the higher the frequency and thus the magnetic field. We also tried to fit our experimental data to the empirical relation

$$\Gamma(T) = \Gamma(\infty) \left[1 + \left(\frac{T - T_N}{T_N}\right)^{-p}\right] \qquad (26)$$

where $\Gamma(T)$ is the full width at half maximum for temperature T and $\Gamma(\infty)$ is the experimental line width of our IMPATT source as observed at high temperatures. At these temperatures, the paramagnetic line width (Λ) is much smaller than $\Gamma(\infty)$. In Eq. 26, C and p are disposable parameters and T_N is the Néel temperature. In Fig. 11, we show $\ln[(\Gamma(T) - \Gamma(\infty))/\Gamma(\infty)]$ as a function of $\ln[(T-T_N)/T_N]$ obtained for Rb_2MnCl_4 with various frequencies. The data clearly show the above mentioned drastic decrease of the line width for higher frequencies(higher magnetic fields !). Furthermore, we encounter for most of the data no straight line but a considerable curvature. Only our X-band data (9.2 GHz) form a straight line and agree well with the data obtained for K_2MnF_4 and Rb_2MnF_4 at 24 GHz (de Wijn, 1973b) (cf. Fig.11. The validity of Eq. 26 may be questionable, on the other hand, for frequencies near and above 100 GHz or magnetic fields of about 4 T. For Rb_2MnCl_4, this is already close to the phase boundary between antiferromagnetic and spin flop phase. If a relation like Eq. 26 is used we ought to insert the magnetic field dependent critical temperature $T_c(H)$ at which the material

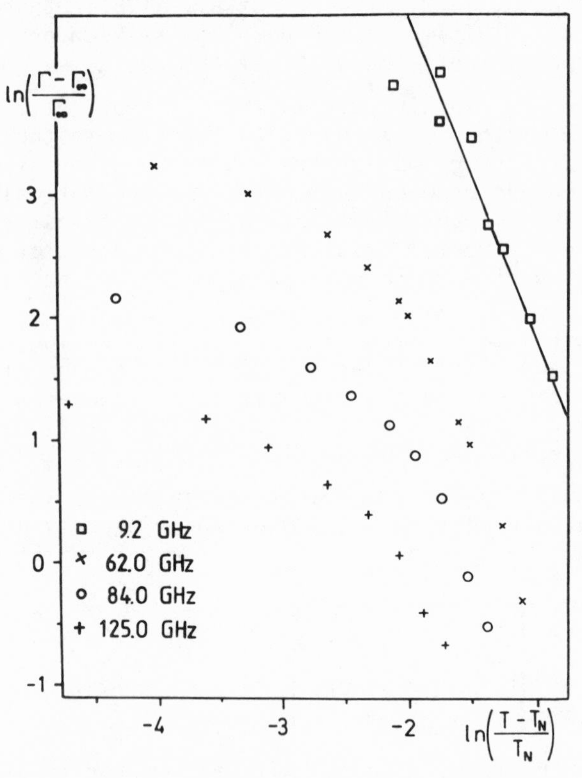

Figure 11. Normalized paramagnetic line width $[\Gamma(T) - \Gamma(\infty)]/\Gamma(\infty)$ of Rb_2MnCl_4 versus normalized temperature $(T-T_N)/T_N$ for various frequencies. $\Gamma(T)$ is the full width at half maximum. The full line represents the corresponding data for K_2MnF_4 (de Wijn et al., 1973b) obtained at 24 GHz.

orders antiferromagnetically when a magnetic field H is applied (cf. Fig. 21 in section 4.3.2). However, the magnetic phase diagram of Rb_2MnCl_4 is still to be determined, for example by means of neutron diffraction. With the knowledge of the magnetic phase diagram, i.e. $T_c(H)$, we hope to perform a more complete analysis of our line width data and to achieve a better understanding of these results.

4.2. Ferromagnetic resonance in K_2CuF_4

As an example for a perovskite-type layer structure with ferromagnetic ordering we have studied K_2CuF_4 by means of magnetic resonance in the mm-wave range. This material is, when below its Curie temperature $T_c = 6.25$ K, an easy-plane ferromagnet, and an almost ideal example of a quasi-two-dimensional X-Y model which has been studied by various methods (Hirakawa et al., 1973; Yamada, 1972; Kubo, 1974; Ikeda, 1974; Le Dang et al., 1976; Kubo et al., 1977; Ferré et al., 1979; Hirakawa et al., 1981; Borovik-Romanov et al., 1980; Kleemann et al., 1981). For a ferromagnet below the Curie temperature, there is an internal field $\vec{H}_{intern}$ which will be derived from the energy density of the system. Since the in-plane anisotropy field is negligibly small for our purposes i.e. $(2-5)\cdot 10^{-4}$T, and only observable at extremely low temperatures (T 3.5 K) (Yamazaki et al., 1981), there is practically no spin canting in K_2CuF_4 at temperatures above 4.2 K (Kleemann et al., 1981). Therefore, we are allowed to treat it as a tetragonal ferromagnetic material with one Cu^{2+} ion per unit cell, even though there are two inequivalent Cu^{2+} sites in the actual K_2CuF_4 structure due to a Jahn-Teller distorsion of the copper fluorene octahedra (Kleemann et al., 1981; Khomski et al., 1973; Haegele et al., 1974; Hidaka et al., 1979; Herdtweck et al., 1981). In Fig. 12, we show this distorsion schematically for a section through the basal plane of the lattice as compared to the regular octahedra of K_2MnF_4.

Describing the lattice approximately with one Cu^{2+} per unit cell, we need not introduce two sublattice magnetizations $\vec{M}_a$ and $\vec{M}_b$ corresponding to the spins at the two inequivalent Cu^{2+} sites but are allowed to consider only "the" magnetization $\vec{M}$. Then the macroscopic enery density U_{FM}, i.e. the $\vec{q} = 0$ part of the spin Hamiltonian with the exhange interaction and the anisotropy energy, may be written as follows

$$U_{FM} = - \frac{H_E}{2M_o} \vec{M}\cdot\vec{M} + \frac{H_A}{2M_o} (M^y)^2 \tag{27}$$

In Eq. 27, H_E $(= 2zSJ/g\mu_B)$ is the "internal" field resulting from the dominating intralayer exhange interaction between nearest neighbors as characterized by the exchange parameter J. The other quantities determining the exchange field are the number of nearest neighbors ($z = 4$), the spin quantum number ($S = 1/2$ for Cu^{2+}), the g- factor

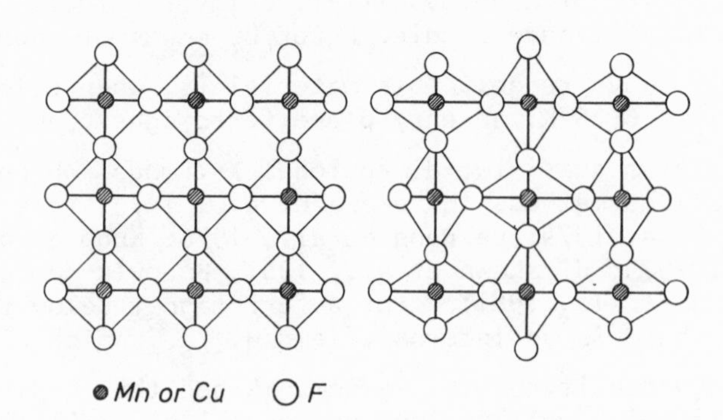

Figure 12. Cross section through a perovskite-type layer with transition metal-fluorene octahedra: regular octahedra as in the case of K_2MnF_4 (to the left) and octahedra distorted by the Jahn-Teller effect as in the case of K_2CuF_4 (to the right).

and Bohr's magneton μ_B. The rather small interlayer exhange interaction (Yamazaki, 1974) can safely be neglected in our context of magnetic resonance. In Eq. 27, H_A is the out-of-plane anisotropy field resulting only from anisotropic exchange interactions and from magnetic dipole-dipole interactions (Kleemann et al., 1981; Yamazaki, 1974, 1976; Tsuru et al., 1976) since crystal field effects are absent in S = 1/2 systems like Cu^{2+}. M_o is the saturation magnetization which is a function of temperature. $M_o(T) - M_o(T=0)$ varies according to a $T^{3/2}$ power law (Hirakawa et al., 1973). From the energy density U_{FM} (Eq. 24), the internal field $\vec{H}_{intern}$ is derived by means of a differentiation

$$\vec{H}_{intern} = - \left(\frac{\partial U_{FM}}{\partial M^x}, \frac{\partial U_{FM}}{\partial M^y}, \frac{\partial U_{FM}}{\partial M^z} \right) \tag{28}$$

Following the literature (Kubo, 1974; Yamazaki, 1974, 1976; Tsuru et al., 1976), the z-axis and the x-axis are chosen to be in the easy plane perpendicular to the crystallographic c-axis while the y-axis is chosen parallel to the hard axis, i.e. the crystallographic c-axis. The external field is assumed to be applied in the easy plane. For a ferromagnet, we have to include demagnetization fields (Keffer, 1966)

$$\vec{H}_o = [H_o \sin\psi - N^x M_o \sin\Theta \ , 0, H_o \cos\psi - N^z M_o \cos\Theta] \quad (29)$$

where ψ and Θ are the angles between the z-axis and the applied static field and the static magnetization, respectively. N^x and N^z are the x- and z-components of the demagnetization tensor of the sample. Including the oscillatory or precessing components, the magnetization reads

$$\vec{M} = [M_o \sin\Theta + m^x, m^y, M_o \cos\Theta + m^z] \quad (30)$$

With $\vec{H}_{intern}$ from Eq. 28 and $\vec{H}_o$ from Eq. 29,a time independent torque is obtained in the equations of motion (cf. Eq. 1) which must vanish for equilibrium. This leads to the equilibrium condition:

$$H_o M_o (\sin\psi \cos\Theta - \cos\psi \sin\Theta) - (N^x - N^z) M_o^2 \sin\Theta \cos\Theta = 0 \quad (31)$$

From Eq. 31, we obtain $\psi = \Theta$ if $N^x = N^z$. For $N^x \neq N^z$, on the other hand, we observe, in general, a refraction at the sample boundary; i.e. ψ and Θ are different, and so are the directions of the applied field inside and outside the sample. With the help of the equilibrium condition Eq. 31, the equations of motion for the oscillating or precessing components of the magnetization may finally be written as follows (cf. Eq. 1)

$$\dot{m}^x = \gamma[H_1 m^y - M_o h^y] \cos\Theta - \Lambda m^x$$

$$\dot{m}^y = \gamma[(H_2 m^z - M_o h^z) \sin\Theta - (H_2 m^x - M_o h^x) \cos\Theta] - \Lambda m^y \quad (32)$$

$$\dot{m}^z = - \gamma[H_1 m^y - M_o h^x] \sin\Theta - \Lambda m^z$$

with

$$H_1 = H_o - N^z M_o \cos^2\Theta - N^x M_o \sin^2\Theta + H_A$$

$$H_2 = H_o - N^z M_o \cos^2\Theta - N^x M_o \sin^2\Theta$$

$$\gamma = g_\perp \mu_\beta / \hbar$$

Resulting from crystal field effects, the g-tensor is anisotropic for K_2CuF_4. With our choice of the coordinate system, we have $g_x = g_z = g_\perp$ and $g_y = g_\parallel$. For reasons of more transparency of these considerations we use only $g_\perp$ in Eq. 32 which is appropiate for H_1, H_2, h^x and h^z. We then have toscale h^y with a factor $g_\parallel / g_\perp$ for numerical calculations. For our further considerations, we recall that m^x, m^y, m^z, h^x, h^y, and h^z are proportional to $e^{i\omega t}$. Replacing the time derivative by the factor $i\omega$, we obtain from Eq. 31 an inhomogeneous system of linear equations for m^x, m^y, and m^z the solution of which yields the dynamic magnetic susceptibility (cf. Eq. 4)

$$\underline{\underline{\chi}} = \begin{bmatrix} \frac{\gamma^2 M_o H_1 \cos^2\Theta}{\Omega_o^2 - (\omega - i\Lambda)^2} & -i\frac{(\omega - i\Lambda)\gamma M_o \cos\Theta}{\Omega_o^2 - (\omega - i\Lambda)^2} & -\frac{\gamma^2 M_o H_1 \sin\Theta\cos\Theta}{\Omega_o^2 - (\omega - i\Lambda)^2} \\ +i\frac{(\omega - i\Lambda)\gamma M_o \cos\Theta}{\Omega_o^2 - (\omega - i\Lambda)^2} & \frac{\gamma^2 M_o H_2}{\Omega_o^2 - (\omega - i\Lambda)^2} & -i\frac{(\omega - i\Lambda)\gamma M_o \sin\Theta}{\Omega_o^2 - (\omega - i\Lambda)^2} \\ \frac{\gamma^2 M_o H_1 \sin\Theta\cos\Theta}{\Omega_o^2 - (\omega - i\Lambda)^2} & +i\frac{(\omega - i\Lambda)\gamma M_o \sin\Theta}{\Omega_o^2 - (\omega - i\Lambda)^2} & \frac{\gamma^2 M_o H_1 \sin^2\Theta}{\Omega_o^2 - (\omega - i\Lambda)^2} \end{bmatrix} \quad (33)$$

where

$$\Omega_o^2 = \gamma^2 H_1 H_2$$

After solving the equations of motions (Eq. 32) for the dynamical susceptibility (Eq. 33), we have to include Maxwell's equations into our considerations in order to be able to describe electomagnetic propagation effects. For layer materials like K_2CuF_4, it is most advantageous to choose the wavevector $\vec{q}$ of the incident radiation parallel to the crystallographic c-axis, i.e. parallel to the y-axis of our coordinate system,

$$\vec{q} = (0, q, 0) \quad (34)$$

As our sample is a thin platelet with its faces perpendicular to the y-axis, we obtain for the demagnetization factors in this geometry

$$N^x = N^z = 0 \qquad N^y = +1 \quad (35)$$

For the dielectric properties, we again assume uniaxial (tetragonal) symmetry:

$$\varepsilon_{xx} = \varepsilon_{zz} = \varepsilon_{\perp} \qquad \varepsilon_{yy} = \varepsilon_{\parallel} \tag{36}$$

With specifications given in Eqs. 34 - 36, the following homogeneous system of linear equations for the components of $\vec{h}$ (cf. Eq. 7) is obtained

$$\begin{aligned} &\left[\frac{(n-ik)^2}{\varepsilon_{\perp}} - (1+\chi^{xx})\right] h^x - \chi^{xy} h^y - \chi^{xz} h^z = 0 \\ &- \chi^{yx} h^x - (1+\chi^{yy}) h^y - \chi^{yz} h^z = 0 \\ &-\chi^{zx} h^x - \chi^{zy} h^y + \left[\frac{(n-ik)^2}{\varepsilon_{\perp}} - (1+\chi^{zz})\right] h^z = 0 \end{aligned} \tag{37}$$

The secular equation of the above systems of equations has two solutions. After inserting the expression for $\underline{\underline{\chi}}$ (Eq.33) into Eq. 37 and after some lengthy algebra, the solutions are

$$\begin{aligned} (n - ik)_1 &= \varepsilon_{\perp}\left[1 + \frac{\omega_o^2 M_o/H_o}{\omega_o^2 - (\omega - i\Lambda)^2}\right. \\ (n - ik)_2 &= \varepsilon_{\perp} \end{aligned} \tag{38}$$

where the resonance frequency ω_o for $H_o \perp c$ under the conditions Eqs. 34 - 36 reads as follows

$$\omega_o^2 = \Omega_o^2 + \gamma^2 M_o H_2 = \gamma^2 (H_o + H_A + M_o) H_o \tag{39}$$

The eigenvectors of the system Eq. 37 are for the two modes with $(n-ik)$ given by Eq. 38

$$\begin{aligned} \vec{h}_1 &= [h_1\cos\Theta,\ 0,\ -h_1\sin\Theta] \\ \vec{h}_2 &= [h_2\sin\Theta,\ 0,\ h_2\cos\Theta\ ] \end{aligned} \tag{40}$$

The results (Eq. 38 and 40) show that the two modes are in this case a magnetic field dependent resonant mode with the oscillatory field $\vec{h}$ perpendicular to the static magnetization M_o (and to the wave vector $\vec{q}$), and a nonmagnetic mode with the oscillating field parallel to the static magnetization. For the calculation of the power transmission (cf. Eq. 12 and Fig. 8), the direction of polarization is assumed to form an angle ψ with the z-axis

$$\vec{h}_o = (h_o \sin\psi\ ,\ 0,\ h_o \cos\psi) \tag{41}$$

Then the amplitudes $\vec{h}_1$ and $\vec{h}_2$ of modes 1 and 2, respectively, are as follows (cf. Eq. 40)

$$\begin{aligned} \vec{h}_1 &= h_o \sin(\psi - \Theta)\ [\cos\Theta, 0, -\sin\Theta] \\ \vec{h}_2 &= h_o \cos(\psi - \Theta)\ [\sin\Theta, 0, \cos\Theta] \end{aligned} \tag{42}$$

The Fresnel coefficients have to be calculated with $\varepsilon_{eff} = \varepsilon_{\perp}$ (cf. Eq. 10). If the analyzer is set parallel to the direction of polarization of the incident radiation, we obtain for the power transmission of the sample (cf. Eq. 12)

$$T_{\parallel} = \left| \sin^2(\psi - \Theta)\ \tilde{t}_1 + \cos^2\ (\psi\ - \Theta)\ \tilde{t}_2 \right|^2 \tag{43}$$

where $\tilde{t}_1$ and $\tilde{t}_2$ are again the amplitude transmission of the sample as given by Eq. 9. If, on the other hand, the analyzer is set perpendicular to the direction of polarization of the incident radiation, the power transmission is

$$T_{\perp} = \sin^2\ (\ \psi - \Theta)\ \cos^2\ (\ \psi - \Theta)\ \left| \tilde{t}_1 - \tilde{t}_2 \right|^2 \tag{44}$$

After these more theoretical considerations it seems appropiate to present and discuss our experimental data on K_2CuF_4 obtained for $H_o \perp c$. In Fig. 13, the measured intensity $I(B)$ is shown as function of applied magnetic field for analyzer parallel or perpendicular to the direction of incident radiation ($I_{\parallel}(B)$ or $I_{\perp}(B)$, resp.), normalized with $I_{\parallel}(B=0)$ (Kullmann et al., to be published). The direction of polarization of the radiation was adjusted perpendicular to the direction of the static field by rotating the whole IMPATT source. Nominally, this would mean $\psi - \Theta = \pi/2$, $\sin(\psi\ - \Theta) = 1$, $\cos(\ \psi - \Theta) = 0$ and, thus, $T\ = 0$ and no contribution of $\tilde{t}_2$ to $T_{\parallel}$ in Eqs. 43 and 44. Since the direction of polarization was not aligned perfectly at right angle to the static field and since the analyzer efficiency is not 100%, there was observed a large signal $I_{\parallel}(B)/\ I_{\parallel}(B=0)$ as expected and a small signal $I_{\perp}(B)/I_{\parallel}\ (B=0)$ due to

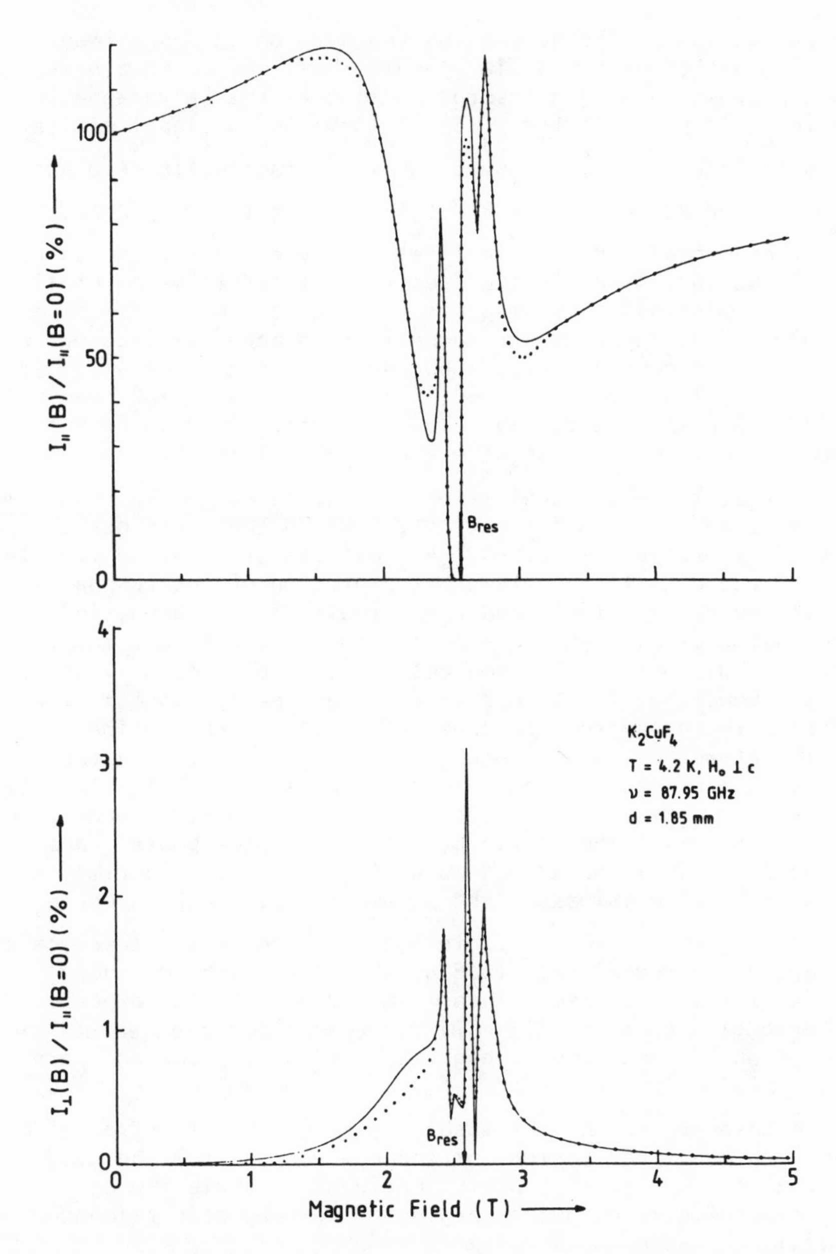

Figure 13. Normalized transmitted intensity of K_2CuF_4 near ferromagnetic resonance versus magnetic field: $I_{\parallel}(B)/I_{\parallel}(B=0)$ with the analyzer parallel and $I_{\perp}(B)/I_{\parallel}(B=0)$ with the analyzer perpendicular to the direction of polarization of the incident radiation, experimental (.....) and calculated (———).

the imperfections. The surprising features of the experimental data, however, are the relatively large number of lines. In a naive way, one would expect a single absorption line at the ferromagnetic resonance (B_{res} in Fig. 13) for $I_{\parallel}(B)/I_{\parallel}(B=0)$ and a signal similar to the middle trace ($H_{\perp}$) in Figs. 9 and 10 (paramagnetic resonance of MAMC at 80 and 220 K) for $I_{\perp}(B)/I_{\parallel}(B=0)$ near the resonance. When these experimental curves were first observed the origin was not clear for so many lines in the spectra. At first, we carefully checked the possibility of magnetostatic modes which give rise to additional lines in microwave magnetic resonance spectroscopy with the sample in a cavity (Auld, 1960; Borovik-Romanov et al., 1980; Walker, 1957, 1958; Dillon, 1960; Fletcher et al., 1960; Damon et al., 1961; Sparks et al., 1969; Yukawa et al., 1974). From the value of the static magnetization at 4.2 K ($M_o \approx 0.112$ T) it is easily verified that the additional lines in Fig. 13 cannot be due to magneto-static modes. Similar estimates can further rule out the possibility of hyperfine splittings. All effects originating from large powers of the oscillating field $\vec{h}$ (saturation of the resonance, etc.) (Bloembergen et al., 1952; Anderson et al., 1955; Damon, 1953, Suhl, 1956;Clogston et al.,1956) are also not possible in our case. Although our home-made IMPATT sources transmit at a power level of 20-30 mW the power level at the sample is reduced to about 1 mW due to the losses in the system and thus too small for saturation effects, etc. Therefore, we had to look for an origin other than the above-mentioned one in order to explain the line structure in Fig. 13. The only origins found by us are the electromagnetic propagation effects. This is demonstrated by calculating the power transmission by means of Eqs. 41 and 42 and comparing it to the experimental data (cf.Fig. 13). The values of the parameters of K_2CuF_4 used in the calculation are compiled in Table 3. The agreement between calculated and experimental data in Fig. 13 is astonishingly good, at least where the line positions are concerned. In the experimental data, some of the maxima and minima are somewhat smeared out due to imperfections of the sample, e.g. not perfectly plane and parallel faces. This is true also for $I_{\perp}(B)$ with the value $\psi - \Theta = 85°$ in order to take account of the misalignment of the direction of polarization and the imperfections of the analyzer. From the good agreement in Fig. 13 it is again concluded that electromagnetic propagation effects cannot be neglected for magnetic resonance in the millimeter-wave range.

Let us now discuss the details of the propagation effects in order to elucidate the origin for the line structure in a more specific way. Our considerations about the power transmission (cf. Eqs. 41 and 42) show that for $H_o \perp c$ and for $\psi - \Theta \approx \pi/2$, interference between the two modes play a negligible role in this case. In a ferromagnet however, the oscillator strength of the

TABLE 3: Values of some parameter for K_2CuF_4

gyromagnetic ratio	$\gamma_\perp$ = 31.96 GHz/T ($g_\perp$ = 2.284) (a,b) $\gamma_\parallel$ = 29.22 GHz/T ($g_\parallel$ = 2.088) (a,b)
dielectric constant	$\varepsilon_\perp = \varepsilon'_\perp - i\varepsilon''_\perp$ $\varepsilon'_\perp$ = 8.29 $\varepsilon''_\perp$ = 0.01 (c)
saturation magnetization	M_o = 0.112 T ≙ 3.58 GHz (4.2 K) (b,d)
anisotropy field	H_A = 0.205 T ≙ 6.55 GHz (4.2 K) (b,e)
damping constant	Λ = 0.01 GHz ≙ 3.10^{-4} T (T = 4.2 K)
sample thickness	d = 1.85 mm
polarization angle	$\phi - \theta \approx 85°$

(a) H.Yamasaki (1974).

(b) W. Kullmann et al. (to be published).

(c) K. Strobel and R. Geick (to be published).

(d) K. Hirakawa and H. Ikeda (1973)

(e) H. Yamasaki (1974)

resonance, i.e. the numerator in the expressions for dynamic susceptibility (cf. Eqs. 4,14, and 33), is much larger than in a paramagnet since the saturation magnetization M_o (= 112 mT) of the ferromagnet K_2CuF_4 is about 35 times larger than the paramagnetic magnetization of MAMC (= 3 mT) at 60 K. That means a $\sqrt{35}$ times larger influence on the optical constants n and k in the case of K_2CuF_4 and that we need not look for temperatures where Λ becomes sufficiently small that χ becomes of order unity. Of course, the damping constant Λ is, in the case of K_2CuF_4, a fit parameter, as in the case of MAMC, and has no simple relation to an actually observed line width. For K_2CuF_4, χ exceeds unity by far for all temperatures where the magnetization is sufficiently large. In order to demostrate these effects, we have plotted in Fig. 14 the calculated values of the refractive index n and of the absorption coefficient k of K_2CuF_4 as a function of applied magnetic field (B in Fig. 14) and show for comparison the transmission $I_{\parallel}(B)/I_{\parallel}(B=0)$, experimental and calculated. The calculated curve for the refractive index n shows that it varies between 0 and 15 in the neighborhood of the resonance field (B_{res}) while its "normal value" outside the resonance is about 3. This variation causes a number of maxima and minima in the transmission due to interferences for reasons of multiple reflections in the sample. As is well known, we obtain maxima or minima of order N or N' if $2dn = N\lambda$ or $2dn = (N'-1/2)\lambda$, respectively. The corresponding order numbers are indicated in Fig. 14 and it is clearly seen that all structures in the transmission curve are explained by the interference effects, except for the transmission minimum near B_{res} which is due to the usual resonance absorption. In this respect it is worth mentioning that the transmission is practically zero in the range 2.5 to 2.6 T where n is nearly zero and k is large, and that the actual resonance occurs at the high, field edge of this range near 2.6 T. This broad transmission maximum is an experimental evidence for the "gap" in the dispersion curve which is obtained when $c^2q^2/\omega^2 = (n - ik)_1$ (cf. Eq. 38) is brought to the form $\omega = f(k)$. Neglecting the damping in these considerations ($\Lambda = 0$, $\varepsilon'' = 0$), we obtain from

$$\frac{c^2q^2}{\omega^2} = \varepsilon_\perp \left[1 + \frac{\omega_o^2 M_o/H_o}{\omega_o^2 - \omega^2}\right] \tag{45}$$

the dispersion relation (K_2CuF_4, $H_o \perp c$)

$$\omega^2 = \frac{1}{2}\left[\omega_o^2\left(1 + \frac{M_o}{H_o}\right) + \frac{c^2q^2}{\varepsilon_\perp} \pm \sqrt{\left[\omega_o^2\left(1 + \frac{M_o}{H_o}\right) + \frac{c^2q^2}{\varepsilon_\perp}\right]^2 - 4\omega_o^2\frac{c^2q^2}{\varepsilon_\perp}}\right] \tag{46}$$

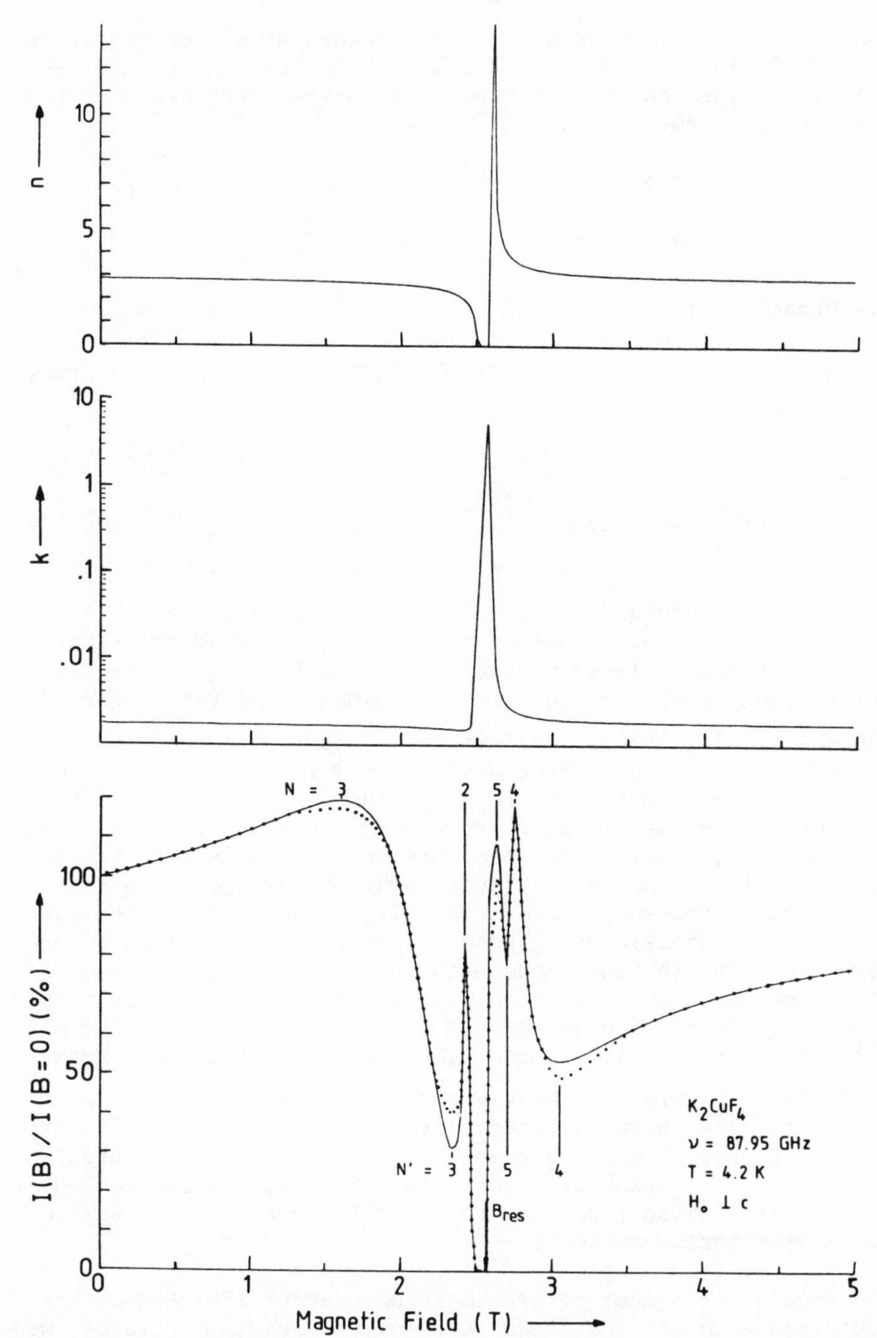

Figure 14: Comparison of the refractive index n, the absorption coefficient k and the normalized transmitted intensity I(B)/I(B=0) for K_2CuF_4 near ferromagnetic resonance: Experimental (.....) and calculated (———).

As shown in Fig. 15, there are three branches of dispersion curves $\omega(q)$ if we include the photon-like branch $\omega = cq/\varepsilon$ (cf. Eq. 38) of the other, the nonresonant mode. The asymptotic values of the frequencies of these branches are

$$\begin{aligned}
&\text{lower branch: small } q: \omega \approx cq/\sqrt{\varepsilon_\perp(1 + M_o/H_o)} && \text{(EM-wave)}\\
&\qquad\qquad\text{large } q: \omega = \omega_o = \gamma\sqrt{H_o(H_o+H_A+M_o)} && \text{(spin-wave)}\\
&\text{middle branch: all } q: \omega = cq/\sqrt{\varepsilon_\perp} && \text{(EM-wave)}\\
&\text{upper branch: small } q: \omega = \omega_o\sqrt{1 + M_o/H_o} && \text{("longitudinal")}\\
&\qquad\qquad\text{large } q: \omega \approx cq/\sqrt{\varepsilon_\perp} && \text{(EM-wave)}
\end{aligned} \tag{47}$$

Here again, small and large values of q mean $q \ll \omega_o\sqrt{\varepsilon_\perp}/c$ and $1/a \gg q \gg \omega_o\sqrt{\varepsilon_\perp}/c$, respectively. The lower branch starts as an electromagnetic wave or photon (EM-wave in Eq. 47) for small q and passes into a spin wave branch (magnon) for large q. Please note that there is no perceptible spin wave dispersion for $q \parallel c$ (the case we are discussing here) in the quasi two-dimensional ferromagnet K_2CuF_4 (Funahashi et al., 1976). That is the reason why no bending of the lower branch up to higher frequencies resulting from the exchange interaction term proportional to q^2 is shown, in contrast to similar graphs for three-dimensional ferromagnets (Auld, 1960; Pincus, 1962; Akhiezer et al., 1968). The upper branch in Fig. 15 (cf. Eq. 47) starts at small q from the frequency, which corresponds to the longitudinal mode frequency (Born et al., 1954) and this branch passes into an electromagnetic wave (EM-wave) or photon-like branch for large q. For the resonant mode with $(n - ik)_\uparrow$ which corresponds to the lower and upper branch in Fig. 15, there is a gap or forbidden zone between these two branches. The width of this gap is $\Delta\omega = \omega_o(\sqrt{1 + M_o/H_o} - 1) \triangleq 1.88$ GHz = 0.06 T which is the range between 2.54 and 2.60 T where n is nearly zero and k is relatively large (cf. fig. 14) and where the transmission $I_\uparrow(B)$ is very small (cf. Figs. 13 and 14). Thus we are able to explain our complicated transmission spectra by means of propagation effects, and our conclusion is that these play an important role for ferromagnetic resonance in the millimeter-wave range.

From a great number of transmission spectra like Figs. 13 and 14, obtained with various frequencies at various temperatures, we have extracted the magnetic field at resonance B_{res} in order to obtain information about the properties of K_2CuF_4 at relatively high magnetic fields. Figs. 16 and 17 show the results obtained

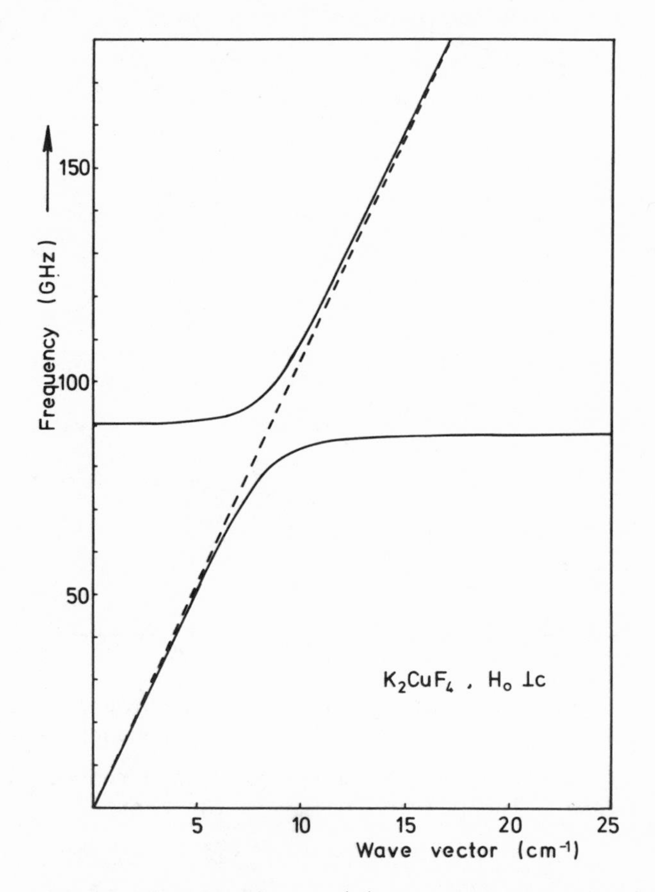

Figure 15: Dispersion relations ω(q) for K_2CuF_4 at 4.2 K with the external field perpendicular to the c-axis and the wavevector q parallel to it. For further details and for values of the parameters see text.

for $H_o \perp c$ and $H_o \parallel c$, respectively. In other words, these investigations have not only been performed for the case $H_o \perp c$, which has been discussed in detail in the preceding paragraphs, but also for $H_o \parallel c$. When treating the more general problem of the external field applied not only in the basal plane but in arbitrary directions, it is no longer suitable and justified to use $\gamma = g_\perp \mu_B/\hbar$

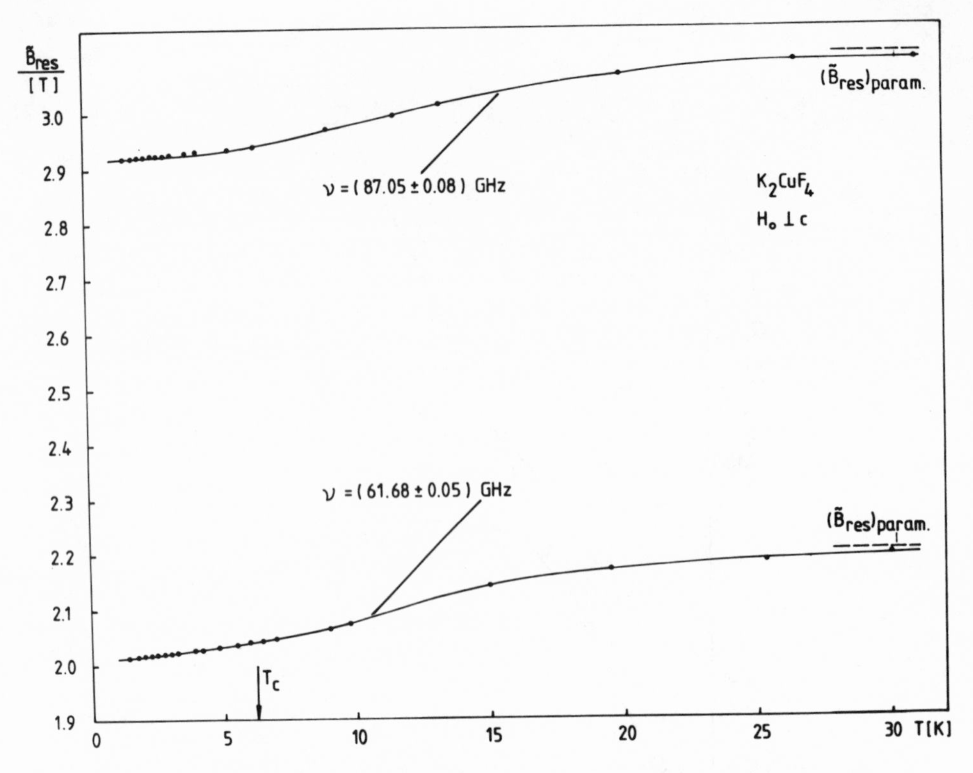

Figure 16: Temperature variation of the ferromagnetic resonance of K_2CuF_4 : magnetic field at resonance $\tilde{B}_{res}$ for two frequencies in the millimeter wave range. The field is applied perpendicular to the c-axis.

in the equations of motion Eq. 32. For reasons of more clarity and in order to retain a simple form of the equations of motion, it is advisable to define artificial fields (Yamazaki, 1974) $\tilde{H}$ and $\tilde{B}$ in the more general case

$$\tilde{H}^{\alpha} = (g_{\alpha\alpha}/2.00)\ H^{\alpha} \qquad \alpha = x,y,z \qquad (48)$$
$$\tilde{B}^{\alpha} = (g_{\alpha\alpha}/2.00)\ B^{\alpha}$$

That means the components of the actual fields are scaled to g=2.00 by multiplying with the appropiate factor $(g_{\alpha\alpha}/2.00)$. For $\tilde{H}$ (and/or $\tilde{B}$), we are allowed to use a unique factor $\gamma = 2.00\mu_B/\hbar$ in the equation of motions similar to Eq. 32. They read with the internal field $\vec{H}_{intern}$ (cf. Eq. 28) as derived from the energy density U_{FM}

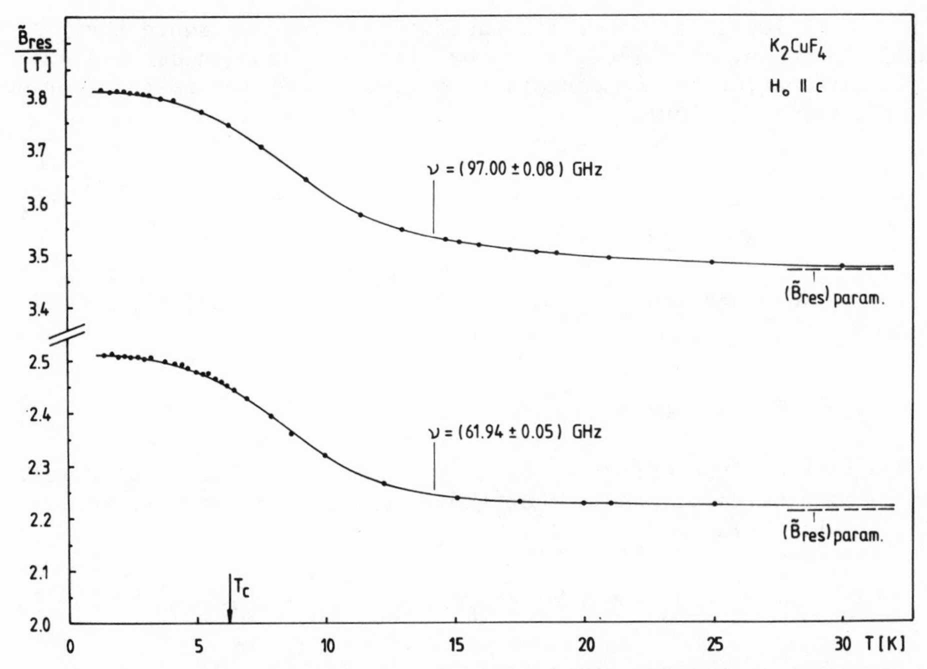

Figure 17: Temperature variation of the ferromagnetic resonance of K_2CuF_4 : magnetic field at resonance B_{res} for two frequencies in the millimeter wave range. The field is applied parallel to the c-axis.

(cf. Eq. 27) and with external field $\vec{H}_o$ parallel to the c-axis, i.e. the y-axis (cf. Eq. 1),

$$\dot{m}^x = -\gamma\ [(\tilde{H}_o - N^y\tilde{M}_o - \tilde{H}_A)\ m^z - M_o\ \tilde{h}^z] - \Lambda m^x$$
$$\dot{m}^z = \gamma\ [(\tilde{H}_o - N^y\tilde{M}_o - \tilde{H}_A)\ m^x - M_o\ \tilde{h}^x] - \Lambda\ m^z \tag{49}$$

where $\gamma = 2.00\mu_B/\hbar$, and $\tilde{H}_o$, $\tilde{H}_A$, $\tilde{h}^x$, etc., are the fields scaled according to Eq. 48. As in Eq. 32, we have included the demagnetization field in Eq. 49. The solution of the inhomogeneous system of linear equations yields two circularly polarized modes in this case

$$m^\pm = (m^x \pm im^z) = \chi^\pm h^\pm = \chi^\pm(h^x \pm ih^z) \tag{50}$$

with

$$\chi^\pm = \gamma M_o/[\Omega_o^2 - (\omega - i\Lambda)^2]$$

and

$$\Omega_o = \gamma(\tilde{H}_o - N^y\ \tilde{M}_o - \tilde{H}_A)$$

For the inclusion of Maxwell's equations, we assume again the conditions Eqs. 34 - 36. Then, the resulting homogeneous system of equations for the components of $\vec{h}$ (cf. Eq. 7) may be transformed to the following form

$$[(n - ik)^2 - \varepsilon_\perp(1 + \chi^\pm)]\, h^\pm = 0 \qquad (51)$$
$$h^y = 0$$

Its solutions are, apart from $h^y = 0$, again two circularly polarized modes with

$$(n - ik)_\pm = \varepsilon_\perp(1 + \chi^\pm) \qquad (52)$$

and with the eigenvectors $h^\pm$ as given in Eq. 50. The final result of these considerations is that the resonance frequency is for $H_o \parallel c$ with $k \parallel c$ and $N^y = 1$

$$\Omega_o = \gamma\,(\tilde{H}_o - \tilde{M}_o - \tilde{H}_A) = (g_\parallel \mu_B/\hbar)(H_o - M_o - H_A) \qquad (53)$$

as compared to the resonance frequency ω_o (cf. Eq. 39) for $H_o \perp c$. These considerations explain why the magnetic field at resonance $\tilde{B}_{res}$ is smaller than the paramagnetic value in Fig. 16 for $H_o \perp c$. Paramagnetic value in this context means that $\tilde{B}_{res}$ corresponds to $\omega = \gamma\tilde{H}_o$, or that the internal fields (H_A, M_o) are zero. We remember that the experiments are performed with constant frequency and that the external field H_o is varied to obtain the resonance. Under these conditions, it follows from Eq. 39 ($H_o \perp c$) that resonance occurs at smaller values of $\tilde{B}_{res}$ for $H_A \neq 0$, $M_o \neq 0$ than for the case $H_A = 0$, $M_o = 0$. Fig. 17 shows, on the other hand, that the resonance field $\tilde{B}_{res}$ for $H_o \perp c$ is larger than the paramagnetic value in agreement with Eq. 53 (for constant frequency). The basic difference between the two configurations leading to the results discussed above is that for $H_o \perp c$ the magnetization and thus the internal field are parallel to H_o since the magnetization is allowed to rotate almost freely in the basal plane due to the smallness of the inplane anisotropy field (Yamasaki et al., 1981). In this situation, the external field for resonance has to be reduced when the internal field is "switched on" by lowering the temperature. For $H_o \parallel c$ on the other hand, the external field is at first perpendicular to the magnetization, and a certain strength of the external field is necessary so that the energy density from the external field may overcome the anisotropy energy, and so that it is energetically more favorable for the magnetization to rotate towards the direction of the external field. For these reasons, the "switching on" of the internal field leads to a higher external one for resonance in this case.

Now let us turn to the temperature variation of the resonance field $\tilde{B}_{res}$ for the two cases under consideration ($H_o \perp c$ and $H_o \parallel c$) as exhibited by the curves in Figs. 16 and 17. We encounter a smooth variation from the paramagnetic value of $\tilde{B}_{res}$ ($\hat{=} \gamma H_o$, cf. Eq. 16) to the ferromagnetic value of $\tilde{B}_{res}$ at low temperatures. This smooth variation extends over a wide temperature range up to temperatures far above the Curie temperature T_c. The external field already forces the system at these temperatures to a high degree of order. But this will certainly not be a long range order, and the correlation length will stay finite. These special high field properties will be discussed in more detail in a forthcoming publication (Kullmann et al., to be published). But the values of $\tilde{B}_{res}$ at low temperatures and the values for H_A and M_o extracted from them agree very well with corresponding data obtained by other groups with considerably lower frequencies, i.e. at lower fields. However, the comparison has to be performed at low temperatures ($T \ll T_c$). For K_2CuF_4, unfortunately, we obtain the combination $A = H_A + M_o$ from the experimentally determined $\tilde{B}_{res}$ of both configurations $H_o \perp c$ and $H_o \parallel c$ (cf. Eqs. 39 and 53), respectively. Thus we are unable to provide experimental values for the two quantities, namely the anisotropy H_A and the magnetization M_o, separately. The temperature variation of the normalized quantity $A(T)/A(0)$ and the extrapolation of our data to zero temperature is shown in Fig. 18. For comparison, there are also given the corresponding data obtained with lower fields, i.e. at lower frequencies. For H_A, we have used the values of H_A reported by Yamazaki (1974), and values of $M_o(T)$ have been calculated by means of the $T^{3/2}$-law (Hirakawa et al., 1973). The value of $M_o(0)$ has been calculated by means of the structural data for K_2CuF_4 (Hirakawa et al., 1973) neglecting any zero-point spin-deviation (Schröder et al., 1980). With $M_o(0) = 0.121$ T, we obtain from our experimental data $H_A(0) = 0.222$ T from the extrapolated value $A(0) = 0.343$ T. This value for H_A agrees quite well with that reported by Yamazaki (1974) for zero temperature. From Fig. 18, it is clearly seen that our experimental data decrease much less with increasing temperature than the data of Yamazaki (1974)

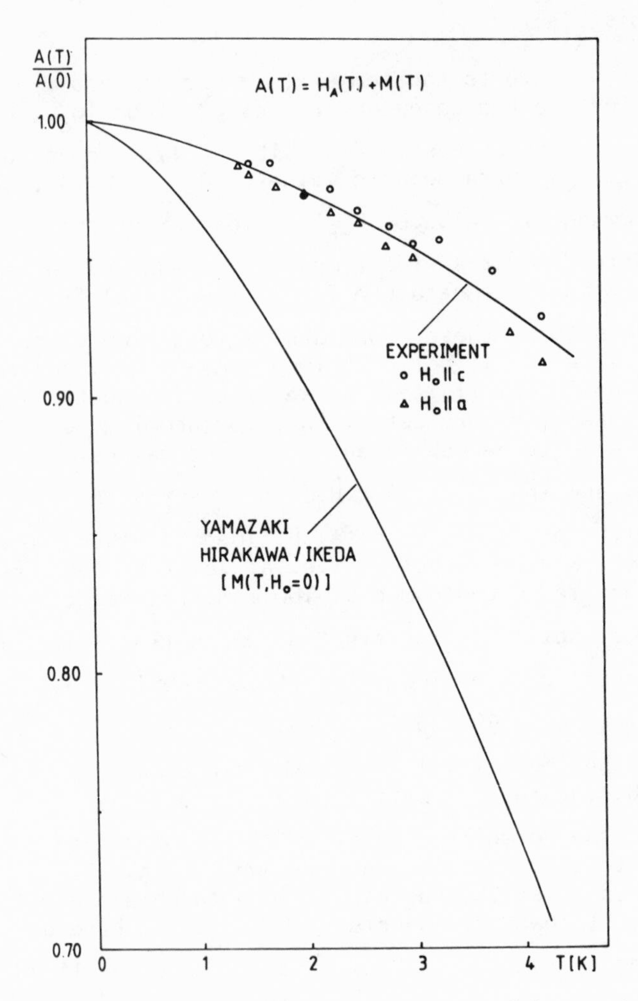

Figure 18: Temperature variation of $A(T) = H_A(T) + M_o(T)$ as obtained by the present investigation in the millimeter wave range for $H_o \parallel c$ and $H_o \perp c$, i.e. $H_o \parallel a$, and as reported for lower external fields by Yamazaki (1974) [$H_A(T)$] and by Hirakawa and Ikeda (1973) [$M_o(T)$].

and Hirakawa and Ikeda (1973). Therefore, our values for T = 4.2 K ($A = 0.316$ T, $M_o = 0.112$ T, $H_A = 0.205$ T, cf. Table 3) are not comparable to their data. Similar observations, i.e. an increase of the magnetization with increasing external field, have been reported for fields up to about 0.15 T (Hirakawa et al., 1979, 1981). In our case, the effects are more drastic as we apply fields up to 4 T. Moreover, we observe analogous effects for the anisotropy field. On the whole, we may conclude that magnetic resonance in the millimeter wave range is well suited to provide accurate information about the internal fields of the system, though the measured temperature variation is strongly infuenced by the application of relatively high magnetic fields (Kullmann et al.).

4.3. Magnetic Resonances in Antiferromagnets

4.3.1. General Considerations

As already mentioned in the introduction and demonstrated with examples in Table 1, there exist a great number of materials with antiferromagnetic order among the perovskite-type layer structures, mostly manganese compounds. In these quasi-two-dimensional uniaxial antiferromagnets, the value of the exchange field is similar to that of manganese compounds which are ordinary three-dimensional antiferromagnets, e.g. $H_E = 54.0$ T for MnF_2 (Manohar et al., 1972) and $H_E = 87.1$ T (Schröder et al., 1980) for Rb_2MnCl_4. But the anisotropy field is much smaller for the quasi-two-dimensional antiferromagnets, namely $H_A = 0.88$ T for MnF_2 (Manohar et al., 1972) and $H_A = 0.172$ T for Rb_2MnCl_4 (Schröder et al., 1980). Therefore, the zero-field antiferromagnetic resonance frequency is considerably diminished, $\nu = \gamma\sqrt{2H_EH_A} = 273$ GHz for MnF_2 and $\nu = 153$ GHz for Rb_2MnCl_4. Moreover, the magnetic ions are less densely packed in the layer structures and, accordingly, the sublattice magnetization is smaller, e.g. $M_o = 0.741$ T for MnF_2 (Manohar et al., 1972) and $M_o = 0.141$ for Rb_2MnCl_4 (Schröder et al., 1980). On the basis of these values, it is evident that magnetic resonance in the millimeter-wave range is very well suited for the investigation of antiferromagnetic perovskite-type layer materials. It is also advantageous that high magnetic fields (up to 10 T) are at our disposal. As is well known, uniaxial easy-axis antiferromagnets undergo a transition from the antiferromagnetic to the spin-flop phase if a sufficiently large external magnetic field is applied parallel to the easy axis (Keffer, 1966). The value of the critical field for this transition ($H_c \approx \sqrt{2\,H_EH_A}$) is easily attained by means of a superconducting solenoid if the zero-field antiferromagnetic resonance frequency falls into the millimeter-wave range. For example, $\nu = 126$ GHz corresponds to $H_c = 4.5$ T. In other words, the experimental apparatus described in Section 2 enables us not only to observe the antiferromagnetic resonance (AFMR) but also the magnetic resonance in the spin-flop regime (denoted spin-flop resonance, abbreviated SFR) in the materials under consideration.

In context with the discussion of the paramagnetic line width in Section 4.1, it was already mentioned that we have investigated Rb_2MnCl_4 as an example of uniaxial antiferromagnets among the perovskite-type layer structures. Its structure (Seifert et al., 1965) is the K_2NiF_4 structure I4/mmm (D_{4h}^{17}) and, in the antiferromagnetic order, the spins are aligned parallel or antiparallel to the crystalline c-axis (Schröder et al., 1980; Epstein et al., 1970). From the point of view of their magnetic structures, the materials with molecular ions like $(C_nH_{2n+1}NH_3)_2MnCl_4$ and $NH_3(CH_2)_mNH_3MnCl_4$ are very similar to Rb_2MnCl_4 (cf. Fig. 2). But the structure of these materials at low temperatures is I4/mmm only in an approximative way when the orientation of the molecules and the canting of the manganese-chlorine octahedra is neglected. The actual low-temperature structures of some of the compounds under consideration are compiled in Table 4 where the average Mn-Mn distance within the layers and the layer separation are also given. Moreover, the materials with molecular ions of the type $(C_nH_{2n+1}NH_3)_2MnCl_4$, the monoammonium compounds, usually exhibit a sequence of structural phase transitions (Knorr et al., 1974; Chapuis et al., 1975; Heger et al., 1975, 1976). For diammonium compounds, i.e. the type $NH_3(CH_2)_mNH_3MnCl_4$, the tendency for structural phase transitions is less pronounced. At room temperature, they usually already have a structure the symmetry of which is considerably less than tetragonal. The reason for the different behavior of monoammonium and diammonium compounds is that the organic ions are bonded in a different way to the perovskite-type layers consisting of manganese chlorine octahedra, as shown in Fig. 19. The monoammoium ions are linked with their NH_3-group to the next perovskite-type layer by hydrogen bridging and form double layers with the CH_3-groups facing each other. The diammonium ions, on the other hand, are linked with their two NH_3-groups to two neighboring perovskite-type layers by hydrogen bridges, thus forming a more stable lattice, as in the case of the monoammonium compounds. Magnetically, the essential difference between the two types of compounds is that the manganese ions form, approximately, a simple tetragonal lattice in the case of the diammonium compounds, and a body-centered tetragonal lattice in the case of the monoammonium compounds (cf. Fig.2). In both types of perovskite-type layer structures with organic ions, the magnetic moments are aligned parallel to the c-axis, i.e. perpendicular to the layers (Van Amstel et al., 1972; Arend et al., 1976; Nosselt et al., 1976; Groenendijk et al., 1979). However, most of the compounds

TABLE 4: Structures of some manganese compounds

	Structure at low temperatures	Structure parameters	$\bar{a}$ (a)	$\bar{c}$ (b)
		Rb_2MnCl_4		
	I4/mmm (D_{4h}^{17})	a = 5.05 Å, c = 16.18 Å (c)	5.05 Å	8.09 Å
		$(C_nH_{2n+1}NH_3)_2MnCl_4$		
n = 1	$P2_1/b$ (C_{2h}^{5})	a = 7.133 Å, b = 7.253 Å, c = 19.35 Å, β = 92.17° (d,h)	5.09 Å	9.68 Å
n = 2	Pbca (D_{2h}^{15})	a = 7.325 Å, b = 7.151 Å, c = 22.04 Å (f-h)	5.12 Å	11.02 Å
n = 3	Pbca (D_{2h}^{15})	a = 7.45 Å, b = 21.41 Å, c = 25.54 Å (f)	5.16 Å	12.77 Å

TABLE 4: Structures of some manganese compounds (continued)

	Structure at low temperatures	Structure parameters	$\bar{a}$(a)	$\bar{c}$(b)
		$NH_3(CH_2)_mNH_3MnCl_4$		
m = 2	$P2_1/b$ (C^5_{2h})	a = 7.130 Å, b = 7.192 Å, c = 8.61 Å, β = 92.68° (i,l)	5.06 Å	8.61 Å
m = 4	$P2_1/b$ (C^5_{2h})	a = 7.177 Å, b = 7.307 Å, c = 10.77 Å, β = 92.67° (i,k,m)	5.12 Å	10.77 Å
m = 5	orthorhombic	a = 7.152 Å, b = 7.360 Å, c = 23.99 Å	5.13 Å	12.00 Å

(a) $\bar{a}$ = average Mn-Mn distance within the layers (cf.Fig.2)

(b) $\bar{c}$ = separation of adjacent layers (cf. Fig. 2)

(c) H.J. Seifert and F.W. Koknat (1965)

(d) K. Knorr et al. (1974)

(e) G. Chapuis et al. (1976)

(f) W. Depmeieret al. (1977)

(g) W. Depmeier (1977)

(h) G. Chapuis (1977)

(i) K. Tichy and H. Arend (1975)

(k) H. Arend et al. (1976)

(l) K. Tichy et al. (1978)

(m) K. Tichy et al. (1980)

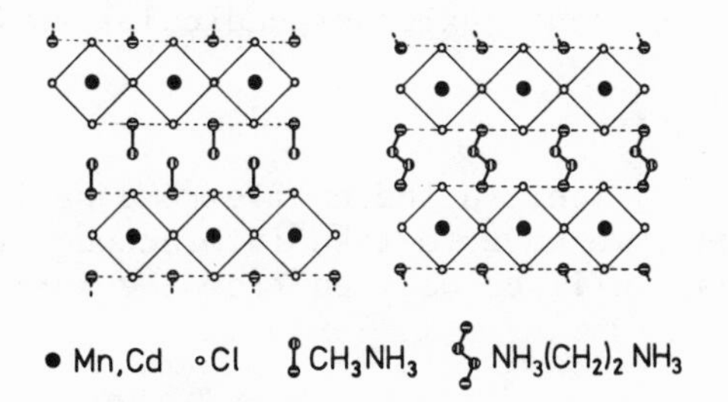

Figure 19: Schematic diagram of the hydrogen-bonding of the molecular ions to the transition metal-chlorine octahedra, in the case of monoammonium compounds $(C_nH_{2n+1}NH_3)_2MnCl_4$ (to the left) and diammonium compounds $NH_3(CH_2)_mNH_3MnCl_4$ (to the right).

under consideration show a small spin canting (Gerstein et al., 1974 a,b) [≈ 0.05° for $(C_3H_7NH_3)_2MnCl_4$ (Groenendijk et al., 1979)] which is ascribed to the antisymmetric Dzialoshinski-Moriya interaction (Dzialoshinski 1957, 1958; Moriya, 1960). The resulting small ferromagnetic moment has been detected in susceptibility measurements from a sharp peak at the transition temperature. Due to the smallness of these effects, we shall neglect them in the discussion of our magnetic resonance results and not add a term of the form$(H_D/M_o)\cdot(\vec{M}_a \times \vec{M}_b)$ to the energy density U_{AFM} of the antiferromagnet. For $(C_3H_7NH_3)_2MnCl_4$ for example, the ratio of the Dzialoshinski field H_D to the exhange field H_E is $H_D/H_E \approx 2.10^{-3}$ (Groenendijk, 1979).

According to the values of the characteristic magnetic quantities as given in the first paragraph of this section, especially that of M_o, electromagnetic propagation effects will not be very important for AFMR and SFR studies of quasi-two-dimensional antiferromagnets. Nevertheless, we shall include the corresponding considerations here in order to use the results for the discussion and analysis of our experimental data. On the other hand, the results will help us to understand why the propagation effects are not so important for the antiferromagnets under consideration. The following discussion will be limited to the case where the external field is applied parallel to the easy axis, i.e. the c-axis. We choose the coordinate system in such a way that the z-axis is parallel to the c-axis and the x- and y-axis perpendicular to it. Then, we have

$$\vec{H}_o = (0, 0, H_o). \tag{54}$$

In the ordered state, i.e. in the antiferromagnetic as well as in the spin-flop phase, we have to take into account the internal field $\vec{H}_{intern}$, which again will be derived from the energy density of the system

$$U_{AFM} = \frac{H_E}{M_o} \vec{M}_a \vec{M}_b - \frac{1}{2} \frac{H_A}{M_o} [(M_a^z)^2 + (M_b^z)^2] + \frac{1}{2} \frac{\hat{H}_A}{M_o} [(M_a^x M_a^y)^2 + (M_b^x M_b^y)^2] \tag{55}$$

In Eq. 55, H_E is the exchange field originating from the dominating intralayer exchange interaction between reighbors which enforces the antiferromagnetic order. Exchange interactions between further neighbors within the layers and between adjacent layers are also very small for the antiferromagnetic perovskite-type layer materials and will be neglected. H_A is the out-of-plane anisotropy field which causes the alignment of the spins parallel or antiparallel to the c-axis. H_A is the probably much smaller in-plane anisotropy field, which is effective only when the spins are flopped into the basal plane. M_o is the temperature-dependent sublattice magnetization, and $\vec{M}_a$ and $\vec{M}_b$ are the magnetizations of the sublattices denoted by subscript a and b, respectively. The superscripts x, y, z denote their respective Cartesian components. In order to include the possibility that the spins flop from their orientation along the c-axis to an orientation perpendicular to it we have to introduce the angles θ_a and θ_b between the directions of $\vec{M}_a$ and $\vec{M}_b$, respectively, and the c-axis. If we include the oscillating or precessing components of $\vec{M}_a$ and $\vec{M}_b$ we obtain

$$\begin{aligned} \vec{M}_a &= [\ (M_o + \Delta M)\sin\theta_a + \tilde{m}_a\cos\theta_a,\ m_a^y,\ (M_o + \Delta M)\cos\theta_a - \tilde{m}_a\sin\theta_a] \\ \vec{M}_b &= [-(M_o - \Delta M)\sin\theta_b + \tilde{m}_b\cos\theta_b,\ m_b^y, -(M_o - \Delta M)\cos\theta_b - \tilde{m}_b\sin\theta_b] \end{aligned} \tag{56}$$

In Eq. 56, $2\Delta M = \chi_\parallel H_o$ is the strongly temperature dependent magnetization caused by H_o parallel to the c-axis ($\chi_\parallel \to 0$ for $T \to 0$). In analogy to the case of a ferromagnet (cf. Eq. 28), the internal fields acting on sublattice a and b are derived by differentiation of the energy density U_{AFM} (cf. Eq. 55)

$$\vec{H}_{a,\ intern} = -\left(\frac{\partial U_{AFM}}{\partial M_a^x}, \frac{\partial U_{AFM}}{\partial M_a^y}, \frac{\partial U_{AFM}}{\partial M_a^z} \right)$$

$$\vec{H}_{b,\ intern} = -\left(\frac{\partial U_{AFM}}{\partial M_b^x}, \frac{\partial U_{AFM}}{\partial M_b^y}, \frac{\partial U_{AFM}}{\partial M_b^z} \right) \tag{57}$$

With the internal fields obtained by means of Eq. 57 and with $\vec{H}_o = (0, 0, H_o)$, i.e. the effective fields $\vec{H}_{a,eff}$ and $\vec{H}_{b,\ eff}$ acting on sublattices a and b, respectively, time-independent torques are obtained in the equations of motion (cf. Eq. 1') which must vanish in equilibrium. This leads to the following equilibrium conditions

$$H_E\ M_o\ \sin(\Theta_a - \Theta_b) - H_A\ M_o\ \sin\Theta_a\ \cos\Theta_b - (M_o + \Delta M)\ H_o\ \sin\Theta_a = 0$$

$$-H_E\ M_o\ \sin(\Theta_a - \Theta_b) - H_A\ M_o\ \sin\Theta_b\ \cos\Theta_b + (M_o + \Delta M)\ H_o\ \sin\Theta_b = 0 \tag{58}$$

We introduce now a common polar angle Θ for the two sublattices and a canting angle ϕ which results from the influence of the external field after the spins have been flopped

$$\Theta_a = \Theta - \phi \qquad\qquad \Theta_b = \Theta + \phi \tag{59}$$

Please note that the angles have been defined here in accordance with the literature, e.g. with the review by F. Keffer (1966), but with a different meaning, especially for ϕ, in comparison to the angles Θ and ϕ defined by means of Eq. 30 and Eq. 41, respectively, in the considerations for a ferromagnet in section 4.2. By inserting Eq. 59 into Eq. 58 and by forming the difference and the sum of the two equations Eq. 58 we obtain a theoretical expression (Keffer, 1966) for the perpendicular susceptibility $\chi_\perp$

$$\sin\phi = \frac{H_o \sin\Theta}{2H_E + H_A \cos 2\Theta}$$

or

$$\chi_\perp = \frac{2\ M_o\ \sin\phi}{H_o \sin\Theta} = \frac{2\ M_o}{2\ H_E + H_A \cos\ 2\Theta} \tag{60}$$

and the well known equilibrium condition (Keffer, 1966) for an uniaxial antiferromagnet

$$\frac{1}{2}\left(\chi_{\perp}-\chi_{\parallel}\right) H_o^2 \sin 2\Theta_o = M_o H_A \sin 2\Theta_o \qquad (61)$$

The equilibrium conditions Eqs. 58 or 60 and 61 ensure only that the time independent torque, i.e. the torque resulting from the static components of the fields, vanishes. In addition to these conditions, the energy density has to be at a minimum for a stable equilibrium. Taking this further condition into account, we arrive finally at the solutions of the equilibrium conditions

$$\Theta = 0 \quad \text{for } \frac{1}{2}\left(\chi_{\perp}-\chi_{\parallel}\right) H_o^2 < M_o H_A \quad \text{(antiferromagnetic phase)}$$

and (61)

$$\Theta = \frac{\pi}{2} \quad \text{for } \frac{1}{2}\left(\chi_{\perp}-\chi_{\parallel}\right) H_o^2 > M_o H_A \quad \text{(spin-flop regime)}$$

If we put $\frac{1}{2}\left(\chi_{\perp}-\chi_{\parallel}\right) H_o^2 = M_o H_A$ and use some further simplifications as $\chi_{\perp} \approx M_o/H_E$ and $\chi_{\parallel} \approx 0$, we obtain the approximate value of the critical field for the spin-flop transition ($H_c \approx \sqrt{2H_E H_A}$). In the following considerations, we keep Θ as a variable but remember that only the values $\Theta = 0$ (antiferromagnetic phase) and $\Theta = \pi/2$ (spin-flop phase) are meaningful. In this way, the following equations apply to both phases depending on the value of Θ. Next we have to include the oscillating or precessing components of $\vec{M}_a$ and $\vec{M}_b$ and also the oscillating field $\vec{h}$. Then, we obtain the equations of motions for the oscillating components with the help of Eqs. 57, 60 and 61 as follows

$$\begin{aligned}
\dot{\tilde{m}}_a &= \gamma[(H_1 + H_3)\, m_a^y + (H_E + H_4)\, m_b^y - (M_o + \Delta M)\, h^y] - \Lambda\, \tilde{m}_a \\
\dot{m}_a^y &= -\gamma[(H_2 + H_3)\, \tilde{m}_a + (H_E + H_4)\, \tilde{m}_b - (M_o + \Delta M)\, \tilde{h}_a] - \Lambda\, m_a^y \\
\dot{\tilde{m}}_b &= -\gamma[(H_1 - H_3)\, m_b^y + (H_E - H_4)\, m_a^y - (M_o - \Delta M)\, h^y] - \Lambda\, \tilde{m}_b \\
\dot{m}_b^y &= \gamma[(H_2 - H_3)\, \tilde{m}_b + (H_E - H_4)\, \tilde{m}_a - (M_o - \Delta M)\, \tilde{h}_b] - \Lambda\, m_b^y
\end{aligned} \qquad (62)$$

where

$$H_1 = H_E + H_A \cos^2\Theta + \hat{H}_A \sin^2\Theta \qquad H_3 = H_o \,(1 - \chi_{\parallel}/2\chi_{\perp})$$

$$H_2 = H_E + H_A \cos 2\Theta \qquad H_4 = \chi_{\parallel} H_o/2\chi_{\perp}$$

$$\tilde{h}_{a,b} = h^x \cos(\Theta \pm \phi) - h^z \sin \Theta \pm \phi)$$

$$\gamma = g\mu_B/\hbar \quad \text{(isotropic)}$$

The solution of the relatively complicated system of equations Eq. 62 is most easily explained using a matrix notation. After replacing the time derivatives once more by the factor $i\omega$ ($\tilde{m}_a$, m_a^y, $\tilde{m}_b$, m_b^y, h^x, h^y, $h^z \sim e^{i\omega t}$), the system may be brought to the form

$$(\overleftrightarrow{D}) \begin{pmatrix} \tilde{m}_a \\ m_a^y \\ \tilde{m}_b \\ m_b^y \end{pmatrix} = (\overleftrightarrow{B}) \begin{pmatrix} h^x \\ h^y \\ h^z \end{pmatrix} \tag{62'}$$

In Eq. 62', $\overleftrightarrow{D}$ and $\overleftrightarrow{B}$ are 4 x 4 and 3 x 4 matrices, respectively, the elements of which can be obtained from Eq. 62. The solution of Eq. 62' involves the inverse $\overleftrightarrow{D}^{-1}$ of matrix $\overleftrightarrow{D}$

$$\begin{pmatrix} \tilde{m}_a \\ m_a^y \\ \tilde{m}_b \\ m_b^y \end{pmatrix} = (\overleftrightarrow{D}^{-1})\,(\overleftrightarrow{B}) \begin{pmatrix} h^x \\ h^y \\ h^z \end{pmatrix} \tag{63}$$

It is of some advantage to write the determinant Δ of $\overleftrightarrow{D}$ which has to be evaluated for forming $\overleftrightarrow{D}^{-1}$ in the following form

$$\Delta = [(\omega - i\Lambda)^2 - \omega_1^2]\,[(\omega - i\Lambda\,) - \omega_2^2\,] \tag{64}$$

where ω_1 and ω_2 are the roots of the corresponding secular equation.

Now the time-dependent magnetization components m^x, m^y and m^z are related to those of the two sublattices by

$$\begin{pmatrix} m^x \\ m^y \\ m^z \end{pmatrix} = (\overleftrightarrow{T}) \begin{pmatrix} \tilde{m}_a \\ m_a^y \\ \tilde{m}_b \\ m_b^y \end{pmatrix} \tag{65}$$

where T is the following 4 x 3 matrix (cf. Eq. 56)

$$\overleftrightarrow{T} = \begin{pmatrix} \cos(\Theta-\phi) & 0 & \cos(\Theta\ \phi) & 0 \\ 0 & 1 & 0 & 1 \\ -\sin(\Theta\ \phi) & 0 & -\sin(\Theta\ \phi) & 0 \end{pmatrix} \tag{66}$$

Finally, we obtain the dynamic susceptibility (cf. Eq. 4) in matrix notation

$$\vec{m} = \overleftrightarrow{\chi}\ \vec{h} = (\overleftrightarrow{T})\ (\overleftrightarrow{D}^{-1})(\overleftrightarrow{B})\ \vec{h} \tag{67}$$

For the inclusion of Maxwell's equations into these considerations, we assume again the wavevector $\vec{q}$ parallel to the c-axis (cf. Eq. 16) and uniaxial symmetry for the dielectric properties (cf. Eq. 15). The latter seems questionable for the materials with organic ions with a lattice symmetry lower than tetragonal at low temperatures. In practice, however, the samples of orthorhombic or monoclinic symmetry usually have a relatively large number of all types of possible domains, and the measured dielectric constants are average values which exhibit a fictitious tetragonal symmetry. Under these assumptions, we obtain for the components of $\vec{h}$ a set of equations analogous to Eq.37. We shall not give these equations in detail here since they are very similar to those derived in Sections 4.1 and 4.2 and rather present only the results. For the antiferromagnetic phase ($\Theta = 0$, $\phi = 0$), the two modes are again circularly polarized

$$h^{\pm} = h^x \pm ih^y \qquad (h^z = 0)$$

with propagation constants

$$\begin{aligned}(n - ik)_{\pm}^2 &= \varepsilon_{\perp}(1 + \chi^{\pm}) \\ &= \varepsilon_{\perp}\left(1 + \frac{\chi_{\perp}\omega_o^2 \pm 2\Delta\omega(\omega - i\Lambda \pm \omega_{\parallel})}{\omega_o^2 - (\omega\ \ i\Lambda \pm \omega_{\parallel})^2}\right)\end{aligned} \tag{68}$$

where

$$\omega_o = \gamma \sqrt{2 H_E H_A}$$

$$\omega_\parallel = \gamma H_o (1 - \chi_\parallel / 2\chi_\perp)$$

$$\Delta\omega = \frac{1}{2} \gamma \chi_\parallel H_o$$

$$\chi_\perp = 2 M_o/(2H_E + H_A) \approx M_o/H_E$$

Eq. 68 reflects the well-known fact that the two resonance frequencies are for an uniaxial antiferromagnet with H_o applied parallel to the easy axis

$$\omega = \omega_o \pm \omega_\parallel = \gamma[\sqrt{2H_E H_A} \pm H_o (1 - \chi_\parallel/2\chi_\perp)] \tag{69}$$

For an antiferromagnet also, it is possible to bring Eq. 68 to the form $\omega = f(q)$ for the discussion of the polariton effects. For $H_o=0$ and at low temperatures ($\chi_\parallel \approx 0$) the resonance frequency is $\omega = \omega_o$. The corresponding "longitudinal" frequency (cf. Eqs. 25 and 47) is is $\omega = \omega_o \sqrt{1 + \chi_\perp}$ and, accordingly, there is also in this case a gap between the dispersion branches (cf. Fig. 15). With the values for MAMC (cf. Table 5), the width of the gap is $\Delta\omega = \omega_o[\sqrt{1+\chi_\perp} - 1] = 74$ MHz which is much smaller than the corresponding quantity for MnF_2 where $\Delta\omega \approx 3$ GHz (= 0.1 cm^{-1}). This means that for MAMC (and the other materials under consideration) electromagnetic propagation effects will become visible in the spectra only at very low temperatures where Λ is sufficiently small (cf. Section 4.1). For the spin-flop phase [$\Theta = \pi/2$, $\sin\phi = H_o / (2 H_E - H_A)$], we obtain two elliptically polarized modes with the same resonance frequency

$$\omega_1^2 \approx \gamma^2 (H_o^2 - 2H_E H_A) \tag{70}$$

The other resonance frequency for the spin-flop phase

$$\omega_2^2 = \gamma^2 \, 2 H_E \hat{H}_A (1 - H_o^2/4H_E^2) \tag{71}$$

can only be observed with the wavevector perpendicular to the c-axis and with a tunable source since the quadratic Zeeman effect is extremely small for this case. Please note that ω_2^2 (Eq. 71) involves not the out-of-plane anisotropy field H_A but the in-plane anisotropy field $\hat{H}_A$.

TABLE 5: Magnetic properties of some manganese compounds

	T_N (K)	$\sqrt{2H_EH_A}$ [a] (T)	H_c [b] (T)	H_E [c] (T)	H_A(AFMR) [d] (T)	$H_A(H_c)$ [e] (T)	M_o [f] (T)
Rb_2MnCl_4							
	56.0 [g,h]	5.47 [g]	5.60 [i]	87.1 [g]	0.172	0.180	0.141
$(C_nH_{2n+1}NH_3)_2MnCl_4$							
n = 1	45.3 [g,k,l]	3.65 [g]	3.60 [k]	71.5 [g]	0.093	0.091	0.117
n = 2	43.1 [m,n]	3.18 [p]	3.10 [m,n]	68.5 [n]	0.074	0.70	0.101
n = 3	39.2 [n]	1.98 [p]	1.63 [n]	66.3 [n]	0.030	0.020	0.086
$NH_3(CH_2)_mNH_3MnCl_4$							
m = 2	40-45 [q]	3.61 [p]	3.50 [q]				0.132
m = 4	40-45 [q]	3.29 [p]	3.50 [q]				0.103
m = 5	40-45 [q]	3.11 [p]	≈ 3.20 [q]				0.192

[a] zero field AFMR frequency at 4.2 K from magnetic resonance measurements

[(b)] Spin-flop critical field at low temperatures

[(c)] exchange field $H_E = 2zSJ/g\mu_B$ ($z = 4$, $S = 2.5$, $g = 2.000$)

[(d)] anisotropy field, calculated from the results of AFMR measurements

[(e)] anisotropy field, calculated from the value of the spin-flop critical field H_c

[(f)] sublattice magnetization calculated from the structural data compiled in Table 4.
Corrections like zero-point spin deviation (de Wijn et al., 1973a; Schröder et al., 1980) have not been taken into account

[(g)] B. Schröder et al. (1980)

[(h)] A. Epstein et al. (1970)

[(i)] N.V.Fedoseeva et al. (1978)

[(k)] W.D. Van Amstel and L. J. de Jongh (1972)

[(l)] B.C. Gerstein et al. (1974b)

[(m)] J. Nösselt et al. (1976)

[(n)] H.A. Groenendijk et al. (1979)

[(p)] present investigation

[(q)] H. Arend et al. (1976)

4.3.2 Results of AFMR and SFR in Rb_2MnCl_4

It is a well-known fact that the order parameter of an antiferromagnet, namely the sublattice magnetization M_o, the exhange field H_E and the anisotropy field H_A are temperature dependent. Therefore, also the zero-field AFMR frequency $\omega = \gamma\sqrt{2 H_E H_A}$ (cf. Eq. 69) is temperature dependent (de Wijn et al., 1973; Shröder et al., 1980) and approaches zero when the temperature approaches the Néel temperature from below. When magnetic resonance experiments are performed with various frequencies and at various temperatures, the properties of the investigated materials are explored as a function of temperature T and external magnetic field. Fig. 20 is an attempt to illustrate the situation in a schematic way. The three coordinate axes are the two-state variables, i.e. the temperature T and the magnetic field H applied parallel to the easy axis, and $h\nu$, which is an abbreviation for the energy or frequency of the $\vec{q} = 0$ magnetic excitation. The bottom, i.e. the H-T-plane, shows the phase diagram of a uniaxial antiferromagnet (Keffer, 1966) with the antiferromagnetic (AF), spin-flop (SF) and paramagnetic phases (PM). The behavior of the system at the phase boundaries and near the bicritical point are problems of current interest (Fisher, 1975a,b; Kosterlitz et al., 1976; Huber et al., 1976; Aharony, 1977). A number of experimental data are already available for MnF_2 (King et al., 1976, 1979; Fisher et al., 1980) and $GdAlO_3$ (Rohrer, 1975; Rohrer et al., 1977), but less is known about layer materials like Rb_2MnCl_4. In the $h\nu$-T-plane to the left in Fig. 20, the temperature variation of the AFMR frequency $\nu \approx \gamma\sqrt{2 H_E H_A}$ for H=0 is shown schematically, as well as its vanishing at the Néel temperature T_N. In the $h\nu$-H-plane to the right in Fig. 20, the magnetic field dependence of the AFMR frequencies is shown for $T \to 0$ (cf. Eqs. 69-71).

In the scheme of Fig. 20, a magnetic resonance experiment corresponds to the determination of the curves of the $\vec{q} = 0$ excitations as a function of T and H in a "horizontal" plane $h\nu$ = const. where ν is the fixed frequency of the IMPATT oscillator. In the antiferromagnetic phase, two types of curves are to be expected. If the chosen oscillator frequency is below the AFMR frequency for H = 0 and T = 0, the magnetic field necessary for resonance will decrease, and the curve will meet the plane H = 0 at a temperature where the zero-field AFMR frequency is equal to the oscillator frequency (see "horizontal" lines in Fig. 20). Then the resonance field will increase rapidly and finally pass into the value for paramagnetic resonance. In fig. 21, the experimental results are shown which were obtained for Rb_2MnCl_4 with a number of frequencies between 9.2 and 248 GHz as a function of temperature (Strobel et al., 1981; Reusch et al., 1981). The graphs of the resonance

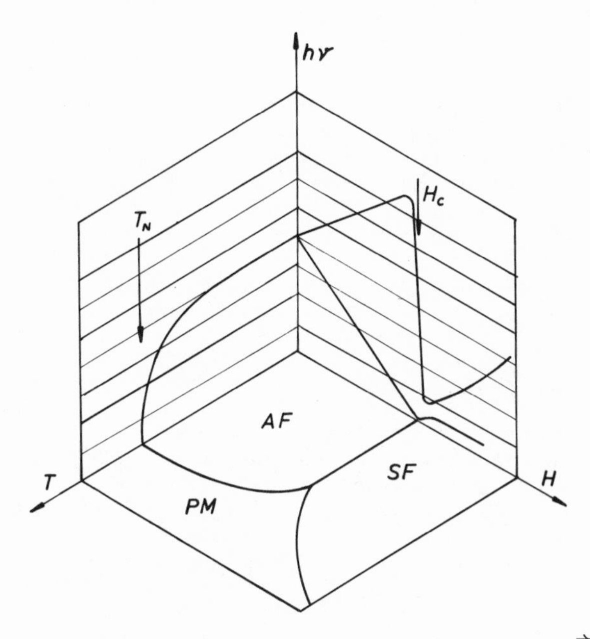

Figure 20: Schematic diagram of the energy hν of the $\vec{q} = 0$ magnetic excitations of a uniaxial antiferromagnet as a function of temperature and of magnetic field H parallel to the c-axis in the antiferromagnetic (AF), spin-flop (SF) and paramagnetic (PM) phases.

field B_{res} for 62, 95 and 125 GHz are good examples of the type discussed above with a zero of B_{res} at a certain temperature. The second type of curve to be expected from Fig. 20 in the antiferromagnetic phase occurs when the oscillator frequency is chosen to be larger than the zero-field AFMR frequency for H= 0 and T = 0 (157 GHz). Then, the magnetic field at resonance will increase with increasing temperature and end at the phase boundary. Good examples for this type are the graphs of B_{res} in Rb_2MnCl_4 for 172 GHz and

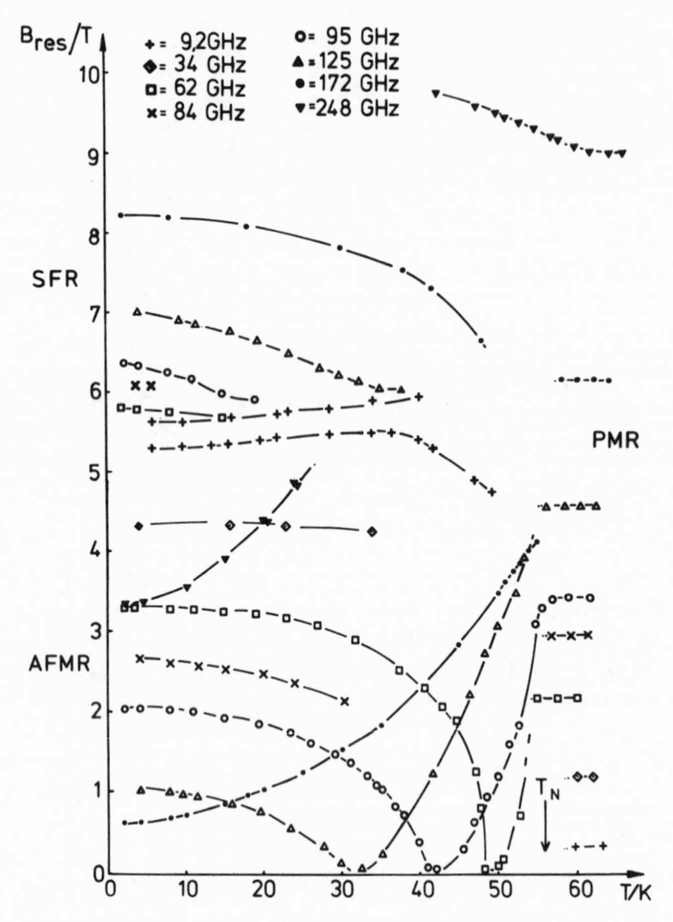

Figure 21: Magnetic resonance of Rb_2MnCl_4 (magnetic field B_{res} at resonance) in the antiferromagnetic (AFMR; B < 5.6 T), in the spin-flop (SFR, B > 5.6 T) and in the paramagnetic state (PMR, T > T_N).

248 GHz in Fig.21. As already mentioned, only one of the two resonance frequencies in the spin-flop phase (cf. Eqs. 70 and 71) can be observed by magnetic resonance. The observable mode will show a slight decrease of B_{res} with increasing temperature (cf. Figs. 20 and 21). Only for high frequencies (ν = 172 GHz and 248 GHz), the field at resonance B_{res} seems to pass from its value in the spin-flop phase to the paramagnetic one at the phase boundary.

Some special comments are necessary for the results at 9.2 GHz (X-band). For this frequency, rather high fields near the spin-flop critical field are needed to reach the resonance condition. Now, the transmission spectra show two resonance lines at temperatures below the Néel temperature (see Fig. 22). What looks like a line splitting are actually two resonances, one in the antiferromagnetic phase and the other in the spin-flop regime. This is easily seen when a small value for ω is inserted and Eqs. 69 and 70 are solved for H_o. The resulting H_o is slightly below $H_c \approx \sqrt{2 H_E H_A}$ for the antiferromagnetic and slightly above H_c in the spin-flop phase. Thus, the X-band resonance measurements at high fields are very useful to obtain an upper and a lower limit for the spin-flop line in the phase diagram (cf. Fig. 20). Moreover, it is very likely that the spin-flop bicritical pint is close to 5.7 T and 40 K where the upper X-band resonance fades off and the lower resonance curve shows a rather abrupt bend.

It was already mentioned that the line widths of magnetic resonance show a typical broadening towards and seem to diverge at the Néel temperature. Fig. 23 shows the line width of the AFMR for $T < T_N$ and of the paramagnetic resonance for $T > T_N$ obtained with various frequencies. The results exhibit that the line broadening is more drastic on the antiferromagnetic side. In Section 4.1, we have already discussed the concept of plotting the logarithm of the line width as a function of the logarithm of the reduced temperature $|T - T_c| / T_c$ in search for characteristic and eventually critical properties of the system. But as already pointed out, this concept will only lead to meaningful results if the phase boundaries have been determined for Rb_2MnCl_4, e.g. by means of neutron diffraction and if for T_c, not simply T_N for H = 0, but T_c (H) is inserted in accordance with resonance field obtained experimentally. In other words, a more complete analysis of our data obtained by magnetic resonance will be possible when the above-mentioned experiments have been performed. Already, the data presented in Figs. 21-23 demonstrate that magnetic resonance in the millimeter-wave range is a useful tool for investigating antiferromagnets.

4.3.3 Dependence of the AFMR frequency on the layer separation

Perovskite-type layer structures with molecular ions offer the

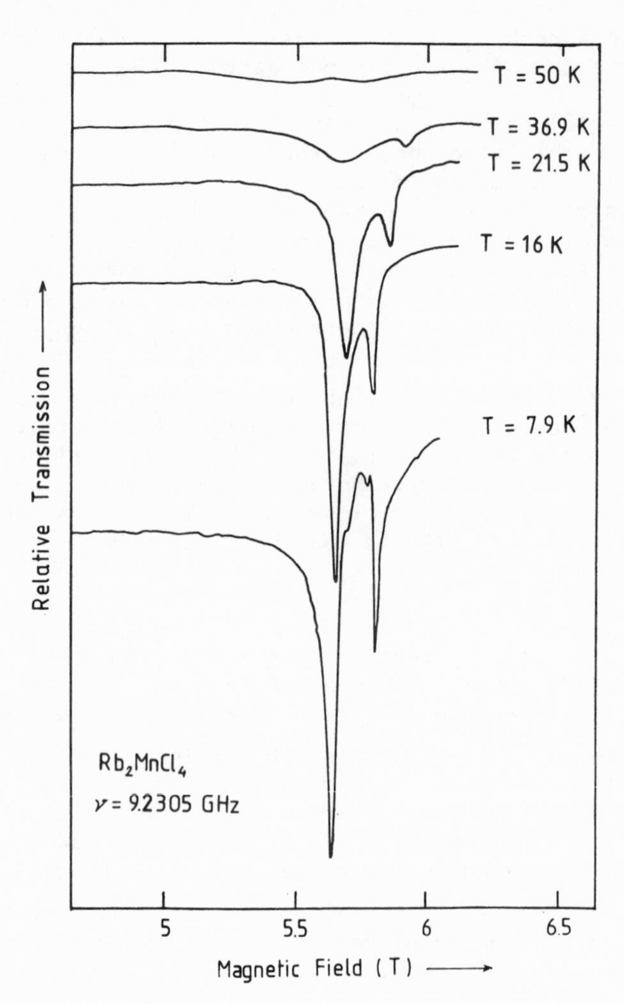

Figure 22: Magnetic resonance lines (relative transmission versus magnetic field) of Rb_2MnCl_4 for microwaves (X-band frequencies) at various temperatures. The line to the left is due to antiferromagnetic resonance (AFMR), that to the right due to spin-flop resonance (SFR).

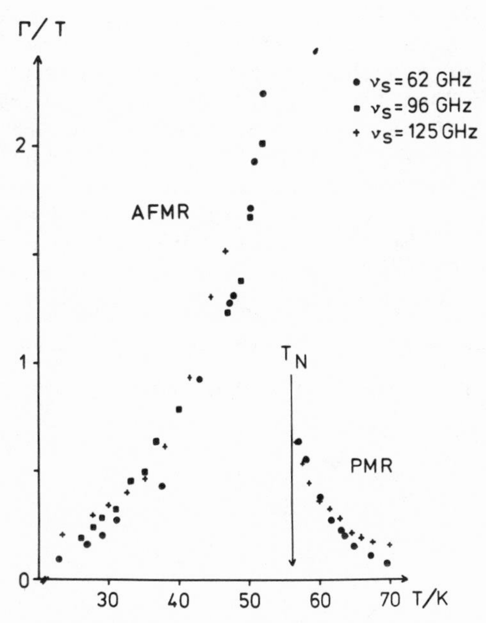

Figure 23: Line width Γ (full width at half maximum) of antiferromagnetic (AFMR) and paramagnetic resonance (PMR) in Rb_2MnCl_4 for various frequencies.

possibility of studying their magnetic properties as a function of layer separation which can be controlled by means of the length of the chain molecules (cf. Table 4). We have investigated the AFMR of three compounds of the type $(C_nH_{2n+1}NH_3)_2MnCl_4$ (n = 1, 2, 3) and of three compounds of the type $NH_3(CH_2)_mNH_3MnCl_4$ (m = 2, 4, 5) at 4.2 K (Strobel et al., 1981; Reusch et al., 1981), in addition to Rb_2MnCl_4. The results obtained with various frequencies are shown in Figs. 24 and 25. In order to obtain graphically a relatively reliable value for the zero-field AFMR frequency, we have plotted the lower Zeeman branch for the external field parallel to the easy axis (cf. Eq. 69) for positive values of the magnetic field B (in Tesla) and the upper branch versus negative values of B. In other words, positive and negative values of B in Figs. 24 and 25 do not mean a reversal of the direction of the field which has no effect in the case of an antiferromagnet. It is merely a trick of a useful graphical presentation of the two Zeeman branches of the AFMR, and the zero-field AFMR frequency is simply obtained as the intersection of the straight line through the experimental points for a particular material and the ν-axis at H = 0. And we note from Fig. 24 that this intersection, and thus the quantity $\omega = \gamma \sqrt{2H_E H_A}$, decreases with increasing layer separation from 153 GHz for Rb_2MnCl_4 to 55.5 GHz for $(C_3H_7NH_3)_2MnCl_4$. This decrease of about a factor of three is

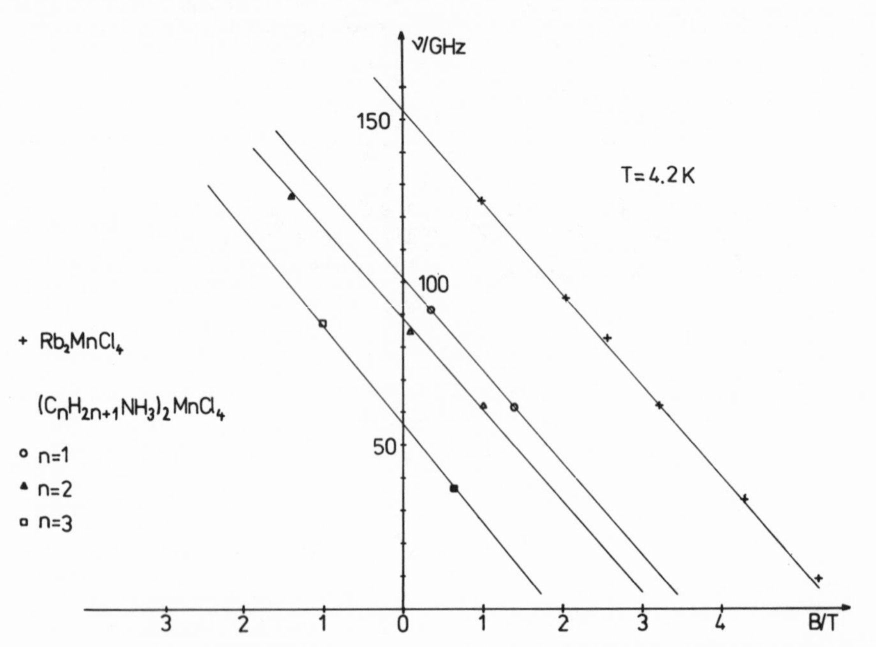

Figure 24: Antiferromagnetic resonance of Rb_2MnCl_4 and of monoammonium compounds ($C_nH_{2n+1}NH_3)_2MnCl_4$: oscillator frequency ν versus magnetic field B. Positive and negative values of the field refer to the lower and the upper spin wave branch (see text).

quite drastic in comparison to the increase of the layer separation from 8.09 Å in Rb_2MnCl_4 to 12.77 Å in $(C_3H_7NH_3)_2MnCl_4$ (cf. Table 4). The corresponding values of $\omega = \gamma\sqrt{2H_E H_A}$ of the diammonium compounds of about 95 GHz are also considerably lower than that of Rb_2MnCl_4 but do not depend so strongly on the chain length.

The zero-field AFMR frequency $\omega = \gamma\sqrt{2H_E H_A}$ as obtained for a number of compounds (cf. Figs. 24 and 25) contains the product of the exchange field H_E and the anisotropy H_A. The question is which one of these quantities causes the decrease for larger layer separation. It will be noted from Fig. 2 and Table 4 that the increase of layer separation is accompanied in these compounds by a slight increase of the Mn-Mn distance, which certainly will also influence the values of H_E. If one of the quantities is known from other experiments, we are able to evaluate the other from the AFMR results and to discuss our results in more detail. For the monoammonium

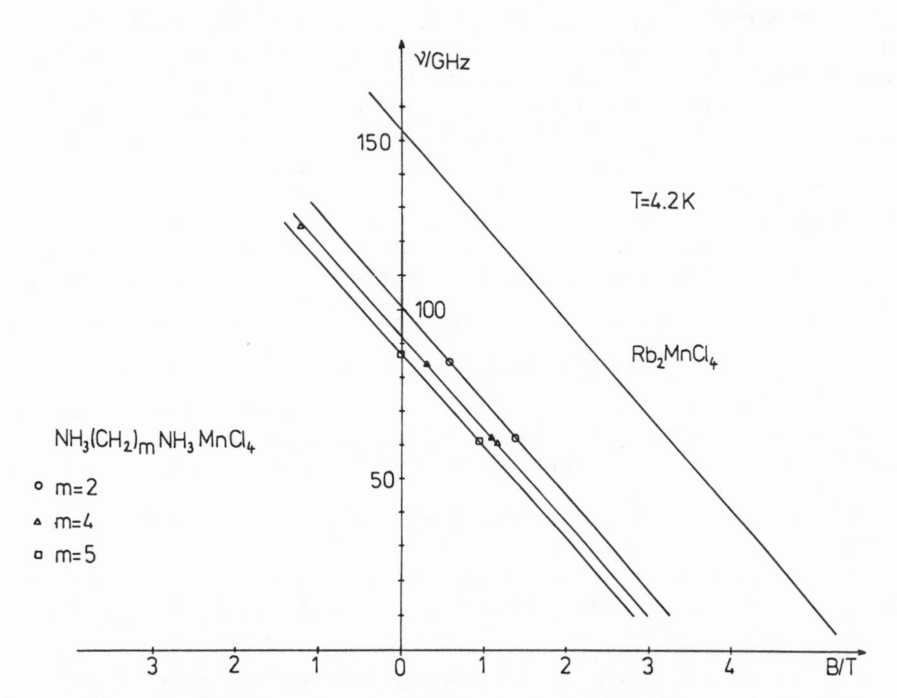

Figure 25: Antiferromagnetic resonance of Rb_2MnCl_4 and of diammonium compounds $NH_3(CH_2)_mNH_3MnCl_4$: oscillator frequency ν versus magnetic field B. Positive and negative values of the field refer to the lower and the upper spin wave branch (see text).

compounds at least, values of the exchange field are available in the literature (Schröder et al., 1980; Groenendijk et al., 1979). In Table 5, we have compiled all the data on the magnetic properties of the compounds under consideration, including the Néel temperature T_N, the spin critical field H_c and the sublattice magnetization M_o. The latter has been calculated as a first approximation from the structural data (see Table 4) neglecting all corrections like zero-point spin deviation which are not negligible in quasi-two-dimensional antiferromagnets. From the zero-field AFMR frequency $\sqrt{2 H_E H_A}$, the anisotropy field H_A (AFMR) has been evaluated where values for H_E are available. From $H_c \approx \sqrt{2 H_E H_A}$, a second value for the anisotropy field was calculated. However, we believe that more accurate values for $\sqrt{2 H_E H_A}$ are obtained by means of magnetic resonance than from the spin - flop transition, which is often smeared out and not sharp in systems with small anisotropy. The data for Rb_2MnCl_4 and for the monoammonium compounds in Table 5 show that the Néel temperature T_N, the exhange field H_E and the sublattice magnetization M_o exhibit about the same relative decrease with increasing layer separation. In a simple, naive picture, one could say that the decrease of the sublattice magnetization reflects the decrease of the density of manganese ions according to the increasing layer separation. The "molecular" field H_E and thus also the Néel temperature vary in accordance with the sublattice magnetization M_o. Of course, this picture is too simple and we must not neglect the dependence of the exchange field H_E on the Mn-Mn distance $\bar{a}$. It was found by de Jongh and Block (de Jongh et al.,1975) from experimental data and theoretical considerations that H_E varies according to a $(\bar{a})^{-12}$ power law. However, the decrease of H_E in our compounds (cf. Table 5) is in most cases larger and can only partly be explained with a $(\bar{a})^{-12}$ power law. More elaborate theoretical considerations on the actual superexchange paths would probably provide a better understanding of the variation of H_E in these compounds.

Now let us turn to the variation of the anisotropy field as a function of layer separation. The data in Table 5 clearly shows that the anisotropy field H_A decreases much more rapidly than the other quantities discussed so far from Rb_2MnCl_4 to $(C_3H_7NH_3)_2MnCl_4$. This means that the drastic decrease of the zero-field ARMR frequency (cf. Fig. 24 and Table 5) originates mainly from that of the anisotropy field. Thus the data in Table 5 demonstrate that the perovskite-type layer materials approach more and more an ideal two-dimensional Heisenberg system with respect to their properties extracted from the magnetic excitations. Of course, the two-dimensional Heisenberg model would not lead to a three-dimensional long-range magnetic order and, in that respect, other more Ising-like properties of the system are essential. Less is known for the diammonium compounds, especially no values for the exchange field. Therefore, we cannot perform a similar analysis for the magnetic resonance data. We note from Table 5 that the values of the zero-field AFMR frequency from resonance experiments are again close to the values for H_c reported in the literature.

ACKNOWLEDGEMENTS

We are much indebted to Dr. W. Haydl, Institut für Angewandte Festkörperphysik in Freiburg i. Br. (Fed. Rep. Germany), for kindly providing the Schottky barrier and Gunn diodes to us. Financial support of this work by the Deutsche Forschungsgemeinschaft is gratefully acknowledged.

REFERENCES

Abelés, F., 1972, Optical Properties of Solids, 24 ff, North Holland Publishing Company - Amsterdam

Ahanory, A., 1977, Physica 86-88 B : 545

Akhiezer, A. I., Baryakhtar, V. G., and Peletminski, S. V., 1968, "Spin Waves", North Holland Amsterdam

Anderson, P. W., and Suhl, H., 1955, Phys. Rev. 100 : 1788

Arend, H., Tichý, K., Baberschke, K., and Rys, F., 1976, Solid State Commun., 8:999

Auld, B.A., 1960, J. Appl. Phys., 31:1642

Birgeneau, R. J., Guggenheim, H. J., and Shirane, G., 1970 a, Phys. Rev. B 1:2211

Birgeneau, R.J., de Rosa, F., and Guggenheim, H. J., 1970 b, Solid State Commun., 8:13

Birgeneau, R. J., Guggenheim, H. J., and Shirane, G., 1973, Phys. Rev., 8:304

Bloembergen, N., and Damon, R.W., 1952, Phys. Rev., 85:699

Born, M., and Huang, K., 1954, "Dynamical Theory of Crystal Lattices", Oxford University Press, New York.

Born, M., and Wolf, E., 1975, "Principles of Optics", 36 ff, Pergamon Press, Oxford.

Borovik-Romanov, A. S., Kreines, N.M., Laiho, R., Levola, T., and Zhotikov, G., 1980, J.Phys. C: Solid State Physics, 13:879.

Bose, S. M., Foo, E-Ni, and Zuniga, M. A., 1975, Phys. Rev.,B 12:3855.

Breed, D. T., 1967, Physica (Utrecht), 37:35.

Chapuis, G., 1977, Phys. Stat. Sol., (a) 43:203.

Chapuis, G., Arend, H., and Kind, R., 1975, Phys. Stat. Sol., (a) 31:449.

Chapuis, G., Kind, R., and Arend, H., 1976, Phys. Stat. Sol., (a) 36:285.

Clogston, A.M., Suhl, H., Walker, L.R., and Anderson, P.W., 1956, Phys. Rev., 101:903.

Czerny, M., and Turner, A. F., 1930, Zeitschrift f. Physik, 61:792.

Damon, R. W., 1953, Rev. Mod. Phys., 25:239 .

Damon, R. W., and Eshbach, J. R., 1961, J.Phys. Chem. Sol., 19:308.

Day, P., 1979, Accounts of Chemical Research, 12:236.

de Jongh, L. J., and Block, R., 1975, Physica, 79 B : 568.

de Jongh, L. J., and Miedema, A. R., 1974, Adv. Phys., 23:1.

Depmeier, W., 1977, Acta Cryst., B 33:3713.

Depmeier, W., Felsche, J., and Wildermuth, G., 1977, J. Solid State Chem., 21:57.

de Wijn, H. W., Walker, L. R., Davis, J. L., and Guggenheim, H.J., 1972, Solid State Commun., 11:803.

de Wijn, H. W., Walker, L. R., and Walstedt, R. E., 1973 a, Phys. Rev., B 8:285.

de Wijn, H. W., Walker, L. R., Geschwind, S., and Guggenheim, H. J., 1973 b, Phys. Rev., B 8:299.

Dillon, J. F., 1960, J. Appl. Phys., 31:1605.

Dzialoshinski, I.E., 1957, J. Exptl. Theor. Phys.(USSR), 32: 1547 Engl. Translation: 1957, Soviet Physics JETP, 5:1259.

Dzialoshinski, I. E., 1958, J. Phys. Chem.Sol.,4:241

Epstein, A., Gurewitz, E., Makovsky,J., and Shaked, H., 1970, Phys. Rev., B 2:3703.

Fedoseeva, N.V., Spevakova, I. P., Bazhan, A. N., and Beznosikov, B.V. 1978, Fiz. Tverd. Tela (Leningrad) 20:2776.
English Translation: 1978, Soviet Physics Solid State, 20:1600.

Ferré, J., Regis, M., Farge, Y., and Kleemann, W., 1979, J. Phys. C: Solid State Physics, 12: 2671.

Fisher, M. E., 1975 a, Phys. Rev. Letters, 34:1634.

Fisher, M. E., 1975 b, AIP Conference Proceedings, 24: 273.

Fisher, M. E., Chen, J. H., and Au-Yang, H., 1980, J. Phys. C: Solid State Physics , 13: L 459.

Fletcher, R.C., and Kittel, C., 1960, Phys. Rev., 120:2004.

Funahashi, S., Moussa, F., and Steiner, M., 1976, Solid State Commun., 18 : 443.

Gerstein, B. C., Chow, Chee, Caputo, R., and Willett, R. D., 1974 a, AIP Conference Proceedings, 24:361.

Gerstein, B. C., Chang, K., and Willett, R. D., 1974 b, J. Chem. Physics, 60 : 3454.

Groenendijk, H.A., van Duyneveldt, A. J., and Willett, R. D., 1979, Physica, 98 B:53.

Haegele, R., and Babel, D., 1974, Zeitschrift anorg. allg. Chemie, 409 : 11.

Heger, G., Mullen, D., and Knorr, K., 1975, Phys. Stat. Sol., (a) 31:455.

Heger, G., Mullen, D., and Knorr, K., 1976, Phys. Stat. Sol.,(a) 35:627.

Herdtweck, E., and Babel, D., 1981, Zeitschrift anorg. allg. Chemie, 474:113.

Hidaka, M., and Walker, P.J., 1979, Solid State Commun., 31:383.

Hirakawa, K., and Ikeda, H., 1973, J. Phys. Soc., Japan, 35:1328.

Hirakawa, K., and Ubukoshi, K., 1981, J. Phys. Soc., Japan, 50:1909.

Hirakawa, K., and Yoshizawa, H., 1979, J.Phys. Soc., Japan, 47:368.

Hirakawa, K.,Shirane, G., and Axe, J.D., 1981, J. Phys. Soc., Japan, 50:787

Holah, G., 1980, Int. Journal Infrared and Millimeter Waves, 1:225 and 235.

Hönl, H., Maue, A. W., and Westphal, K., 1966, "Theorie der Beugung" in "Encyclopaedia of Physics" (S. Flügge ed.) XXV/1 : 218, Springer, Heidelberg.

Houghton, J. T., and Smith, S. D., 1966, "Infrared Physics" chapter 5: 174 ff, University Press, Oxford.

Huber, D.L., and Raghavan, R., 1976, Phys. Rev., B 14:4068.

Hughes Aircraft Company, Electron Dynamics Division, Product information.

Hutchings, M. T., Fair, M. J., Day, P., and Walker, P. J., 1976, J. Phys. C:Solid State Phys., 9: L55.

Ikeda, H., 1974, J. Phys. Soc., Japan, 37: 660.

Keffer, F., 1966, "Spin Waves" in " Encyclopaedia of Physics". (S. Flugge ed.) Vol. XVIII/2 : 1, Springer, Heidelberg.

Khomski, D. I., and Kugel, K. I., 1973, Solid State Commun., 13: 763.

King, A. R., and Rohrer, H., 1976, AIP Conference Proceedings, 29:420.

King, A. R., and Rohrer, H., 1979, Phys. Rev., B 19 : 5864.

Kleemann, W., Ferre, J., and Schafer, F.J., 1981, J. Phys. C: Solid State Physics, 14: 4463.

Kneubühl, F., and Affolter, E., 1979, "Infrared and Submillimeter-Wave Waveguides" in " Infrared and Millimeter Waves" (K. J. Button ed.) Vol. 1, Academic Press, New York.

Knorr, K., Jahn, I. R., and Heger, G., 1974, Solid State Commun., 15 : 221

Kosterlitz, J. M., Nelson, D. R., and Fisher, M.E., 1976, Phys. Rev., B 13: 412.

Kubo, H., 1974, J. Phys. Soc., Japan, 36: 675.

Kubo, H., Machida, Y., and Uryu, N., 1977, J. Phys. Soc. Japan,43: 459

Kullman, W., Grieb, T., Strobel, K., Geick, R., to be published.

Kuno, H.J., 1979, "IMPATT Devices for Generation of Millimeter Waves", in "Infrared and Millimeter Waves" (K.J.Button ed.) Vol. 1, Academic Press, New York.

Lamarre, J. M., Coron, N., Courtin, R., Dambier, G., and Charra, M., 1981, Int. Journal Infrared and Millimeter Waves, 2: 273

Le Dang, K., and Veillett, P., 1976, Phys. Rev., B 13: 1919.

Manohar, C., and Venkataraman, G., 1972, Phys. Rev., B 5: 1993.

Mermin, N.D., and Wagner, H., 1966, Phys. Rev. Letters, 17: 1133.

Mitsuishi, A., Otsuka, Y., Fujita, S., and Yoshinaga, H., 1963, Jap. J. Appl. Phys., 2: 574.

Moriya, T., 1960, Phys. Rev., 120: 91.

Nösselt, J., Heger, G., and Moser, R., 1976, Progress Report KFK 2357, Teilinstitut nukleare Festkörperphysik des IAK, Kernforschungszentrum Karlsruhe, 81

Pincus, P., 1962, J. Appl. Phys. , 33: 553

Renk, K.F., and Genzel, L., 1962, Appl. Optics., 1: 643.

Reusch, W., Zehendner, B., Strobel, K., and Geick, R., 1981, Digest of the Sixth International Conference on Infrared and Millimeter Waves, Miami Beach, paper T-2-1 (IEEE Cat. No. 81 CH-1645-1 MTT).

Rohrer, H., 1975, Phys. Rev. Letters, 34:1638.

Rohrer, H. and Gerber, Ch., 1977, Phys. Rev. Letters, 38: 909.

Sarmento, E.F., and Tilley, D. R., 1976 a, J. Phys. C: Solid State Phys., 9 : 2943.

Sarmento, E.F., and Tilley, D. R., 1976 b, J. Phys. C: Solid State Phys., 10: 795.

Sarmento, E.F., and Tilley, D.R., 1976 c, J. Phys. C:Solid State Phys., 10: 4209.

Schneider, M. V., and Phillips, T. G., 1981, Int. Journal Infrared and Millimeter Waves, 2: 15.

Schröder, B., Strobel, K., Wagner, G., and Geick, R., 1978, Infrared Physics, 18: 893.

Schröder, B., Wagner, V., Lehner, N., Kesharwani, K. M., and Geick, R., 1980, Phys. Stat. Sol., (b) 97: 501.

Seifert, H. J., and Koknat, F. W., 1965, Zeitschrift anorg. allg. Chemie, 341: 269.

Skalyo, J., Shirane, G., Birgeneau, R. J., and Guggenheim, H. J., 1969, Phys. Rev. Letters, 23: 1394.

Slichter, C. P., 1978, "Principles of Magnetic Resonance", in "Solid State Sciences, Vol. 1, Springer, Heidelberg.

Sparks, M., Titmann, B. R., Mee, J. E., and Newkirk, C., 1969, J. Appl. Phys., 40: 1518.

Storey, J. M. V., Watson, D. M., and Townes, C. H., 1980, Int. Journal Infrared and Millimeter Waves, 1: 15

Strobel, K., and Geick, R., 1981, Physica, 108 B: 951

Strobel, K., and Geick, R., J. Phys. C: Solid State Physics : to be published.

Strobel, K., Frank, J., Rosshirt, K., and Geick, R., 1980, Int. Journal Infrared and Millimeter Waves, 1: 295.

Suhl, H., 1956, Phys. Rev., 101: 1437.

Tichý, K., and Arend, H., 1975, Report AF-SSP-84, Institut für Reaktortechnik der Eidgen. Technischen Hoschschule Zürich, CH-5303 Würenlingen, Switzerland.

Tichý, K., Benes, J., Halg, W., and Arend, H., 1978, Acta Cryst., B 34: 2970.

Tichý, K., Benes, J., Kind, R., and Arend, H., 1980, Acta Cryst., B-36: 1355.

Tsuru, K., and Uryu, N., 1976, J.Phys. Soc. Jap., 41: 804.

Ulrich, R., 1968, Applied Optics, 1987

Ulrich, R., 1967, Infrared Physics, 7: 37 and 65

Ulrich, R., Renk, K. F., and Genzel, L., 1963, IEEE Transactions MTT 11: 363.

van Amstel, W. D., and de Jongh, L. J., 1972, Solid State Commun., 11: 1423.

Vogel, P., and Genzel, L., 1964, Infrared Physics, 4: 257.

Walker, L. R., 1957, Phys. Rev., 105: 390.

Walker, L. R., 1958, J.Appl. Phys., 29: 318.

Yamada, I., 1972, J. Phys. Soc. Japan,33: 979.

Yamazaki, H., 1974, J. Phys. Soc. Japan, 37: 667

Yamazaki, H., 1976, J. Phys. Soc. Japan, 41: 1911

Yamazaki, H., Morishige, Y., and Chikamatsu, M., 1981, J. Phys. Soc. Japan, 50: 2872.

Yokazawa, Y., 1971, J. Phys. Soc. Japan, 31: 1590.

Yukawa, T., and Abe, K., 1974, J. Appl. Phys., 45: 3146.

THE REFLECTION OF OPEN ENDED CIRCULAR WAVEGUIDES

APPLICATION TO NONDESTRUCTIVE MEASUREMENT OF MATERIALS

Fred E. Gardiol, Thomas Sphicopoulos and
Viron Teodoridis

Laboratory of Electromagnetism and Acoustics
Department of Electrical Engineering
Ecole Polytechnique Federale de Lausanne
Chemin de Bellerive 16, CH-1007 Lausanne, Switzerland

ABSTRACT

The measurement of dielectric properties of a material based on the reflection in an open-ended waveguide is a simple and nondestructive technique, which does not require extracting any sample of the material to be measured. It is therefore of interest for the study of moisture content in materials, and also in the field of biomedical applications. The material one wishes to measure is placed on a flat metallic flange which terminates the waveguide, directing the microwave radiation within the material.

In order to determine the relationship between measured quantities such as VSWR and phase with the dielectric properties of the material, a theoretical study was carried out, considering the electromagnetic field near the aperture. Imposing the continuity of the tangential field components yields a dyadic Green's equation, which is solved by numerical techniques. A computer program was set up to carry out the complete resolution of the problem. Results are presented and compared to experimental values. Finally, in order to provide a simpler relationship to the user, a third-order polynomial expansion was developed, which gives a good approximation of the exact values over specified ranges of the parameters.

1. INTRODUCTION

Several recent developments in microwaves and millimeter waves consider the radiation from and to the opening of a metallic waveguide, for various purposes in the fields of materials measurements and of biomedical applications.

In the area of non-destructive measurement of materials, a flanged open-ended waveguide is placed next to the material under test. The reflection factor, within the waveguide, is then measured by means of a standard technique (slotted-line or reflectometer). In this way it is possible to measure materials which must not be disturbed, for instance walls of ancient buildings, pressure-sensitive compounds, or to continuously monitor a fabrication process. The more usual microwave techniques (Altschuler, 1963), in contrast, always required cutting and machining of samples to be placed in suitable test jigs (waveguide or cavity); they are thus destructive and do not provide a response in real time. The new nondestructive methods become particularly interesting with the increasing availability of low-cost, rugged and easy-to-use solid state sources; they provide interesting means to solve in a simple fashion rather difficult problems (Decreton et al., 1974).

However, measuring the reflection factor within the open-ended waveguide is only the experimental part of the process. The final quantity which one wishes to determine is the permittivity of the material, which is a rather complex function of the reflection factor. The relationship between reflection and permittivity can be determined experimentally, by carrying a large number of measurements with well-known calibrated materials and then interpolating. Such an approach, comparative in nature, is limited in practice, as a large range of known materials is necessary. Measurement errors then tend to get compounded in the process, providing in the end rather poor accuracy (running a very large number of measurements can improve accuracy by statistical means: the time needed for the measurement increases accordingly). It is therefore preferable to determine the required relationship by analyzing the electromagnetic fields around the aperture. The basic approach is misleadingly simple: solve Maxwell's equations in both media, then apply boundary conditions at the interface. The detailed development is, however, quite complex. During the process, the field distribution is also determined, showing the material area participating in the measurement.

The field of thermography considers the microwave and millimeter wave radiation of the human body itself, which is related to its equivalent noise temperature, itself a measure of the actual physical temperature. As electromagnetic radiation in those frequency ranges propagates to some extent through living tissues, it is possible to probe for hot spots deep within the body. Such hot spots can correspond to malignant tumors (Gautherie et al., 1979) or to rhumatismal irritations of joints (Edrich et al., 1977). In contrast, infrared rays only show the surface temperature.

In this last application, the signal is not fed from the waveguide towards the outside, but actually originates outside the waveguide, within the biological material, and flows into the waveguide. Very sensitive radiometers are required to detect it. Since reciprocity relationships apply at very small signal amplitudes, the electromagnetic field study described earlier permits here the determination of the depth at which a temperature rise can be detected, together with the shielding effect of intermediate layers.

The diathermy process, nowadays more usually called hyperthermia, consists of heating up some parts of the body for therapeutic purposes, utilizing some form of non-ionizing electromagnetic radiation (Licht, 1965). Some degree of focusing is achieved using microwaves, most often in the 915 and 2450 MHz ISM frequency ranges (Industrial, Scientific and Medical). In the millimeter wave ranges, several bands are in the process of being allocated: 24-24.25 GHz; 61-61.5 GHz; 122-123 GHz; 244-246 GHz (ITU). These will definitely become of interest for particular applications as adequate sources and equipment become available.

Microwave diathermy has been in use since 1947 for the treatment of several kinds of arthritis, sprains and strains, sciatic syndrome, infarctus, inflammatory and dental problems, furuncles and even eye diseases (Licht, 1965).

Applicators used to transfer the power from the source (magnetron) to the patient to be treated were for a long time radiating elements (antennas). As, however, a large part of the power radiated was lost to the environment, being of no use for the actual treatment and creating a hazard for the operators, recent designs, of the contact type, focus the power on the part to be treated (Kantor et al. 1980). They are actually open-ended waveguides, surrounded by loaded chokes designed to prevent leakage and unnecessary exposure. In recent years, hyperthermia has shown significant promise for the treatment of several kinds of cancers (Sterzer et al., 1980).

In the three non-conventional applications of microwaves described, one finds the same kind of element: a section of waveguide, terminated by a flange and placed in close contact with the material to be measured, treated or studied. In all situations, it is necessary to determine the reflection factor in the waveguide. This is actually the unknown one wishes to obtain in the study of materials; in the two medical applications, the reflection should be reduced as much as possible to ensure optimum power transfer. The knowledge of the radiated fields is of paramount importance in hyperthermia, where it is important to accurately deposit a certain dose of heating at a specific location, without disturbing the surrounding medium.

In contrast with the better known applications of microwave (radar and communications), where signals propagate over long distances far from the antennas, we are here considering the immediate vicinity of the aperture. Antenna theory, which considers almost exclusively the far field region of the radiating elements, is therefore of little use for these applications. A detailed analysis of the near fields is therefore necessary to derive the desired formulations.

2. DESCRIPTION OF THE GEOMETRY

A circular metallic waveguide having an inner radius a is terminated in the transverse plane z=0 by a flat metallic flange extending (theoretically) to infinity in the transverse direction (Fig. 1). The metal surfaces are assumed to consist of a perfect electric conductor (pec), over the surface of which the tangential electric components vanish. The waveguide itself is filled with a lossless nonmagnetic dielectric of relative permittivity ε_c. The material measured (or

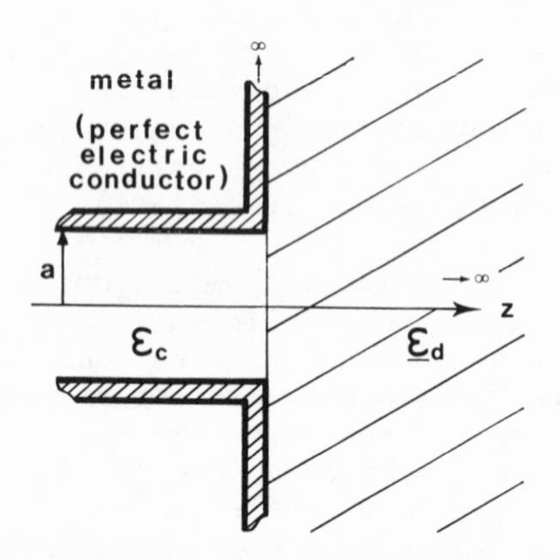

Fig. 1. Geometry considered for the mathematical derivations, circular waveguide of radius a, loaded with a dielectric of permittivity ε_c, terminated by a flat infinite flange in plane z = 0, radiating into an infinite half-space material of complex permittivity $\underline{\varepsilon}_d$.

treated) is assumed to completely fill the right-hand half plane $0 \leq z \leq \infty$; it is homogeneous, isotropic, linear and nonmagnetic, entirely defined by its complex relative permittivity $\underline{\varepsilon}_d$:

$$\underline{\varepsilon}_d = \varepsilon'_d - j\varepsilon''_d \tag{1}$$

An unmodulated sine-wave of frequency $f = \omega/2\pi$ is applied, its time-dependence $\exp(j\omega t)$ being omitted (in the usual fashion) in the following development. Complex notation is used throughout, the actual field being obtained in terms of its phasor equivalent by means of:

$$\vec{E}(t) = \text{Re}\,[\sqrt{2}\,\underline{\vec{E}}\,\exp(j\omega t)] \tag{2}$$

The geometrical conditions listed represent a simplified model of the real situation (Fig. 2); they are a prerequisite for the mathematical development. In practice, they can be relaxed to some extent: since the electromagnetic fields decay rapidly as one moves away from the aperture into the material, the effect of the far edges of the sample soon becomes negligible, particularly so if the material is

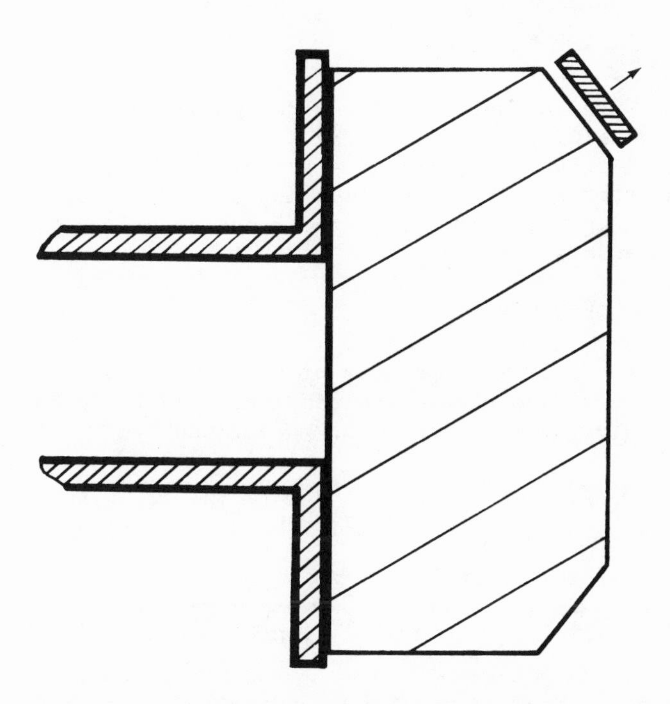

Fig. 2. Geometry encountered in practical applications. Both the flange and the material have finite sizes. Moving a short-circuit plate next to the back side of the material should not perturb the reflection in the waveguide.

lossy. It is easy to check if this requirement is met: moving a piece of metal next to the outside surface should not appreciably perturb the wave reflected into the waveguide.

As the guide materials are lossless, it can support either propagating modes (above cutoff), which have a purely imaginary propagation factor $\underline{\gamma} = j\beta$, or evanescent modes (below cutoff) for which the propagation factor is purely real $\gamma = \alpha$. If several modes can propagate simultaneously, the field distribution in the waveguide remains complex far from the aperture, so that measurement data become practically intractable (multimode operation). For measurement purposes, it is therefore necessary to restrict the frequency range to allow only one mode to propagate (dominant mode). At a reference plane, selected far enough from the aperture, the amplitudes of all evanescent higher-order modes have become negligible, and the reflected dominant mode can be measured by well-known microwave methods (slotted line or reflectometry). This frequency range is located between the cutoff of the dominant TE_{11} mode and that of the first higher-order TM_{01} mode, so that

$$\frac{87,843}{a\sqrt{\varepsilon_c}} < f < \frac{114,754}{a\sqrt{\varepsilon_c}} \tag{3}$$

It may be argued that the TM_{01} mode should not be excited in a structure possessing perfect symmetry. This is of course true for the theoretical situation only; in practice, small mechanical irregularities and material inhomogeneities may suffice to excite other modes; it is necessary to restrict the range accordingly (one may also use mode-suppressors).

3. THEORETICAL DEVELOPMENT

Only the transverse components of the fields (denoted by the subscript t) are needed to ensure the continuity in the plane of the aperture. Within the waveguide, they are given by

$$\underline{\vec{E}}_t = U_0\,[\vec{F}_1(\vec{r}_t)\exp(-j\beta_1 z) + \sum_{n=1}^{\infty}\underline{\Gamma}_n\vec{F}_n(\vec{r}_t)\exp(+\underline{\gamma}_n z)] \tag{4}$$

$$\underline{\vec{H}}_t = U_0\,[\underline{Y}_1\vec{W}_1(\vec{r}_t)\exp(-j\beta_1 z) - \sum_{n=1}^{\infty}\underline{\Gamma}_n\underline{Y}_n\vec{W}_n(\vec{r}_t)\exp(+\underline{\gamma}_n z)] \tag{5}$$

where U_0 is the equivalent amplitude of the incident wave, $\underline{\Gamma}_1$ is the reflection factor of the dominant mode, $\underline{\Gamma}_n$ ($n \neq 1$) are the relative amplitudes of the higher-order evanescent modes, while $\vec{F}_n(\vec{r}_t)$, $\underline{\gamma}_n$ and $\underline{Y}_n$ are, respectively the transverse dependence function of the electric field, the propagation factor and the wave admittance of mode n, while

$$\vec{W}_n(\vec{r}_t) = \vec{e}_z \times \vec{F}_n(\vec{r}_t) \tag{6}$$

The index n, which specifies the waveguide mode, is selected according to their successive appearance over the frequency range. Thus, n = 1 corresponds to the dominant mode (propagating), n = 2 to the first higher-order mode theoretically excited within the structure, and so on.

The transverse mode functions, which are chosen as real quantities, form a complete orthonormal set for the geometry and the excitation considered, satisfying the normalization relation

$$\int_{S'} \vec{F}_m \cdot \vec{F}_n \, dA' = \langle \vec{F}_m, \vec{F}_n \rangle = \delta_{mn} = \begin{cases} 1 & \text{if } m = n \\ 0 & \text{if } m \neq n \end{cases} \tag{7}$$

The integration is carried out over the entire waveguide cross-section S'. This integration will be denoted, from now on, by the bracket notation $\langle a,b \rangle$, as defined in (7).

The electromagnetic fields radiated into the material through the waveguide opening are evaluated by means of aperture theory (Lewin, 1951; Harrington, 1961; Galejs, 1969). The opening, as seen from the material side ($z > 0$) is replaced, for the calculations, by a perfectly conducting metallic wall on which circulates an equivalent magnetic surface current $\vec{\underline{M}}_e$ which is proportional to the actual tangential electric field in the aperture (Fig. 3):

$$\vec{\underline{M}}_e = - \vec{e}_z \times \vec{\underline{E}} \tag{8}$$

This magnetic surface current creates a Hertz magnetic potential in the whole right-hand side half space, given by

$$\vec{\underline{\Lambda}} = \frac{1}{j2\pi\mu_0\omega} \int_{S'} \frac{\exp(-j\underline{k}_d R)}{R} \vec{\underline{M}}_e \, dA' \tag{9}$$

where $\underline{k}_d = \omega\sqrt{\varepsilon_0 \varepsilon_d \mu_0}$ is the complex wave number in the material, R is the distance from the magnetic current source to the observer and the integration is carried out over the complete waveguide opening S'.

The electric and the magnetic fields are in turn provided by:

$$\vec{\underline{E}} = -j\omega\mu_0 \, \nabla \times \vec{\underline{\Lambda}} \tag{10}$$

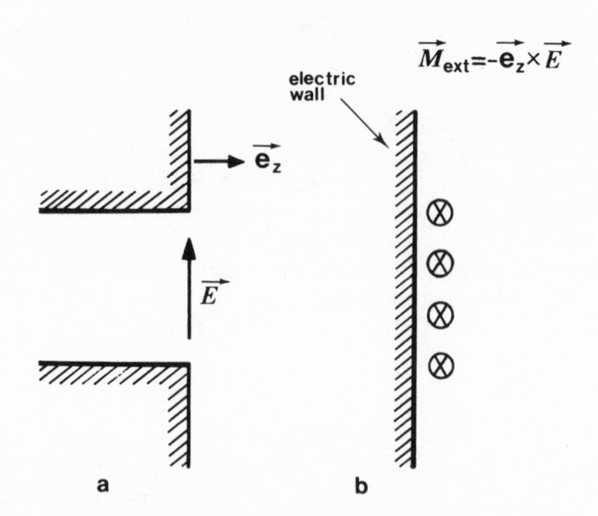

Fig. 3. (a) Aperture in a conducting plane connected to a waveguide; (b) equivalent source considered for the mathematical development of the radiation into the material.

$$\underline{\vec{H}} = \nabla \times \nabla \times \underline{\vec{\Lambda}} = (\underline{k}_d^2 + \nabla\nabla\cdot)\underline{\vec{\Lambda}} \tag{11}$$

The relationship for the magnetic field can be expressed in a simplified form, in terms of the external dyadic Green's function:

$$\underline{\vec{H}} = \int_{S'} \underline{\bar{\bar{G}}}\cdot\vec{M}_e(\vec{r}\,')\,dA' \tag{12}$$

where $\underline{\vec{M}_e} = -\vec{e}_z \times \underline{\vec{E}_t}$ is the equivalent magnetic current in the plane of the aperture, previously defined in (8). The dyadic Green's function for this particular problem can be extracted from (8 - 11), yielding:

$$\underline{\bar{\bar{G}}} = \frac{1}{j2\pi\omega\mu_0}\,(\bar{\bar{I}}\,k_d^2 + \nabla\nabla\cdot)\,\frac{\exp(-jk_dR)}{R} \tag{13}$$

where $\bar{\bar{I}}$ is the two-dimensional unit dyadic.

The boundary conditions, which must be satisfied in the z = 0 transverse plane of the aperture are:

The continuity of the tangential component of the electric field. As this component is actually taken as the source of the radiation in the right-hand half space, this condition is automatically satisfied.

- In the absence of electric surface currents in the plane of the aperture, the tangential components of the magnetic field, obtained by letting $z = 0$ in (5) and in (12) are continuous. This provides a relationship which permits determination of the unknowns of the problem, i.e. the reflection factors $\underline{\Gamma}_n$.

Several methods are available to solve this problem. While the principle itself is rather straightforward, i.e. applying the boundary conditions, the mathematical resolution is rather involved, due to the complexity of the Green's function expression (13):

- Due to the presence of the complex term $\exp(-j\underline{k}_d R)/R$ in the integrand, the integration process just cannot be carried out analytically. The integral must therefore be evaluated by means of numerical techniques.

- When the source and the observer coincide, i.e. when $R = 0$, the integrand becomes singular. This is not an essential singularity, in the sense that the integral itself remains defined. Nevertheless, adequate precautions must be taken, the singularity must be extracted before the application of numerical integration procedures (Gex-Fabry et al., 1979).

- The integration is followed by a double differentiation. This operation must also be carried out numerically. The process is unfortunately quite sensitive to noise, so that even minor rounding errors tend to build up, producing rather large instabilities in the computation process. An alternate way would be to introduce the $\nabla\nabla\cdot$ operator under the integral sign, and carry out the differentiation before the integration. In this case, the singularity becomes a third-order pole, much more difficult to extract [this is the well-known singular behavior of dyadic Green's functions (Collin, 1981)].

While previous problems of a similar nature could be solved by means of a point-matching technique (Gex-Fabry et al., 1979; Mosig et al., 1981), applying the boundary condition for the magnetic field on a discrete number of concentric rings across the waveguide aperture, this approach was not found suitable in the present situation. It appeared preferable to find a way to eliminate the double differentiation, and thus remove the above-mentioned instability problem. This can actually be done by using somewhat more complicated mathematical formulations, of the moment method type. Among the various such techniques available, the method of characteristic modes (Harrington et al., 1971a) was selected, as it appeared to provide optimal convergence and accuracy in the calculations.

In order to use this method, the continuity condition must be expressed in terms of Green's functions on both sides, external (defined in 13) and internal. Proceeding in a similar manner as previously, but this time inside the waveguide, we consider now a waveguide terminated by a short-circuit plate on which circulates a surface current $\underline{\vec{M}}_{int}$ (Fig. 4). The magnetic field in the plane of the aperture, on the waveguide side, is thus given by:

$$\underline{\vec{H}}_{int} = 2\underline{\vec{H}}_1 + \int_{S'} \underline{\bar{\bar{B}}}\cdot\underline{\vec{M}}_{int}(\vec{r}')\,dA' \tag{14}$$

where

$$\underline{\vec{H}}_1 = U_0\,\underline{Y}_1\,\vec{W}_1(\vec{r}_t) \tag{15}$$

is the transverse magnetic field of the incident wave in the waveguide, given by (5), and the internal magnetic surface current is given by $\underline{\vec{M}}_{int} = \vec{e}_z \times \underline{\vec{E}}_t = \underline{\vec{M}}$.

If a real short circuit were placed in the plane of the aperture, the magnetic surface current $\underline{\vec{M}}$ would vanish, and the total magnetic field in this plane would be twice that of the incident wave, as given by (14). It is also apparent that, when replacing $\underline{\vec{M}}$ in (14) by its expansion over the waveguide modes, as given in (4), and then identifying with the transverse magnetic field in the plane of the aperture,

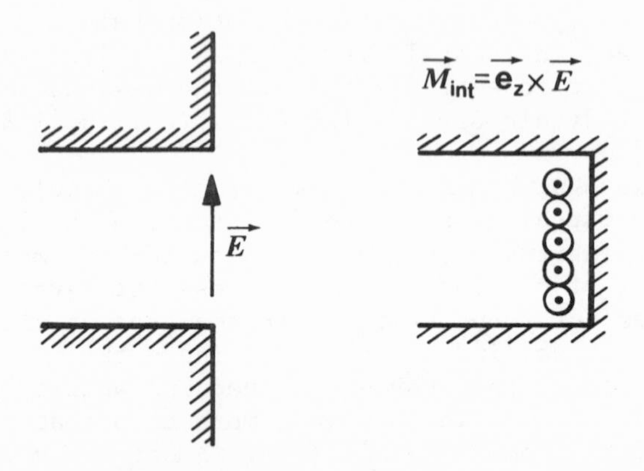

Fig. 4. (a) Aperture in a conductive plane connected to a waveguide. (b) Equivalent source considered for the mathematical development of the reflection into the waveguide.

given by (5), one obtains:

$$\int_{S'} \underline{\bar{\bar{B}}} \cdot \vec{W}_n \, dA' = -\underline{Y}_n \vec{W}_n \tag{16}$$

A more complete derivation of the internal dyadic Green's function, presented in Appendix A, yields (in the plane of the aperture):

$$\bar{\bar{B}}(\vec{r},\vec{r}\,') = -\sum_{n=1}^{\infty} \underline{Y}_n \vec{W}_n(\vec{r}_t) \, \vec{W}_n'(\vec{r}_t\,') \tag{17}$$

Replacing then $\underline{M}_e$ by $-\underline{M}$ in (12), which specifies the continuity of the tangential electric field, and then applying the continuity relationship for the magnetic field in the plane of the aperture one obtains:

$$\vec{\underline{H}}_1 = -\frac{1}{2} \int_{S'} (\underline{\bar{\bar{G}}} + \underline{\bar{\bar{B}}}) \cdot \vec{\underline{M}} \, dA' = \underline{y}_{op}(\vec{\underline{M}}) = g_{op}(\vec{\underline{M}}) + jb_{op}(\vec{\underline{M}}) \tag{18}$$

This constitutes the general formulation of the problem in terms of dyadic Green's functions (Magnetic Field Integral Equation, MFIE). It is then necessary to solve (18) to determine first the unknown magnetic current density $\vec{\underline{M}}$ which, in turn, permits calculation of the reflection factor of the dominant mode $\underline{\Gamma}_1$ (Section 4.).

4. METHOD OF CHARACTERISTIC MODES

The relationship (18), which links the magnetic surface current $\underline{M}$ (V/m) to the transverse magnetic field $\vec{\underline{H}}_1$ (A/m) of the incident wave in the waveguide, defines the complex admittance operator $\underline{y}_{op}$, which has a real part (conductance operator) g_{op} and an imaginary part (susceptance operator) b_{op}. The usual notation has been taken here, even though it may lead to possible confusions with the external and internal Green's functions $\bar{\bar{G}}$ and $\bar{\bar{B}}$. This means that there is no particular link between $\bar{\bar{G}}$ and g_{op} or between $\bar{\bar{B}}$ and b_{op}, but that both operators depend on the sum of both Green's functions. As the system considered is reciprocal, the operator $\underline{y}_{op}$ is symmetrical (Harrington et al., 1971b). A set of base functions $\vec{M}_k$, which correspond to the characteristic modes of the aperture, are derived, and the unknown magnetic current density is then expanded over this set. The eigenvalue equation used to obtain the base functions has the general form:

$$\underline{y}_{op}(\vec{M}_k) = \underline{\nu}_k \, A_{op}(\vec{M}_k) \tag{19}$$

where A_{op} is a real operator which can be selected arbitrarily. For the sake of convenience, A_{op} is here chosen equal to the conductance operator g_{op}, so that

$$g_{op}(M_k) + j\, b_{op}(M_k) = \underline{\nu}_k\, g_{op}(M_k) \qquad (20)$$

It is readily apparent that the eigenvalues of (20) have the form

$$\underline{\nu}_k = 1 + j\, \lambda_k \qquad (21)$$

with

$$b_{op}(\vec{M}_k) = \lambda_k\, g_{op}(\vec{M}_k) \qquad (22)$$

Since both operators b_{op} and g_{op} are real and symmetrical, the eigenvalues, λ_k, and the eigenvectors, $\vec{M}_k$, are real. As these eigenvalues are defined by a homogeneous linear equation (22), their amplitudes are arbitrary. They are here normalized by the imposition of an orthogonality condition:

$$< \vec{M}_m, g_{op}(\vec{M}_k) > = \delta_{mk}\, P_{kk} = P_{kk} \qquad (23)$$

To simplify matters, the power level P_{kk} is set equal to 1 Watt. The bracket notation, which defines the "scalar product", was first introduced in (7). It follows directly from (22) and (23) that

$$< \vec{M}_m, b_{op}(\vec{M}_k) > = \lambda_k\, \delta_{mk}\, P_{kk} \qquad (24)$$

The unknown magnetic current density $\underline{\vec{M}}'$ can now be expanded over the set of the eigenvectors $\vec{M}_k$

$$\underline{\vec{M}}' = \sum_k \underline{a}_k\, \vec{M}_k \qquad (25)$$

Replacing $\underline{\vec{M}}'$ by its series expansion in (18), dot-multiplying by $\vec{M}_m$ and integrating over the waveguide cross-section yields:

$$\sum_k \underline{a}_k < \vec{M}_m, \underline{y}_{op}(\vec{M}_k) > - < \vec{M}_m, \underline{\vec{H}}_1 > = 0 \qquad (26)$$

The orthogonality conditions (23) and (24) are then applied, so that all the terms of the series but one vanish, yielding

$$\underline{a}_m\, (1 + j\lambda_m)\, P_{mm} = \underline{I}_m \qquad (27)$$

where $\underline{I}_m$ stands for

$$\underline{I}_m = < \vec{M}_m, \vec{H}_1 > \qquad (28)$$

The unknown coefficients $\underline{a}_m$ can thus be determined and, after replacing by their respective values in (25), the expression for the magnetic current density is obtained

$$\underline{\vec{M}}' = \sum_m \frac{\underline{I}_m}{(1 + j\lambda_m) P_{mm}} \vec{M}_m \tag{29}$$

At this point, however, the eigenvalues λ_k and the eigenvectors $\vec{M}_k$ are not yet known, and their determination is by no means straightforward, due to the rather complex nature of the operators. Their values are determined by expanding the eigenvectors $\vec{M}_k$ over the set of the transverse mode functions $\vec{W}_n$, which were defined in (6):

$$\vec{M}_k = \sum_n u_{kn} \vec{W}_n \tag{30}$$

Since both $\vec{M}_k$ and the $\vec{W}_n$ are real, it directly follows that the coefficients u_{kn} are also real quantities.

This series development is then introduced into (22), which is in turn dot-multiplied by $\vec{W}_m$ and integrated over the waveguide cross-section, yielding

$$\sum_n u_{kn} < \vec{W}_m, b_{op}(\vec{W}_n) > = \lambda_k \sum_n u_{kn} < \vec{W}_m, g_{op}(\vec{W}_n)> \tag{31}$$

The bracketed terms of (31) are then written in simplified form as

$$b_{mn} = < \vec{W}_m, b_{op}(\vec{W}_n) > \tag{32}$$

$$g_{mn} = < \vec{W}_m, g_{op}(\vec{W}_n) > \tag{33}$$

In this manner, one obtains a set of linear equations, which can be written in matrix form:

$$\begin{bmatrix} b_{11} & b_{12} & \cdots & b_{1N} \\ b_{21} & b_{22} & & \\ & & & \\ b_{N1} & b_{N2} & & b_{NN} \end{bmatrix} \begin{bmatrix} u_{k1} \\ u_{k2} \\ \\ u_{kN} \end{bmatrix} = \lambda_k \begin{bmatrix} g_{11} & g_{12} & \cdots & g_{1N} \\ g_{21} & g_{22} & & \\ & & & \\ g_{N1} & g_{N2} & & g_{NN} \end{bmatrix} \begin{bmatrix} u_{k1} \\ u_{k2} \\ \\ u_{kN} \end{bmatrix} \tag{34}$$

In (34), the set of equations, which is theoretically infinite, has been truncated, including only the first N equations. The value for N will be determined in the computation process, in order to ensure adequate convergence of the results. Symbolically, the matrix equation (34) is written in the form

$$[b][u]_k = \lambda_k [g][u]_k \tag{35}$$

The g_{mn} and b_{mn} terms which appear in (34) are respectively the real and imaginary parts of the components of the admittance matrix, given by:

$$\underline{y}_{mn} = g_{mn} + jb_{mn} = \langle \vec{W}_m, y_{op}(\vec{W}_n) \rangle$$

$$= -\frac{1}{2} \int_S \int_{S'} \vec{W}_m \cdot (\underline{\bar{\bar{G}}} + \underline{\bar{\bar{B}}}) \cdot \vec{W}_n \; dA' \; dA \tag{36}$$

Replacing $\underline{\bar{\bar{G}}}$ and $\underline{\bar{\bar{B}}}$ by their respective values given by (13) and (17) one obtains:

$$\underline{y}_{mn} = -\frac{1}{2} \int_S \int_{S'} \vec{W}_m \cdot \{ [\bar{\bar{I}} \underline{k}_d^2 + \nabla\nabla] \frac{\exp(-j\underline{k}_d R_0)}{2\pi j \omega \mu_0 R_0} - \sum_k \vec{W}_k \vec{W}_k \} \cdot \vec{W}_n$$

$$dA' \; dA \tag{37}$$

where R_0 is the distance between the two points considered in the integration, respectively in the $\vec{r}$ and $\vec{r}'$ coordinate systems.

A most interesting identity (Dubost et al., 1976) permits simplification of the term which contains derivations:

$$\int_S \int_{S'} \vec{W}_m \cdot \nabla\nabla \cdot \frac{\exp(-j\underline{k}_d R_o)}{2\pi j \omega \mu_o R_o} W_n \; dA \; dA'$$

$$= -\int_S \int_{S'} (\nabla \cdot \vec{W}_m)(\nabla' \cdot \vec{W}_n) \frac{\exp(-j\underline{k}_d R_0)}{2\pi j \omega \mu_0 R_0} \; dA \; dA' \tag{38}$$

The prime (') sign after ∇ specifies operation on the primed coordinates. The integration of the internal dyadic Green's function is self-evident, taking into consideration the orthonormality condition (7)

$$\int_S \int_{S'} \vec{W}_m(\vec{r}_t) \cdot \sum_k (-\underline{Y}_k) \vec{W}_k(\vec{r}_t) \vec{W}_k(\vec{r}_t') \cdot \vec{W}_n(\vec{r}_t') \; dA' dA = -\underline{Y}_n \delta_{mn} \tag{39}$$

Replacing into (37), the admittance term becomes

$$\underline{y}_{mn} = -\frac{1}{2}\int_S\int_{S'}[\vec{W}_m\cdot\vec{W}_n k_{\underline{d}}^2 - (\nabla\cdot\vec{W}_m)(\nabla'\cdot\vec{W}_n)]\frac{\exp(-j\underline{k}_d R_0)}{2\pi j\omega\mu_0 R_0}$$

$$dA'dA + \frac{\underline{Y}_n\delta_{mn}}{2} \qquad (40)$$

Finally the reflection factor $\underline{\Gamma}_1$ for the dominant mode, which is the quantity one wishes to determine in most cases, is obtained from (8), (4) and (6) which yield $\underline{\vec{M}}$ in the plane of the aperture (z=0):

$$\underline{\vec{M}} = \vec{e}_z \times \underline{\vec{E}} = U_0[\vec{W}_1 + \sum_n \underline{\Gamma}_n\vec{W}_n] \qquad (41)$$

Multiplying by $\vec{H}_1$, given by (15), integrating over the waveguide cross-section and taking advantage of the orthonormality condition (7) yields:

$$<\vec{H}_1,\underline{\vec{M}}> = U_0^2Y_1(1 + \underline{\Gamma}_1) \qquad (42)$$

and hence

$$\underline{\Gamma}_1 \frac{<\vec{H}_1,\underline{\vec{M}}>}{U_0^2Y_1} 1 \qquad (43)$$

It must be remembered here that U_0 is quite an arbitrary quantity: as both $\vec{H}_1$ and $\underline{\vec{M}}$ are proportional to U_0, the value chosen does not affect the resulting $\underline{\Gamma}_1$. In practice, U_0 is chosen equal to 1 volt (it is kept in the equations to ensure dimensional consistency).

Replacing respectively $\vec{H}_1$ and $\underline{\vec{M}}$ in (43) by their developments over the $\vec{W}_n$ set of functions, making use respectively of (15) and of (29), we obtain:

$$<\vec{H}_1,\underline{\vec{M}}> = U_0Y_1<\vec{W}_1, \sum_m \frac{\underline{I}_m}{(1+j\lambda_m)P_{mm}} \sum_n u_{mn}\vec{W}_n>$$

$$= U_0Y_1\sum_m \frac{\underline{I}_m u_{m1}}{(1+j\lambda_m)P_{mm}} = (U_0Y_1)^2\sum_m \frac{u_{m1}^2}{(1+j\lambda_m)P_{mm}} \qquad (44)$$

with (23)

$$P_{mm} = <\vec{M}_m, g_{op}(\vec{M}_m)> \qquad (45)$$

Finally, introducing (44) into (43), one obtains:

$$\underline{\Gamma}_1 = Y_1 \sum_m \frac{u_{m1}^2}{(1 + j\lambda_m)P_{mm}} - 1 \tag{46}$$

This relation allows one to determine the reflection coefficient once the eigenvalues λ_m and the first term of the eigenvectors u_{m1} are known. An alternate formulation for the reflection coefficient, giving actually directly the input admittance, can be determined, considering another formulation for the magnetic field in the waveguide, given by:

$$\underline{\vec{H}}_{int} = (1 - \underline{\Gamma}_1)\underline{\vec{H}}_1 + \int_{S'} \underline{\bar{\bar{B}}}_{ho} \cdot \underline{\vec{M}}_{int} \, dA' \tag{47}$$

As can be seen when comparing (47) with the previously used relationship (14), $\underline{\bar{\bar{B}}}_{ho}$ is the part of the internal dyadic Green's function which operates exclusively on the higher-order mode terms [the term n=1 is excluded from the summation in (17)]. The boundary condition (18) then takes the form

$$(1 - \underline{\Gamma}_1)\underline{\vec{H}}_1 = - \int_{S'} (\underline{\bar{\bar{G}}} + \underline{\bar{\bar{B}}}_{ho}) \cdot \underline{\vec{M}} \, dA' \tag{48}$$

Dot-multiplying both sides by the magnetic current $\underline{\vec{M}}$ and integrating over the waveguide cross-section then yields:

$$(1 - \underline{\Gamma}_1) \int_S \underline{\vec{M}} \cdot \underline{\vec{H}}_1 \, dA = - \int_S \int_{S'} \underline{\vec{M}} \cdot (\underline{\bar{\bar{G}}} + \underline{\bar{\bar{B}}}_{ho}) \cdot \underline{\vec{M}} \, dA \, dA' \tag{49}$$

Combining with (42), the relationship for the input admittance is obtained

$$\frac{\underline{Y}}{Y_1} = \frac{1 - \underline{\Gamma}_1}{1 + \underline{\Gamma}_1} = - U_0^2 Y_1 \frac{\int_S \int_{S'} \underline{\vec{M}} \cdot (\underline{\bar{\bar{G}}} + \underline{\bar{\bar{B}}}_{ho}) \cdot \underline{\vec{M}} \, dA \, dA'}{\{\int_S \underline{\vec{M}} \cdot \underline{\vec{H}}_1 \, dA\}^2} \tag{50}$$

While this relationship is inherently more complicated than (46), it presents an interesting advantage, that of being also a variational principle for the admittance. This means that this relationship is insensitive to first-order changes in the magnetic current density. It must be remembered that this last quantity is expressed in terms of an infinite series, which must be truncated when carrying the calculations. The above expression will thus be relatively insensitive to truncation errors, and can thus be expected to converge more rapidly to the true solution. It is also quite appreciated to have several alternative methods available for the calculations, as they provide a way to at least partially check the validity of the developments and the accuracy of the results obtained. Proceeding in the same way as previously, the magnetic field of the dominant waveguide mode $\underline{\vec{H}}_1$ and the magnetic current density $\underline{\vec{M}}$ are replaced by their respective developments, yielding after some simplifications:

$$\frac{\underline{Y}}{Y_1} = \frac{1}{Y_1 \sum_n \frac{u_{n1}^{\star 2}}{(1 + j\lambda_n^{\star}) P_{nn}^{\star}}} \tag{51}$$

where the $\lambda_n^{\star}$, $u_{n1}^{\star}$ and $P_{nn}^{\star}$ are respectively the eigenvalue, eigenvector and normalizing power obtained when the base equation used to define the admittance operator is taken as (49) instead of (18).

5. CIRCULAR WAVEGUIDE EXCITED BY THE TE_{11} MODE

The previous developments were carried out in a rather general form, in order to be applicable to homogeneous waveguides having any arbitrary cross-section. From now on, the particular case of a circular waveguide, defined in Section 2 and excited by the dominant TE_{11} mode, will be considered more specifically.

The dominant TE_{11} mode which feeds the waveguide is spatially degenerate. This means that actually two modes can exist and propagate with the same dispersion characteristic $\beta(\omega)$, their fields having azimuthal dependences in either $\cos\phi$ or $\sin\phi$. As the structure is axially symmetrical and the material isotropic, both modes produce the same reflection; it is thus only necessary to consider one of them. The same azimuthal dependence applies to all the fields excited within the structure; this means that all the higher-order modes near the aperture plane must be of either the TE_{1m} or of the TM_{1m} subsets. In the frequency scale, these modes appear in an alternate fashion, their cutoff frequencies being respectively linked to the extrema or to the zeros of the Bessel function J_1 (Fig. 5). This means that, when using the numbering index n defined at the beginning of Section 3, odd values of n always correspond to TE modes, and even values of n to TM modes. The transverse dependence functions $\vec{F}_n$ and $\vec{W}_n$ for these modes, as well as their propagation factors and wave impedances, are presented in Table I. The first values of the cutoff wavenumbers p_n are presented in Table II. The normalization factors C_n are selected to satisfy the orthonormality condition (7).

Since the problem presents circular symmetry, all the fields and geometrical quantities are expressed in the circular cylindrical coordinate system, which becomes the two-dimensional polar coordinates (ρ,ϕ) and (ρ',ϕ') over the plane of the aperture $(z = 0)$. Figure 6 presents a graphical representation, showing respectively the position

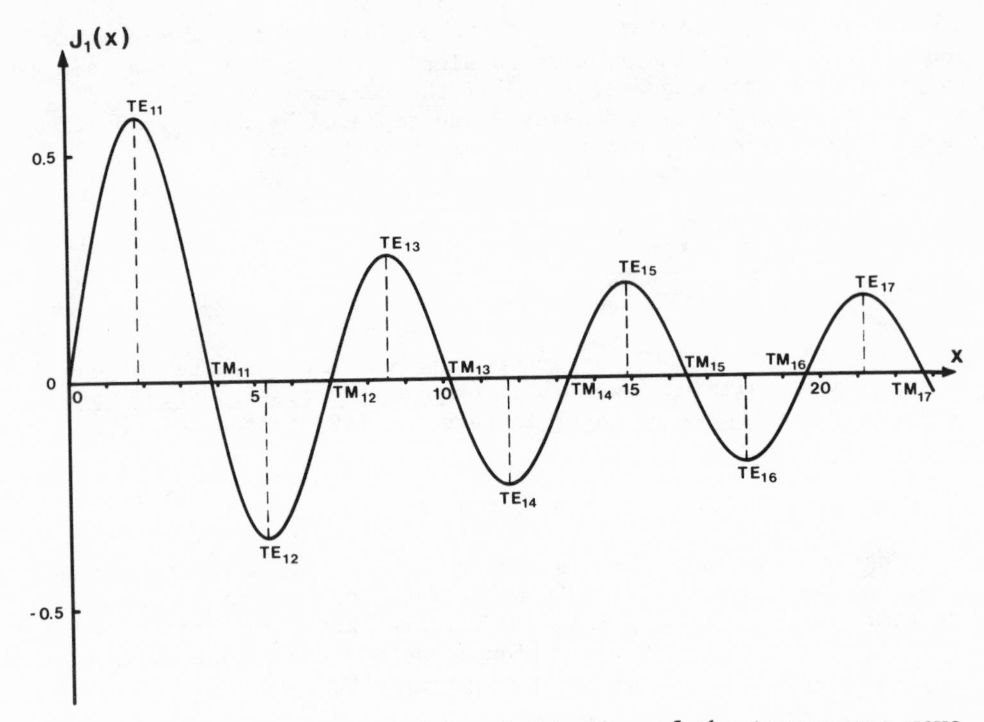

Fig. 5. Bessel function $J_1(x)$ and location of the transverse wave numbers $p_n a$ for TE_{1m} modes (extrema) and TM_{1m} modes (zeros). See also Table I for the numerical values of $p_n a$.

vectors of the source $\vec{r}'$ and of the observer $\vec{r}$, the coordinate systems at both locations and the geometrical relationships between them. The distance R_0, from the source in the $\vec{r}'$ system to the observer in the $\vec{r}$ system is given by

$$R_0 = |\vec{r} - \vec{r}'| = \sqrt{\rho^2 + \rho'^2 - 2\rho\rho'\cos\psi} \tag{52}$$

It is apparent that the angular dependence is a function of the difference between the two angles, denoted here by

$$\psi = \phi - \phi' \tag{53}$$

The sum of the angles is also introduced as a new variable, to simplify further developments

$$\psi' = \phi + \phi' \tag{54}$$

The integration over the azimuthal coordinates is modified to take into account the newly introduced coordinates

Table I. Characteristic Quantities for TE_{1m} AND TM_{1m} MODES*

	TE modes: n odd	TM modes: n even
$\vec{F}_n$	$\vec{e}_z \times \nabla_t \psi_n$	$-\nabla_t \phi_n$
$F_{n\rho}$	$C_n \frac{1}{\rho} J_1(p_n \rho) \sin\phi$	$-C_n p_n J_1'(p_n \rho) \sin\phi$
$F_{n\phi}$	$C_n p_n J_1'(p_n \rho) \cos\phi$	$-C_n \frac{1}{\rho} J_1(p_n \rho) \cos\phi$
$\vec{W}_n = \vec{e}_z \times \vec{F}_n$	$-\nabla_t \psi_n$	$-\vec{e}_z \times \nabla_t \phi_n$
$W_{n\rho}$	$-C_n p_n J_1'(p_n \rho) \cos\phi$	$C_n \frac{1}{\rho} J_1(p_n \rho) \cos\phi$
$W_{n\phi}$	$C_n \frac{1}{\rho} J_1(p_n \rho) \sin\phi$	$-C_n p_n J_1'(p_n \rho) \sin\phi$
transverse potential	$\psi_n = C_n J_1(p_n \rho) \cos\phi$	$\phi_n = C_n J_1(p_n \rho) \sin\phi$
boundary condition	$J_1'(p_n a) = 0$	$J_1(p_n a) = 0$
propagation factor $\underline{\gamma}_n$	if n = 1 $\underline{\gamma}_n = j\beta_1 = j\sqrt{k_c^2 - p_1^2}$ if n ≠ 1 $\gamma_n = \alpha_n = \sqrt{p_n^2 - k_c^2}$	$\gamma_n = \alpha_n = \sqrt{p_n^2 - k_c^2}$
wave impedance $\underline{Y}_n$	$\underline{\gamma}_n / j\omega\mu_0$	$j\omega\varepsilon_0\varepsilon_c/\alpha_n$
normalization factor C_n	$\sqrt{2}\{\sqrt{\pi}[(ap_n)^2-1]^{\frac{1}{2}} J_1(p_n a)\}^{-1}$	$\sqrt{2}\{\sqrt{\pi} a p_n J_1'(p_n a)\}^{-1}$
with:	$\vec{F}_n = \vec{e}_\rho F_{n\rho} + \vec{e}_\phi F_{n\phi}$ $k_c = \omega\sqrt{\varepsilon_0 \varepsilon_c \mu_0}$	$\vec{W}_n = \vec{e}_\rho W_{n\rho} + \vec{e}_\phi W_{n\phi}$ $\nabla_t f = \vec{e}_\rho \frac{\partial f}{\partial \rho} + \vec{e}_\phi \frac{1}{\rho}\frac{\partial f}{\partial \phi}$

*F.E. Gardiol (1981).

Table II. Transverse Wave Numbers p_n of The Modes Excited at the Aperture

n	$p_n \cdot a$	mode
1	1,841	TE_{11}
2	3,832	TM_{11}
3	5,331	TE_{12}
4	7,016	TM_{12}
5	8,536	TE_{13}
6	10,173	TM_{13}
7	11,706	TE_{14}
8	13,324	TM_{14}
9	14,864	TE_{15}
10	16,471	TM_{15}
11	18,016	TE_{16}
12	19,616	TM_{16}
13	21,164	TE_{17}
14	22,760	TM_{17}

$$\int_0^{2\pi}\int_0^{2\pi} A(\phi,\phi')\, d\phi'\, d\phi = \int_0^{2\pi}\int_0^{2\pi} A\,(\psi,\phi)\, d\psi\, d\phi \tag{55}$$

The vector product $\vec{W}_m(\vec{r})\cdot\vec{W}_n(\vec{r}\,')$ which appears in (40) is developed

$$\vec{W}_m\cdot\vec{W}_n = \vec{e}_\rho\cdot\vec{e}_{\rho'} W_{m\rho} W_{n\rho'} + \vec{e}_\rho\cdot\vec{e}_{\phi'} W_{m\rho} W_{n\phi'} + \vec{e}_\phi\cdot\vec{e}_{\rho'} W_{m\phi} W_{n\rho'} + \vec{e}_\phi\cdot\vec{e}_{\phi'} W_{m\phi} W_{n\phi'} \tag{56}$$

The products of the unit vectors are obtained, considering Fig. 6:

$$\vec{e}_\rho\cdot\vec{e}_{\rho'} = \vec{e}_\phi\cdot\vec{e}_{\phi'} = \cos(\phi-\phi') = \cos\psi \tag{57}$$

$$\vec{e}_\rho\cdot\vec{e}_{\phi'} = -\vec{e}_\phi\cdot\vec{e}_{\rho'} = \sin(\phi-\phi') = \sin\psi \tag{58}$$

Replacing into (56) yields:

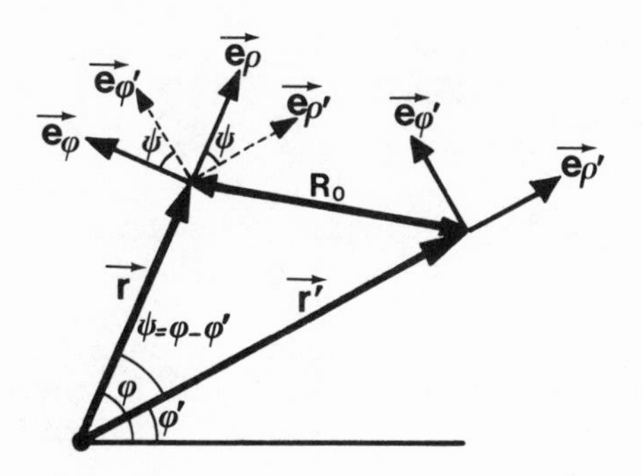

Fig. 6. Definition of the coordinate systems at the source $\vec{r}'$ and at the observer $\vec{r}$ locations.

$$\vec{W}_m \cdot \vec{W}_n = (W_{m\rho} W_{n\rho'} + W_{m\phi} W_{n\phi'}) \cos \psi + (W_{m\rho} W_{n\phi'} - W_{m\phi} W_{n\rho'}) \sin \psi \tag{59}$$

The expression of the transverse functions, given in Table I, are then introduced into (59), leading to a somewhat involved development, which must be carried out for the four possible combinations of modes: TE-TE, TM-TM, TE-TM and TM-TE. The complete derivation for the first case mentioned is presented in detail in Appendix B. During the process, certain simplifications can be carried out. The integration with respect to ϕ (55) can also be carried out at that point, since the multiplying factor appearing in (40) is not a function of this quantity. The terms remaining after the integration yield:

$$\int_0^{2\pi} \vec{W}_m \cdot \vec{W}_n d\phi = \frac{1}{2} \pi K_{mn} \left[(-1)^{m+n} J_0(p_m\rho) J_0(p_n\rho') + J_2(p_m\rho) J_2(p_n\rho') \cos 2\psi\right] \tag{60}$$

with

$$K_{mn} = p_m p_n C_m C_n \tag{61}$$

The divergence terms become, respectively for TE modes (n odd)

$$\nabla \cdot \vec{W}_n = - \nabla \cdot \nabla_t \Psi_n = - \nabla_t^2 \Psi_n = p_n^2 \Psi_n \tag{62}$$

whereas for TM modes,

$$\nabla\cdot\vec{W}_n = -\nabla\cdot(\vec{e}_z \times \nabla_t\Phi_n) = -\nabla\Phi_n\cdot(\nabla \times \vec{e}_z) + \vec{e}_z\cdot(\nabla\times\nabla_t\Phi_n) \equiv 0 \qquad (63)$$

The first term on the right contains the derivative of a constant, the second is the curl of a gradient: both vanish for all TM modes.

The contribution of the TE modes is developed, introducing the transverse mode potentials given in Table I, then carrying out the integration with respect to ϕ as was done previously. The resulting expression is

$$\int_0^{2\pi} (\nabla\cdot\vec{W}_m)(\nabla'\cdot\vec{W}_n)\, d\phi$$

$$= \frac{1}{2}\,[1-(-1)^{mn}]\pi p_m p_n K_{mn} J_1(p_m\rho) J_1(p_n\rho')\cos\psi \qquad (64)$$

The admittance matrix element can thus be expressed in the form

$$y_{mn} = -k_d^2[(-1)^{m+n}\, \underline{T}_{mn}^0 + T_{mn}^2] + [1-(-1)^{mn}] p_m p_n K_{mn} T_{mn}^1 + \frac{1}{2}\, \underline{Y}_m \delta_{mn} \qquad (65)$$

where all the integrals are of the form

$$\underline{T}_{mn}^{\ell} = \frac{K_{mn}}{j8\omega\mu_0} \int_0^a\int_0^a\int_0^{2\pi} J_\ell(p_m\rho) J_\ell(p_n\rho')\cos\ell\psi\, \frac{\exp(-j\underline{k}_d R_0)}{R_0}\, \rho\rho'\, d\psi\, d\rho'\, d\rho \qquad (66)$$

$$\ell = 0,1,2$$

This integral can be further developed to carry out the computations, letting

$$\underline{k}_d = k_d' - jk_d'' \qquad (67)$$

The real and imaginary parts of expression $\underline{T}_{mn}^{\ell} = T_{mn}^{\ell'} + jT_{mn}^{\ell''}$ are then respectively given by:

$$T_{mn}^{\ell'} = -\frac{K_{mn}}{4\omega\mu_0} \int_0^a\int_0^a\int_0^{\pi} J_\ell(p_m\rho) J_\ell(p_n\rho')\cos\ell\psi\, \frac{\exp(-k_d'' R_0)}{R_0}\, \sin(k_d' R_0)\, \rho\rho'\, d\psi\, d\rho'\, d\rho \qquad (68)$$

$$T^{\ell''}_{mn} = - \frac{K_{mn}}{4\omega\mu_0} \int_0^a \int_0^a \int_0^\pi J_\ell(p_m\rho) J_\ell(p_n\rho') \cos\ell\psi \, \frac{\exp(-k''_d R_0)}{R_0}$$

$$\cos(k'_d R_0) \; \rho\rho' \; d\psi \; d\rho' \; d\rho \qquad (69)$$

$$\ell = 0,1,2$$

It is apparent, that when $R_0 = 0$, which happens when the source and observators point coincide for $\rho = \rho'$ and $\psi = 0$, the denominator vanishes. The real part of the integral is not singular, due to the presence of the sine function, which also vanishes under the same conditions. On the other hand, the imaginary part contains a cosine term, which remains finite: the integrand thus possesses a non-essential singularity, which must first be extracted before the computation of the integral by means of numerical methods can be carried out. The procedure used to extract the singularity is outlined in Appendix C.

6. CALCULATION PROCEDURE

A computer program was elaborated, in Fortran IV for operation on the VAX-11 Computer of the Electrical Engineering Department of the Ecole Polytechnique. The specified input data are the waveguide radius a, the frequency f, the dielectric constants of both media ε_c and $\underline{\varepsilon}_d$.

The first step of the procedure is to calculate the functions $\underline{T}^\ell_{mn}$ corresponding to the interactions between the first N modes excited at the aperture. The admittance terms $\underline{y}_{mn}$ are then given by (65), the real and imaginary parts yielding the two N x N square matrices of (34). The eigenvalues and eigenvectors of this equation are extracted and, upon introduction into (46), yield the reflection factor $\underline{\Gamma}_1$. Calculations of $\underline{\Gamma}_1$ with an increasing number of waveguide modes N exhibit a fast convergence towards the true value (Fig. 7). The convergence presents an alternate character for the modulus of $\underline{\Gamma}_1$: this means that successive values give respectively higher and lower bounds bracketing the true value. A good approximation is obtained already with 5 modes. Upper and lower bounds almost coincide after 10 modes. The argument of $\underline{\Gamma}_1$ also exhibits fast convergence.

The alternate approach presented at the end of Section 4, making use of the formulation (50) and (51) based upon the dyadic Green's function equation (47) was also implemented; it converges towards values practically identical with those of the other approach.

Calculated results for lossless dielectrics are presented in Fig. 8, in terms of the normalized frequency $k_c a$, with the ratio

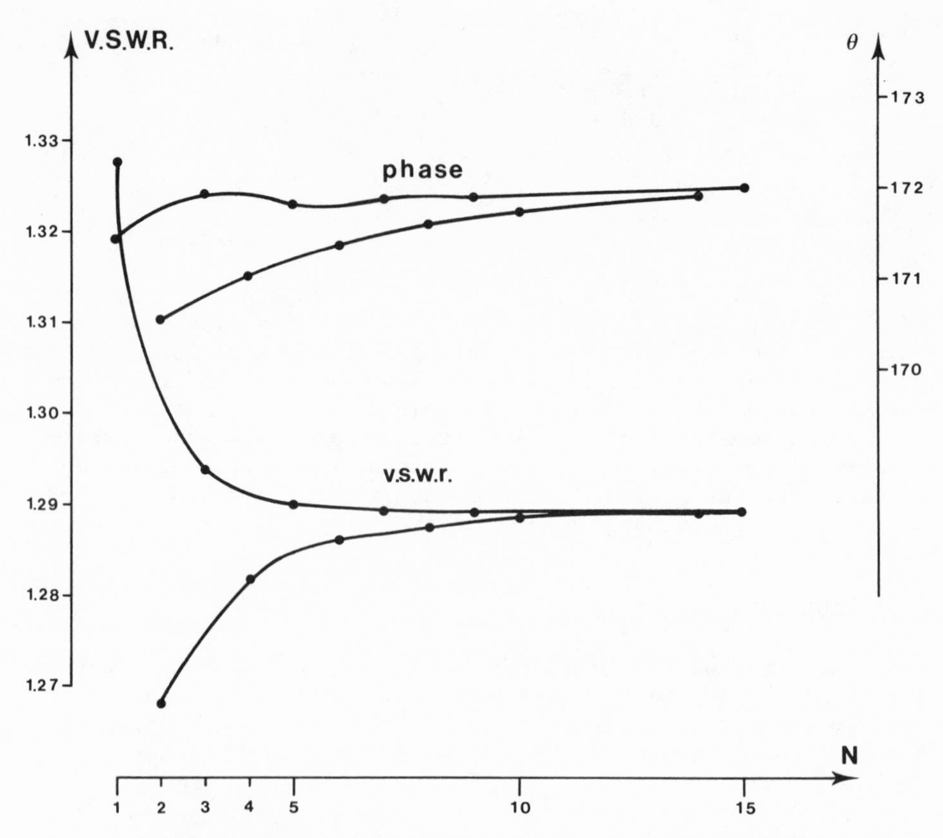

Fig. 7. Convergence of calculated values for VSWR and for $\theta = \arg(\underline{\Gamma}_1)$ as a function of the number N of modes considered in the computations, for $k_c a = 2.17$, homogeneous case.

$\varepsilon_d'/\varepsilon_c$ as parameter (1 to 16). For lossy dielectric, $\underline{\Gamma}_1$ is represented in the polar coordinate system (Smith chart) in Fig. 9, at a fixed normalized frequency $k_c a$.

Calculations with the present program require typically about 2 minutes for one set of data (a, f, ε_c, $\underline{\varepsilon}_d$) considering the interactions between 5 modes; the computation time is roughly proportional to N(N+1). It is dominated by the numerical integration process. It could be reduced by about 50%, without loss of accuracy, by keeping in memory the values of J_ℓ used during the integration process, rather than calculating them each time.

7. EXPERIMENTAL VERIFICATION

The method developed being of a rather intricate mathematical nature, it provides various possible sources of potential errors as

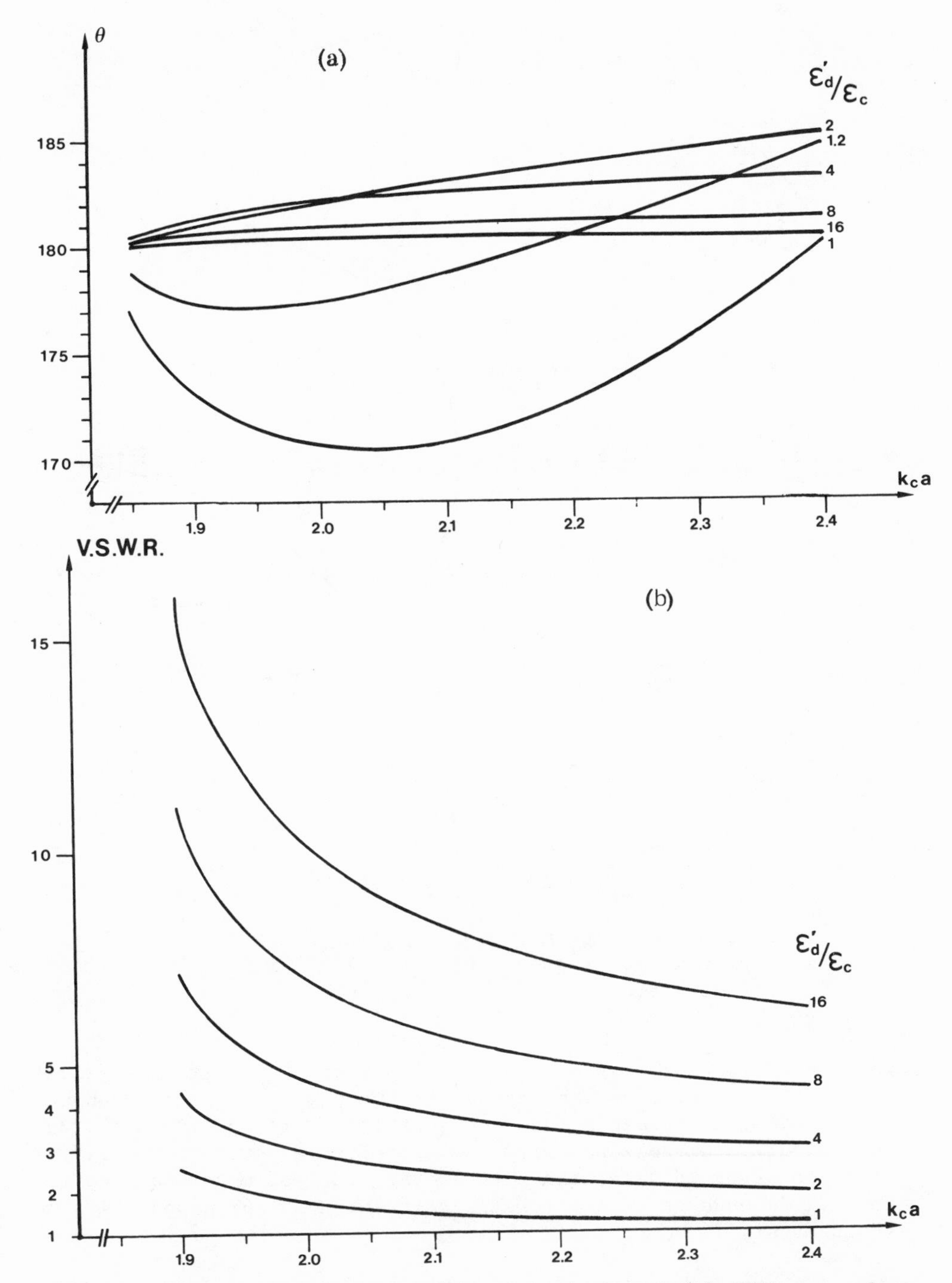

Fig. 8. Calculated values of (a) $\theta = \arg(\underline{\Gamma}_1)$ and (b) VSWR as a function of the normalized frequency $k_c a$ for lossless dielectrics of relative permittivities $\varepsilon_d/\varepsilon_c$ between 1 and 16.

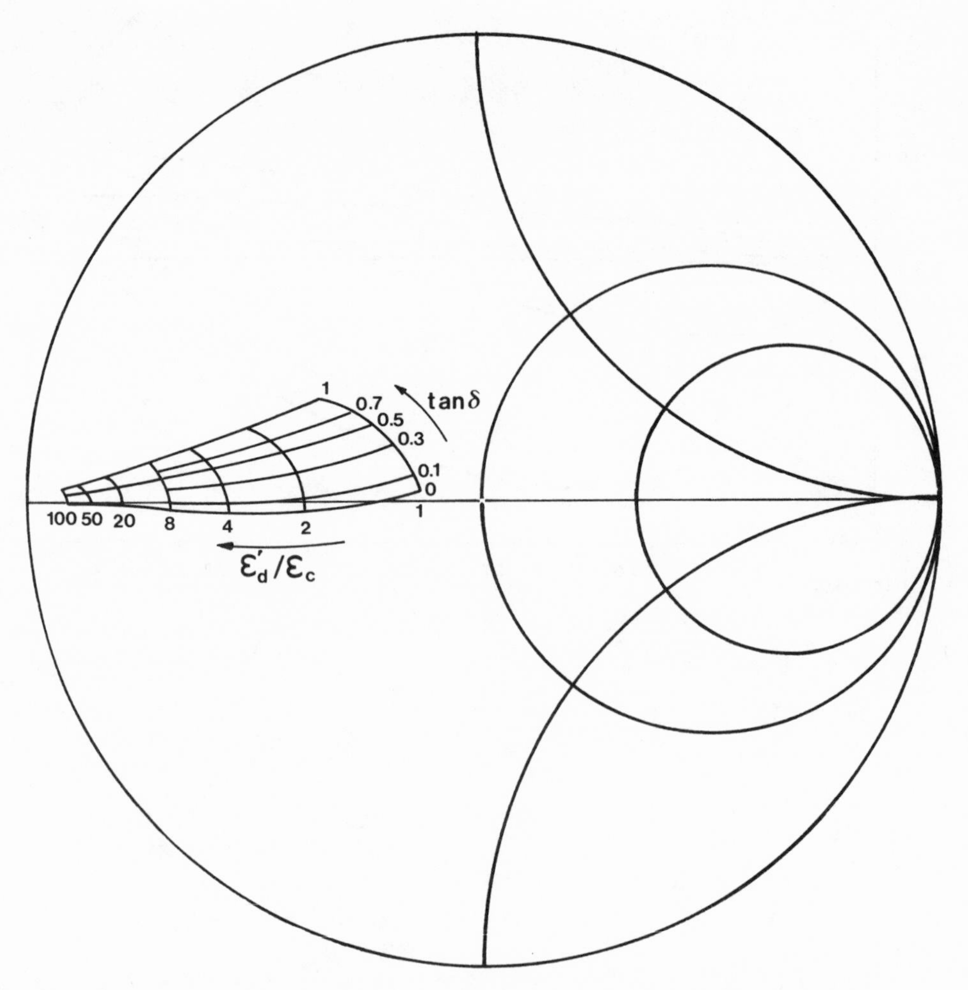

Fig. 9. Plot of reflection factor in polar diagram (Smith chart) at a normalized frequency of $k_c a = 2.15$ with $\varepsilon_d'/\varepsilon_c$ and $\tan\delta$ as parameters.

well as of inaccuracies during the computation. A first verification was made (see Section 6) by comparing two similar approaches based on different dyadic Green's function equations. It was found furthermore highly desirable to carry out a limited experimental verification, in order to ascertain whether the order of magnitude and the general dependence of the results calculated do correspond with measured data.

A slotted line in circular tubing was fabricated at the machine shop of the Department of Electrical Engineering of Ecole Polytechnique. It was fitted into a standard slotted line carriage and tests were made with an empty waveguide ($\varepsilon_c = 1$) radiating into open space ($\varepsilon_d = 1$). Measurements were carried out over the frequency range (3); they are reported and compared with the theoretical curve in Fig. 10.

The agreement between theory and experiment is quite adequate, within the experimental error of the measurement (as the slotted line was simply machined out of a circular metallic tube, it was not possible to end the slot with gradually tapering sections, hence a certain residual VSWR).

8. POLYNOMIAL APPROXIMATION

The rather complicated mathematical procedure developed here can only be implemented on a rather powerful computing machine: it will therefore be of practical use only to a limited number of potential users. Furthermore, it provides the reflection factor $\underline{\Gamma}_1$ in the waveguide (and related quantities $\underline{Y}$, VSWR, phase, etc.) as a function of

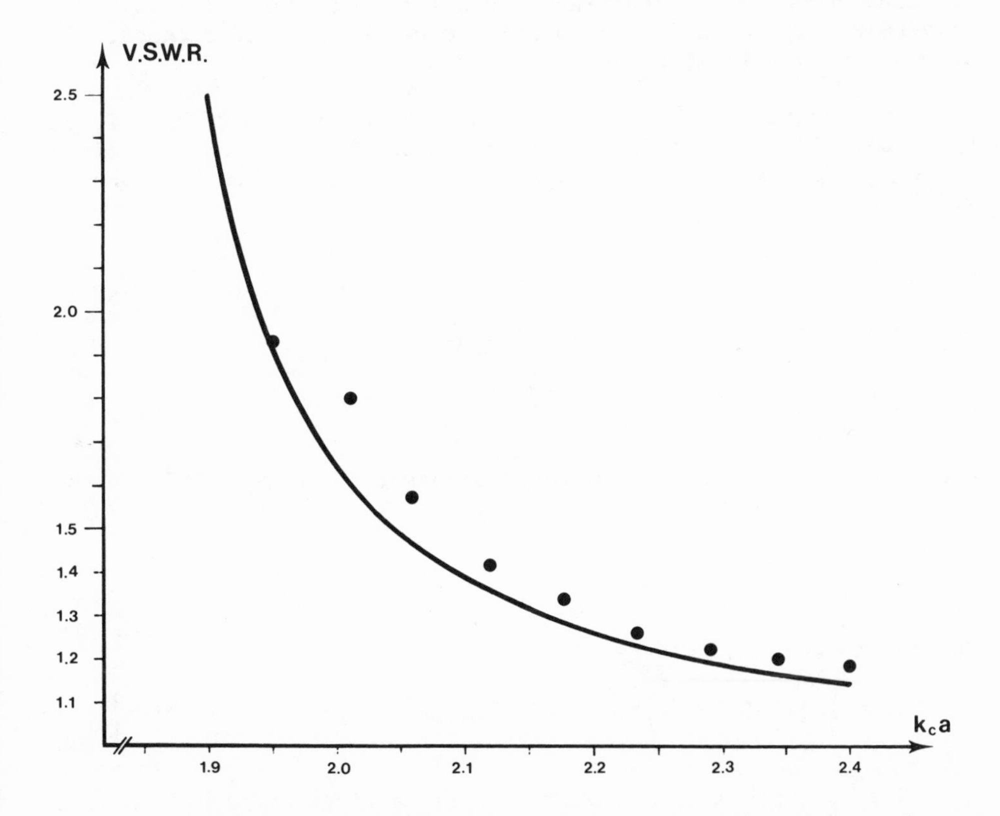

Fig. 10. Comparison of calculated values of VSWR (solid curve) with slotted line measurements (dots) as a function of the normalized frequency $k_c a$ for the homogeneous case $\varepsilon'_d = \varepsilon_c = 1$.

the dielectric constant ratio $\underline{\varepsilon}_d/\varepsilon_c$. The situation encountered in practice is actually the opposite one: measurements yield the VSWR and the phase of the reflection factor, and from them one wishes to deduce $\underline{\varepsilon}_d/\varepsilon_c$.

A first way to do this, at one given frequency, is to plot the point corresponding to the measurements in the Smith chart of Fig. 9 and determine both $\varepsilon_d'/\varepsilon_c$ and $\tan\delta$ from the graphical representation. This approach may be somewhat inaccurate, depending considerably on the skill of the operator.

A second technique, proposed in a previous publication (Decreton et al., 1974), places the complete analysis program within an optimization loop, which searches for the value of $\underline{\varepsilon}_d/\varepsilon_c$, which yields, at the output of the analysis program, the value of $\underline{\Gamma}_1$ obtained from the measurement. This approach is unfortunately quite time-consuming.

A simpler approach, which provides an answer to both problems, was considered here: starting from a set of calculated points, a polynomial approximation was developed using best fit methods. The general outline is given below.

The function to be approximated is here called f(x,y): it can be either $\varepsilon_d'/\varepsilon_c$ or $\tan\delta$, as a function of $x \sim s\cos\theta$ and $y \sim s\sin\theta$ where s = VSWR and θ is the phase of $\underline{\Gamma}_1$. The function is known at a set of points x_i, y_i, where

$$f_i = f(x_i,y_i) \tag{70}$$

An approximation for f is a polynomial f_p:

$$f(x,y) \simeq f_p(x,y) = \sum_n A_n P_n(x,y) \tag{71}$$

where the A_n are the unknown coefficients to be determined and the P_n are functions of the form

$$P_n(x,y) = x^m y^\ell \tag{72}$$

The error function is defined as:

$$\sigma^2 = \sum_i [f_i - f_p(x_i,y_i)]^2 = \sum_i [f_p^2(x_i,y_i) - 2f_p(x_i,y_i)f_i + f_i^2] \tag{73}$$

The best possible approximation is obtained by minimizing the error function, i.e. by setting its derivative with respect to the parameters A_n equal to zero, as follows:

$$\frac{\partial\sigma^2}{\partial A_j} = \sum_i \frac{\partial f_p^2(x_i,y_i)}{\partial A_j} - 2\sum_i f_i \frac{\partial f_p(x_i,y_i)}{\partial A_j}$$

$$= 2\sum_i [f_p(x_i,y_i)-f_i] \frac{\partial f_p(x_i,y_i)}{\partial A_j} = 0 \quad (74)$$

$$\text{for } j = 1,2\ldots N$$

Replacing f_p by its polynomial development, given in (71), one obtains:

$$\sum_n A_n \left[\sum_i P_n(x_i,y_i)P_j(x_i,y_i)\right] = \sum_i f_i P_j(x_i,y_i) \quad (75)$$

$$j = 1,2\ldots N$$

This system provides a matrix equation for the unknown expansion coefficients A_n.

Third order polynomials were found to provide a good approximation over a limited range of parameters. The average relative error is defined by

$$\frac{\sigma}{(\sum_i f_i^2)^{\frac{1}{2}}} = \frac{\{\sum_i[(f_i-f_p(x_i,y_i)]^2\}^{\frac{1}{2}}}{[\sum_i f_i^2]^{\frac{1}{2}}} \quad (76)$$

while the maximum relative error is given by

$$\varepsilon_{max} = \text{Max}\left[\frac{f_i-f_p(x_i,y_i)}{f_i}\right] \quad (77)$$

The development was carried out for normalized values of x and y located between 0 and 1, obtained as follows:

$$x = \frac{s\cos\theta - \min(s\cos\theta)}{\max(s\cos\theta) - \min(s\cos\theta)} \quad (78)$$

$$y = \frac{s\sin\theta - \min(s\sin\theta)}{\max(s\sin\theta) - \min(s\sin\theta)} \quad (79)$$

Polynomial expansions for both $\varepsilon_d'/\varepsilon_c$ and tan δ over various domains were calculated. They are presented in matrix form,

$$f_p(x,y) = [1\ x\ x^2\ x^3]\begin{bmatrix} c_{00} & c_{01} & c_{02} & c_{03} \\ c_{10} & c_{11} & c_{12} & c_{13} \\ c_{20} & c_{21} & c_{22} & c_{23} \\ c_{30} & c_{31} & c_{32} & c_{33} \end{bmatrix}\begin{bmatrix} 1 \\ y \\ y^2 \\ y^3 \end{bmatrix} = \sum_{i=0}^{3}\sum_{j=0}^{3} c_{ij}x^i y^j \quad (80)$$

Table III lists the values of the c_{ij} coefficients, for a range in either $\varepsilon_d'/\varepsilon_c$ or tan δ, together with the

$$\left\{\begin{matrix} \max \\ \min \end{matrix}\left(s \begin{matrix} \cos\theta \\ \sin\theta \end{matrix} \right)\right\}$$

limiting values appearing in (78-79). The average and maximum relative errors, defined in (76) and (77) are also given. It is apparent that smaller domains of definition must be taken for the tan δ expansion than for the one of $\varepsilon_d'/\varepsilon_c$.

An example of the accuracy by the polynomial approximation is given in Fig. 11. The crosses correspond to the values calculated with the complete analysis, the curves to those obtained using the polynomial expansion.

The curves corresponding respectively to ε_d' = constant and to tan δ = constant are obtained directly from (80), letting $f_p(x,y)$ = K (constant) and $x = x_0$ (constant). Defining

$$B_j = \sum_{i=0}^{3} c_{ij}\, x_0^i \quad (81)$$

and replacing into (80) we obtain

$$B_3y^3 + B_2y^2 + B_1y + B_0 = K \quad (82)$$

This third-order equation has three solutions for y, only one of which fits within the domain of the known values of y_i. The calculation process is repeated with different values of x_0, yielding the two families of curves represented in Fig. 11.

The polynomial expansion technique was here applied at a particular normalized frequency $k_ca = 2.15$. It can later on be extended over the complete frequency range, in which case the c_{ij} coefficients become themselves functions of k_ca.

Use of the polynomial expansion brings the problem within the scope of desktop minicomputers and even of some programmable pocket calculators.

Table III. Terms c_{ij} of the Polynomial Expansion (80)

for $\varepsilon_d'/\varepsilon_c$

$1 < \varepsilon_d'/\varepsilon_c < 10$

$0 < \tan\delta < 1$

$\min(s\cos\theta) = -7.6202$

$\max(s\cos\theta) = -1.2836$

$\min(s\sin\theta) = -0.1680$

$\max(s\sin\theta) = 1.3177$

i ↓ \ j →	0	1	2	3
0	15.8130	5.4730	-31.6871	14.7781
1	-26.8562	-6.2975	53.2283	-27.1405
2	15.4451	-8.1283	-13.1256	10.4807
3	-3.4594	9.1683	-8.4915	1.6418

average relative error 0.122%

maximum relative error 0.815%

$10 < \varepsilon_d'/\varepsilon_c < 100$

$0.1 < \tan\delta < 1$

$\min(s\cos\theta) = -30.6045$

$\max(s\cos\theta) = -5.9880$

$\min(s\sin\theta) = -0.4005$

$\max(s\sin\theta) = 1.0313$

i ↓ \ j →	0	1	2	3
0	182.284	-252.280	380.939	-211.017
1	-413.617	1445.818	-2153.813	942.034
2	510.351	-2218.075	2794.431	-942.985
3	-246.241	881.456	-755.365	50.737

average relative error 1.41%

for $\tan\delta$

$1.1 < \varepsilon_d'/\varepsilon_c < 10$

$0.1 < \tan\delta < 1$

$\min(s\cos\theta) = -7.6202$

$\max(s\cos\theta) = -1.4290$

$\min(s\sin\theta) = -3.0854$

$\max(s\sin\theta) = 1.2490$

i ↓ \ j →	0	1	2	3
0	0.0730	1.9670	-2.4026	2.4497
1	-0.2534	-4.9547	14.1166	-10.0428
2	0.9446	9.5698	-25.8385	17.3350
3	-0.7626	-6.4481	14.9202	-10.1229

average relative error 0.716%

maximum relative error 9.043%

$10 < \varepsilon_d'/\varepsilon_c < 100$

$0.1 < \tan\delta < 1$

$\min(s\cos\theta) = -30.6045$

$\max(s\cos\theta) = -5.9880$

$\min(s\sin\theta) = -0.4005$

$\max(s\sin\theta) = 1.0313$

i ↓ \ j →	0	1	2	3
0	0.3706	-4.7611	10.4163	-5.0352
1	-2.4308	37.0375	-68.4883	34.8151
2	5.6738	-73.4825	134.5276	-66.8193
3	-3.9186	42.4623	-76.3359	37.2368

average relative error 3.495%

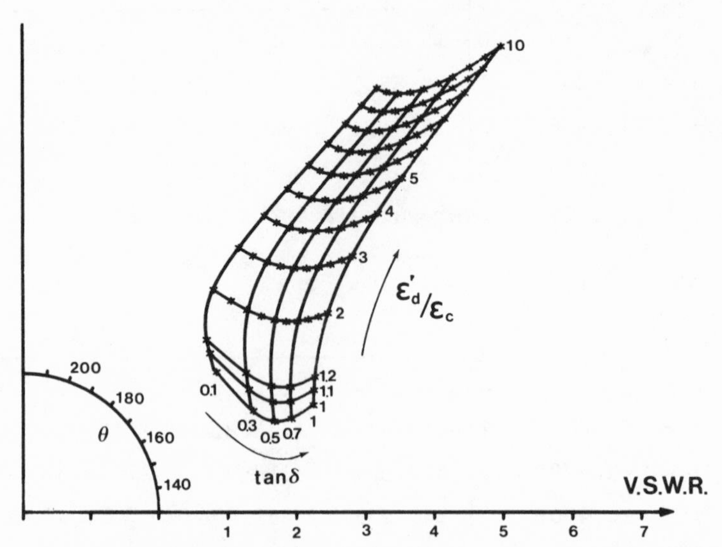

Fig. 11. Comparison of values calculated with the complete analysis program (crosses) and values obtained using third-order polynomial expansion (solid lines) for $k_c a = 2.15$.

9. CONCLUSION

The possibility of using an open-ended circular waveguide terminated by a large flat metallic flange, operating in the dominant TE_{11} mode of propagation to carry out nondestructive measurements of materials has been considered. The material to be measured must have at least one flat face, enabling a good contact with the waveguide probe. For such a technique to be practical, a way had to be found to correlate, if possible by means of simple equations, the measured values (VSWR, phase-shift of reflection factor) with the dielectric permittivity of the material.

To do that, a detailed mathematical analysis of the field distribution, both within the waveguide and outside of the aperture, is carried out. The continuity of the tangential electric and magnetic fields in the plane of the aperture is imposed, yielding a dyadic admittance Green's function formulation. This relationship is developed with the method of characteristic modes. Eigenvalues and eigenvectors of the aperture problem are determined and the reflection factor expressed in terms of these quantities. The expansion over the waveguide modes yields a matrix equation, the terms of which are calculated by means of numerical integration of a product of Bessel functions. During the process, a singularity in the integrand has to be extracted.

A computer program was elaborated; calculated values are presented as a function of complex permittivity and of frequency. Comparison of calculated values with measurement data show good agreement.

The mathematical approach being quite involved and leading to rather lengthy calculations on a large computer, a simplified approximation was elaborated, based on a two-dimensional third-order polynomial approximation. A simple equation is then obtained, easy to program on a microcomputer or a pocket calculator. The material properties ε_d' and tan δ can then easily be derived from measurement data, VSWR and reflection phase-shift θ.

With the advent of a simple, easy to use equation providing a good accuracy over a specified range of parameters, one of the main stumbling blocks encountered when using open-ended waveguides to make nondestructive measurements of materials has been removed.

ACKNOWLEDGMENT

The authors wish to thank J.D. Decotignie and L.G. Bernier for the help provided in the computation, J.F. Zurcher for the design of the slotted line and S. Mamane for preliminary mathematical developments.

This project was supported in part by the Swiss National Science Foundation under successive Grants 2.086-0.78, 2.244-0.79 and 2.657-0.80.

APPENDIX A. DEVELOPMENT OF THE INTERNAL GREEN'S FUNCTION $\underline{\bar{\bar{B}}}$

The transverse magnetic field in the plane of the aperture, on the waveguide side (z < 0) is provided by (30), an integral equation over the magnetic surface current $\underline{\vec{M}}$ in the plane of the aperture, multiplied by the internal dyadic Green's function $\underline{\bar{\bar{B}}}$. This last function represents the field produced by a point source of magnetic current density $\underline{\bar{M}}$ given by

$$\underline{\vec{M}}(\vec{r}) = \underline{\vec{M}}_0 \, \delta(\vec{r} - \vec{r}') \qquad \text{(A1)}$$

At a point $\vec{r}$, the magnetic field $\underline{\vec{H}}$ resulting from the presence of the point source at $\vec{r}'$ is given by the general linear expression:

$$\vec{\underline{H}}(\vec{r}) = \underline{\overline{\overline{B}}}(r,r')\cdot\vec{\underline{M}}_0 \tag{A2}$$

This relationship defines the internal dyadic Green's function $\underline{\overline{\overline{B}}}$. Its value is determined by considering the electromagnetic field solutions, i.e. the waveguide modes (propagating and evanescent), which are the solutions to Maxwell's equations in the presence of boundary conditions (Felsen et al., 1973). Rather lengthy calculations provide the expression:

$$\underline{\overline{\overline{B}}}(\vec{r},\vec{r}') = \frac{1}{j\omega\mu_0}(\nabla x\nabla x\ \vec{e}_z)(\nabla' x\nabla' x\ \vec{e}_z)\underline{S}''(\vec{r},\vec{r}') + k_c^2(\nabla x\ \vec{e}_z)(\nabla' x\ \vec{e}_z)\underline{S}'(\vec{r},\vec{r}') \tag{A3}$$

The $\underline{S}''$ scalar term represents the contribution of the TE modes (n odd). It is a summation over products of the transverse potentials $\underline{\Psi}_n$ for TE modes (Table I).

$$\underline{S}''(\vec{r},\vec{r}') = \frac{1}{2j\omega\mu_0}\sum_{n\ \text{odd}} \underline{\Psi}_n(\rho,\phi)\underline{\Psi}_n^*(\rho',\phi')\underline{Z}_n \exp(-\gamma_n|z-z'|) \tag{A4}$$

In a similar manner, the $\underline{S}'$ term provides the contribution of the TM modes (n even), a summation over their transverse potentials Φ_n:

$$\underline{S}'(\vec{r},\vec{r}') = \frac{1}{2j\omega\varepsilon_0\varepsilon_c}\sum_{n\ \text{even}} \Phi_n(\rho,\phi)\Phi_n^*(\rho',\phi')\underline{Y}_n \exp(-\underline{\gamma}_n|z-z'|) \tag{A5}$$

The wave characteristic impedance $\underline{Z}_n = 1/\underline{Y}_n$ is given in Table I.

When the $\underline{S}''$ and $\underline{S}'$ developments, given respectively by (A4) and (A5), are introduced into (A3) and the derivations are carried out, the transverse vector functions $\vec{W}_n(\rho,\phi)$ given in (6) and (7) appear. One obtains in this manner for TM modes:

$$(\nabla\ x\ \vec{e}_z)\phi_n(\rho,\phi) = \phi_n(\nabla\ x\ \vec{e}_z) + \nabla\phi_n\ x\ \vec{e}_z = \vec{W}_n \tag{A6}$$

It must here be noted that the magnetic surface current $\vec{\underline{M}}$ is purely transverse and that only the transverse part of the magnetic field $\vec{\underline{H}}$ is required to ensure the continuity of the fields. This means that only the transverse components of the internal dyadic Green's function $\underline{\overline{\overline{B}}}$ need to be determined, and one obtains in this way for TE modes:

$$(\nabla\ x\ \nabla\ x\ \vec{e}_z)\psi_n(\rho,\phi)\exp(-\underline{\gamma}_n|z-z'|)\Big|_t$$

$$= \nabla\ x\ (\nabla\ x\ \vec{e}_z)\psi_n(\rho,\phi)\exp(-\underline{\gamma}_n|z-z'|)\Big|_t$$

$$= \nabla \times (\vec{e}_z \times \vec{W}_n)\exp(-\underline{\gamma}_n|z-z'|)\Big|_t$$

$$= \underline{\gamma}_n \text{ sign}(z-z')(\vec{W}_n)\exp(-\underline{\gamma}_n)|z-z'|) \tag{A7}$$

Carrying out the products one then finds:

$$\underline{\bar{\bar{B}}}_t(r,r') = \frac{1}{2}\sum_{\substack{n\\ \text{odd}}} \frac{-\underline{\gamma}_n^2\underline{Z}_n}{-(\omega\mu_0)^2}\vec{W}_n(\rho,\phi)\vec{W}_n'(\rho',\phi')\exp(-\underline{\gamma}_n|z-z'|)$$

$$+\frac{1}{2}\sum_{\substack{n\\ \text{even}}} \frac{k_c^2\,\underline{Y}_n}{-\omega^2\mu_0\varepsilon_0\varepsilon_c}\vec{W}_n(\rho,\phi)\vec{W}_n'(\rho',\phi')\exp(-\underline{\gamma}_n|z-z'|) \tag{A8}$$

Several simplifications are possible at this point:

• for TE modes

$$\frac{\underline{\gamma}_n\underline{Z}_n}{(\omega\mu_0)^2} = \frac{\underline{\gamma}_n^2}{(\omega\mu_0)^2} = \frac{j\omega\mu_0}{\underline{\gamma}_n} = -\frac{\underline{\gamma}_n}{j\omega\mu_0} = -\underline{Y}_n \tag{A9}$$

• for TM modes

$$\frac{k_c^2\,\underline{Y}_n}{-\omega^2\mu_0\varepsilon_0\varepsilon_c} = \frac{\omega^2\mu_0\varepsilon_0\varepsilon_c}{-\omega^2\mu_0\varepsilon_0\varepsilon_c}\underline{Y}_n = -\underline{Y}_n \tag{A10}$$

The internal dyadic Green's function applied to transverse components then takes the simpler form:

$$\underline{\bar{\bar{B}}}_t(\vec{r},\vec{r}') = -\frac{1}{2}\sum_n \underline{Y}_n\vec{W}_n(\rho,\phi)\vec{W}_n'(\rho',\phi')\exp(-\underline{\gamma}_n|z-z'|) \tag{A11}$$

The Green's function obtained in this way applies to the case of a waveguide of infinite extent in both directions. For the semi-infinite situation considered here, this expression must be multiplied by two, in order to take into account the doubling of the current on the short-circuit. This yields the value given in (17).

APPENDIX B. DEVELOPMENT OF $\vec{W}_m \cdot \vec{W}_n$ TERM FOR 2 TE MODES

The starting point of the development is equation (59):

$$\vec{W}_m \cdot \vec{W}_n = (W_{m\rho}W_{n\rho'} + W_{m\phi}W_{n\phi'})\cos\psi + (W_{m\rho}W_{n\phi'} - W_{m\phi}W_{n\rho'})\sin\psi \tag{59}$$

The transverse functions given for TE modes (n odd) in Table I become, making use of properties of the Bessel functions:

$$W_{m\rho} = -C_m p_m J_1'(p_m\rho)\cos\phi = \frac{C_m p_m}{2}\,[-J_0(p_m\rho) + J_2(p_m\rho)]\cos\phi \tag{B1}$$

$$W_{m\phi} = C_m \frac{1}{\rho} J_1(p_m\rho)\sin\phi = \frac{C_m p_m}{2}\,[J_0(p_m\rho) + J_2(p_m\rho)]\sin\phi \tag{B2}$$

The first term of (59) then becomes:

$$\begin{aligned} &W_{m\rho}W_{n\rho'} + W_{m\phi}W_{n\phi'} \\ &\quad = \frac{K_{mn}}{4}\,[J_0(p_m\rho)J_0(p_n\rho') + J_2(p_m\rho)J_2(p_n\rho')]\cos\psi \\ &\quad - \frac{K_{mn}}{4}\,[J_0(p_m\rho)J_2(p_n\rho') + J_2(p_m\rho)J_0(p_n\rho')]\cos\psi' \end{aligned} \tag{B3}$$

where

$$K_{mn} = p_m p_n C_m C_n \tag{61}$$

The second term of (59) then yields:

$$\begin{aligned} &W_{m\rho}W_{n\phi'} - W_{m\phi}W_{n\rho'} \\ &\quad = \frac{K_{mn}}{4}\,[J_0(p_m\rho)J_0(p_n\rho') - J_2(p_m\rho)J_2(p_n\rho')]\sin\psi \\ &\quad - \frac{K_{mn}}{4}\,[J_0(p_m\rho)J_2(p_n\rho') - J_2(p_m\rho)J_0(p_n\rho')]\sin\psi' \end{aligned} \tag{B4}$$

Integrating over ϕ between 0 and 2π, the crossed-terms in J_0J_2 and J_2J_0 all vanish and one obtains for the TE-TE term:

$$\int_0^{2\pi} \vec{W}_m \cdot \vec{W}_n \, d\phi = \frac{\pi K_{mn}}{2}[J_0(p_m\rho)J_0(p_n\rho') + J_2(p_m\rho)J_2(p_n\rho')\cos 2\psi] \tag{B5}$$

The same development is carried out for the other mode combinations. For the TM-TM terms, the same expression (B5) is obtained.

The cross-products between different kinds of modes TE-TM and TM-TE yield a negative sign for the J_0J_0 expression. These results are presented in (60).

APPENDIX C. EXTRACTION OF THE SINGULARITY IN $T_{mn}^{\ell''}$

The technique used to eliminate the singularity which appears in the integrand consists of replacing it by a simple discontinuity, which does not impede the numerical integration process. This technique can be used with integrals of the type

$$I = \int_{S'} \frac{F(\rho',\phi')}{G(\rho',\phi')} \, dA' \tag{C1}$$

for which

$$G(\rho_0,\phi_0) = 0 \tag{C2}$$

with the point ρ_0,ϕ_0 located within the integration domain S'. This integral can be decomposed as follows:

$$I = \int_{S'} \frac{F(\rho',\phi')-F(\rho_0,\phi_0)}{G(\rho',\phi')} \, dA' + F(\rho_0,\phi_0) \int_{S'} \frac{dA'}{G(\rho',\phi')}$$

$$= I_1 + F(\rho_0,\phi_0)\, I_2 \tag{C3}$$

The integrand of I_1 is no longer singular if the zero of the denominator is of first order: it remains finite but discontinuous (Gex-Fabry et al., 1979) in the position $\rho' = \rho_0$ and $\phi' = \phi_0$. The domain of integration then is divided in such a way that the discontinuity remains at the border of the domain. The remaining problem is to find a way to integrate, if possible by analytical means, the integral I_2, whose integrand is still singular.

In the particular problem considered here, the integral which presents a singularity is of the form:

$$\int_0^a\int_0^a\int_0^\pi J_\ell(p_m\rho)J_\ell(p_n\rho')\cos(\ell\psi)\exp(-k_d''R_0)\frac{\cos(k_d'R_0)}{R_0}\,\rho\rho'\,d\psi d\rho' d\rho$$

$$= \int_0^a J_\ell(p_m\rho)\;\rho\;d\rho \left[\int_0^a\int_0^\pi [J_\ell(p_n\rho')\cos(\ell\psi)\right.$$

$$\exp(-k''_d R_0)\frac{\cos(k'_d R_0)}{R_0} - \frac{J_\ell(p_n\rho)}{R_0}]\rho'\, d\psi\, d\rho'$$

$$+ J_\ell(p_n\rho) \int_0^a\int_0^\pi \frac{1}{R_0}\rho'\, d\psi\, d\rho'\Bigg]$$

$$= \int_0^a J_\ell(p_m\rho)\,[I_1 + J_\ell(p_n\rho) I_2]\, \rho\, d\rho \tag{C4}$$

where

$$R_0 = (\rho^2 + \rho'^2 - 2\rho\rho' \cos\psi)^{\frac{1}{2}} \tag{52}$$

The integral I_1 is not singular; it can be integrated numerically. The integration with respect to ψ of I_2 yields (Gradsteyn et al., 1965):

$$\int_0^\pi \frac{d\psi}{\sqrt{\rho^2 + \rho'^2 - 2\rho\rho' \cos\psi}} = \begin{cases} 2/\rho\; K(\rho'/\rho) & \text{for } \rho' < \rho \\ \\ 2/\rho'\; K(\rho/\rho') & \text{for } \rho' > \rho \end{cases} \tag{C5}$$

so that:

$$I_2 = \frac{2}{\rho}\int_0^\rho K\left(\frac{\rho'}{\rho}\right)\rho'\, d\rho + 2\int_\rho^a K\left(\frac{\rho}{\rho'}\right) d\rho' \tag{C6}$$

These integrals have also been evaluated (Gradsteyn et al., 1965); they yield:

$$I_2 = 2a\; E(\rho/a) \tag{C7}$$

where K and E are, respectively, the elliptical integrals of first and of second types.

The integral finally takes the form:

$$\int_0^a J_\ell(p_m\rho)\;[I_1(\rho) + 2aJ_\ell(p_n\rho)E(\rho/a)]\;\rho\, d\rho \tag{C8}$$

with

$$I_1(\rho) = \int_0^a \int_0^\pi [J_\ell(p_n\rho')\cos(\ell\psi)\exp(-k_d''R_0)\cos(k_d'R_0) - J_\ell(p_n\rho)] \frac{\rho'\, d\psi\, d\rho'}{R_0} \qquad \text{(C9)}$$

REFERENCES

Altschuler, H. M., Dielectric constant, in "Handbook of Microwave Measurement", M. Sucher and J. Fox, eds., Vol. II, pp. 495-548, Polytechnic Press, Brooklyn, N.Y. (1963).

Collin, J. R., Dyadic Green's functions, current views on singular behavior, in "Abstracts of XX General Assembly", International Union of Radio Science, 34, Washington, D.C. (1981).

Decreton, M. C., and Gardiol, F. E., Simple nondestructive method for the measurement of complex permittivity, IEEE Trans. Intrum. Meas. IM-23 (4) 434-438 (1974).

Dubost, G., and Zisler, S., "Antennes à large bande", 51, Masson, Paris (1976).

Edrich, J., and Smyth, C. J., Millimeter wave thermography as subcutaneous indicator of joint inflammation, in "Proceedings of the Seventh European Microwave Conference, Copenhagen", pp. 713-717, Microwave Exhibitions and Publishers, London (1977).

Felsen, L. B., and Marcuvitz, N., "Radiation and Scattering of Waves", pp.196-197, Prentice-Hall, Englewood Cliffs, N.J. (1973).

Galejs, J., "Antennas in Inhomogeneous Media", Pergamon Press, Oxford, England (1969).

Gardiol, F. E., "Hyperfréquences", pp. 44-45, Georgi, St.-Saphorin, Switzerland (1981).

Gautherie, M., Edrich, J., Zimmer, R., Guerquin-Kern, J. L., and Robert, J., Millimeter wave thermography, application to breast cancer, J. Microwave Power 14 (3) 123-129 (1979).

Gex-Fabry, M., Mosig, J. R., and Gardiol, F. E., Reflection and radiation of an open-ended circular waveguide: application to nondestructive measurement of materials, Archiv für Elektrotechnik und Übertragungstechnik 33 (12) 473-478 (1979).

Gradsteyn, I. S., and Ryzhik, I. M., "Table of Integrals, Series and Products, 387, 626, Academic Press, New York (1965).

Harrington, R. F., "Time Harmonic Electromagnetic Fields", McGraw-Hill, New York (1961).

Harrington, R. F., and Mautz, R. J., Theory of characteristic modes for conducting bodies, IEEE Trans. Antennas and Propagation AP-19 (5) 622-627 (1971a).

Harrington, R. F., and Mautz, R. J., Computation of characteristic modes for conducting bodies, IEEE Trans. Antennas and Propagation AP-19 (5) 629-639 (1971b).

Kantor, G., and Witters, D. M., A 2450-MHz slab-loaded direct contact applicator with choke, IEEE Trans. Microwave Theory Tech. MTT-28 (12) 1418-1422 (1980).

Lewin, L., "Advanced Theory of Waveguides", Iliffe, London, England (1951).

Licht, S., "Therapeutic Heat and Cold", Waverly Press, Baltimore (1965).

Mosig, J. R., Besson, J. C. E., Gex-Fabry, M., and Gardiol, F. E., Reflection of an open-ended coaxial line and application to nondestructive measurement of materials, IEEE Trans. Instrum. Meas. IM-30 (1) 46-51 (1981).

Sterzer, F., Paglione, R., Nowogrodzi, M., Beck, E., Mendecki, J., Friedenthal, E., and Botstein, C., Microwave apparatus for the treatment of cancer, Microwave Journal 23 (1) 39-44 (1980).

INDEX